The Cultural Politics of Anti-Elitism

This book examines the highly ambivalent implications and effects of anti-elitism. It draws on this theme as a cross-cutting entry point to provide transdisciplinary analysis of current conjunctures and their contradictions, drawing on examples from popular culture and media, politics, fashion, labour and spatial arrangements.

Using the toolboxes of media and discourse analysis, hegemony theory, ethnography, critical social psychology and cultural studies more broadly, the book surveys and theorizes the forms, the implications and the ambiguities and limits of anti-elitist formations in different parts of the world. Anti-elitist sentiments colour the contemporary political conjuncture as much as they shape pop cultural and media trends. Populists, right-wing authoritarian ones and others, direct their anger at cultural, political and, sometimes, economic elites while supporting other elites and creating new ones. At the same time, "elitist" knowledge and expertise, decision-making power and taste regimes are being questioned in societal transformations that are discussed much more positively under headlines such as participation or democratization.

The book brings together a group of international, interdisciplinary case studies in order to better understand the ways in which the battle cry "against the elites" shapes current conjunctures and possible future politics, focusing on themes such as nationalist political discourse in India, Austria, the UK and Hungary, labour struggles and anti-oligarchy rhetoric in Russia, tax-avoiding elites and fiscal imaginaries, working-class agency, Melania Trump as a celebrity narrative in Slovenia, aesthetic codes of the Alt-Right, football hooliganism in Germany, "hipster hate" in German political discourse or the politics of expertise and anti-elite iconography in high fashion internationally. The book is intended for undergraduates, postgraduates and postdoctoral researchers.

Moritz Ege is professor of Cultural Studies/Popular Cultures at the University of Zurich. His publications cover a range of topics in urban ethnography, cultures of social inequality, political dynamics of the popular and conjunctural analysis.

Johannes Springer teaches cultural studies at the Institute of Music at the University of Applied Sciences Osnabrück, Germany. His areas of interest and publications include pop music history, music video studies, labour in creative industries, production of culture perspectives and theories of space and place, stars and fandom.

The Cultural Politics of Anti-Elitism

Edited by Moritz Ege and Johannes Springer

Routledge
Taylor & Francis Group
LONDON AND NEW YORK

First published 2023
by Routledge
4 Park Square, Milton Park, Abingdon, Oxon OX14 4RN

and by Routledge
605 Third Avenue, New York, NY 10158

Routledge is an imprint of the Taylor & Francis Group, an informa business

British Library Cataloguing-in-Publication Data
A catalogue record for this book is available from the British Library

ISBN: 978-0-367-69260-5 (hbk)
ISBN: 978-0-367-69261-2 (pbk)
ISBN: 978-1-003-14115-0 (ebk)

DOI: 10.4324/9781003141150

Typeset in Times New Roman
by SPi Technologies India Pvt Ltd (Straive)

Contents

Acknowledgements

This book is the result of a longer process of exploring anti-elitism in contemporary culture and cultural history at the Department of European Ethnology/Cultural Anthropology at the University of Göttingen. It would not have been possible without the support and co-operation of a number of partners to whom we want to express our gratitude. The project *"Against the elites!" The cultural politics of anti-elitism in the current conjuncture* included a conference at the University of Göttingen and a series of talks on topics ranging from fashion, football, the city/country divide and labour struggles which were co-hosted by organisations such as the *Agrarsoziale Gesellschaft*, the Göttingen branch of the labour union *Ver.di*, the football *Supporter Crew 05* and Martina Glomb and her fashion design students from the University of Applied Sciences and Arts Hannover and the *Zukunftswerkstatt Ihmezentrum*. A group of graduate students and the book's editors also curated "Krisenkino" (crisis cinema) film series at cinema *Lumière* in Göttingen looking at anti-elitism in cinema history, interrogating different varieties of anti-elitist narratives and representations.

Special thanks go to graphic designer Christian Heinz for translating our ideas into a design concept that not only graced our events posters and flyers but also this book's chapter introductions. The overall project goes back to discussions within a working group "thinking (in) conjunctures", where we particularly thank Manuela Bojadžijev, Alexander Gallas and Ove Sutter. We also want to thank our former colleagues at the Institute of Cultural Anthropology/European Ethnology, University of Göttingen, especially Regina Bendix and administrative staff members Esther Lauer and Sabine van Eeckhoutte, for their support throughout the project and events.

The project was made possible by the financial support of the Ministry for Science and Culture of Lower Saxony *(Niedersächsisches Vorab)*. Additional support to render this book Open Access has come from the University of Klagenfurt (Faculty of Cultural Studies), University of Zurich, University of Basel, and University of the Arts/London College of Communication.

Contributors

Atlanta Ina Beyer is a lecturer in Gender and Diversity at Rhine-Waal University of Applied Sciences in Kleve, Germany, and a cultural worker and editor. Her research focuses on queer and queer-feminist movements and subcultures, popular culture and gender studies. She co-edited *Perverse Assemblages. Queering Heteronormativity Inter/Medially* (2017); her dissertation is on music and zines of the queer punk movement since 1985.

Rebecca Bramall is reader in Cultural Politics at the London College of Communication, University of the Arts London. Her research explores the relationship between culture and economy, with a current focus on fiscal imaginaries. Key publications include *The Cultural Politics of Austerity* (2013) and a special issue of the journal *New Formations on Austerity* (2016).

John Clarke is an Emeritus professor in the Faculty of Arts and Social Sciences at the Open University, UK. He is also a Leverhulme Emeritus fellow while working to develop a conjunctural analysis of "Brexit and Beyond", which may eventually result in a book. His most recent publication is a series of conversations, titled *Critical Dialogues: Thinking Together in Turbulent Times*, published in 2019.

Sebastian Dümling is a lecturer at the University of Basel's Department of Cultural Studies and European Ethnology. As a researcher on narrative and discourse, he is primarily interested in how history is represented – particularly in the context of the New Right. Recent publications include "Changing Societies, Changing Narratives. How to talk about 'social change' and be understood", published in *Journal of European Ethnology and Cultural Analysis* (2020).

Moritz Ege is professor of Popular Cultures at the University of Zurich, Department of Social Anthropology and Popular Cultures. His main research interests are urban ethnography, protest movements, the history of popular culture and conjunctural analysis. Recent publications include the co-edited volume *Urban ethics. Conflicts over the good and proper life in cities* (Routledge, 2021).

Sonja Eismann is one of the founders and editors of Missy Magazine, a Berlin-based feminist bi-monthly on politics, pop culture and style. Her research interests lie in the representation of gender in popular culture, feminist theory and activism and the utopian potential of fashion. She teaches courses in Gender, Cultural, Popular Music and Fashion Studies at universities in Germany, Austria and Switzerland.

Richard Gebhardt is a political scientist working in adult education for unions, federal centres for education, parties, etc. His research interests include right-wing populism and extremism in Germany, hooliganism and football fan cultures. He recently edited *Fäuste, Fahnen, Fankulturen : die Rückkehr der Hooligans auf der Straße und im Stadion* (2017) and co-edited *Volksgemeinschaft statt Kapitalismus? Zur sozialen Demagogie der Neonazis* (2009).

Paolo Gerbaudo is a sociologist and political theorist at King's College London. He is the author of a number of articles on social media, populism and political organisation and of the monographs *Tweets and the Streets* (2012), *The Mask and the Flag* (2017), *The Digital Party* (2019) and *The Great Recoil* (2021).

Stefanie Hürtgen is associate professor at the University of Salzburg, Austria, and associate member of the Institute for Social Research in Frankfurt, Germany. Her fields of research include global and European political economy, sociology of work and labour geography. Her recent publications include *Structural Heterogeneity in Europe: The Arrival of an Apparently Developmental Problem in the Global North and the Question of Transnational Solidarity* (2020).

Breda Luthar is professor of Media and Communication Studies at the Faculty of Social Sciences, University of Ljubljana, Slovenia. She has a background in sociology and focuses her research on politics, media and popular culture, class and culture and material/consumer culture in socialism. Her latest publication is *Intimate Media and Technological Nature of Sociality* (2020).

Olga Reznikova is research associate at the Department of Social Anthropology and Popular Cultures at the University of Zurich. Her work focuses on the city, workers' protests, racism and anti-Semitism, employing ethnographic, cultural–analytical and historical methods. Her Ph.D. thesis (2021, Göttingen) was on the topic "Angry Truckers: social protest in Russia." Her latest publication is Russländischer Kolonialismus? Zur Kritik des westlichen antiimperialistischen Weltbildes (Russian colonialism? A critique of the Western anti-imperialist worldview), in: Julian Warner (Hg.), After Europe: Beiträge zur dekolonialen Kritik, Berlin 2021.

Sanam Roohi is Alexander von Humboldt fellow at the Centre for Modern Indian Studies, University of Göttingen, and on the editorial board of

Comparative Migration Studies journal. Her work straddles the themes of critical mobilities, embodied migration infrastructures and transnational resource flows within the India diaspora. She has published in journals, including *Modern Asian Studies, Journal of Contemporary Asia and Ethnic and Migration Studies.*

Brigitta Schmidt-Lauber is head of the department for European Ethnology at the University of Vienna and holds a professorship there. Her research interests lie in urban–rural relationships, the history and cultural analysis of everyday life as well as ethnographic studies Her recent publications include the edited volume *Andere Urbanitäten* (2018, Vienna: Böhlau), a critical view on perceptions of the city.

Alexandra Schwell is a professor of empirical cultural analysis at the University of Klagenfurt, Austria. Her major research interests include political anthropology, popular culture, border studies, ethnographic methods and Europeanization processes. Her recent publications include (2019) *Who's Afraid of the Big, Bad ...?: Populism and the Threatened Border in Austria.*

Johannes Springer teaches cultural studies at the Institute of Music at the University of Applied Sciences Osnabrück, Germany. His areas of interest and publications include pop music history, music video studies, labour in creative industries, production of culture perspectives and theories of space and place, stars and fandom.

Diana Weis studied theatre studies and modern German literature at Freie Universität Berlin and holds a Ph.D. from the University of Hamburg. She is a professor of Fashion Journalism at BSP Business School Berlin. Her research focuses on beauty norms, body image and female beautification practices, especially within digital cultures. In 2020, she published her book *Modebilder. Abschied von Real Life.*

Jens Wietschorke is senior lecturer at the Department of Empirical Cultural Studies and European Ethnology at the University of Munich. His research interests include the historical analysis of class relations, popular culture (with a focus on the first half of the 20th century), urban anthropology and the anthropology of architecture and space. He recently published the book *1920er Jahre. 100 Seiten* (2020).

“AGAINST THE ELITES!”

THE CULTURAL POLITICS OF ANTI-ELITISM BETWEEN POPULISM, POP CULTURE AND EVERYDAY LIFE - AN INTRODUTION

INTRODUCTION

MORITZ EGE AND JOHANNES SPRINGER

1 The cultural politics of anti-elitism between populism, pop culture and everyday life

An introduction

Moritz Ege and Johannes Springer

This is a book about anti-elite rhetoric, narratives, imagery and movements. It asks: What is characteristic of anti-elite articulations, be it in populist politics, pop culture or everyday life more broadly? Which kinds of elites are being imagined, caricatured and criticised by whom, through which media and why? What social actors, parties, movements, artists, subcultures, technologies and milieus are involved in producing, shaping and mediating anti-elite articulations? To what ends and with which results? And how are relationships of power and dominance challenged and reconfigured in that process?

In providing answers to these questions, the chapters most of them were finalized in 2021 (they go back to a series of events, including a conference under the title "Against the elites!" The cultural politics of anti-elitism in the current conjuncture held in 2018), contribute to a socio-cultural analysis of current conjunctures.[1] For this purpose, we do not define who "the elite" *really* are. Instead, we pursue the usages of the term in different contexts and try to understand better what image critics of the elites have of their adversaries. The concept "elite" itself, to us, is a subordinate category of socio-cultural analysis. It can be useful as a heuristic tool and in precisely defined circumstances, but it can also be misleading as an overstretched analytical concept: It allows a dubious self-aggrandisement for those who believe in the existence and rightful claim to power of an elite in the sense of "the select few" (and, usually, consider themselves part of it). For those who decry "the elites", the term may have uses that we would categorise as progressive or reactionary, but it often also leads to questionable slippages, as we will show. It runs the danger of replacing other critical concepts and analyses that are more structurally grounded or more phenomenologically acute.

In focusing on anti-elite articulations of different types, the book also aims at circumventing the "programmatic bias" (Caiani and Padoan, 2020, p.6) of many studies into populism that neglect the spheres of cultural production, ways of life, aesthetics or affects. And, at the same time, it also intends to avoid a mere culturalism that ignores the role of politics and the economic sphere or reduces them to mere cultural dynamics. In this introduction, we will sketch out the overall concerns of the book and present the argument that anti-elitism is a crucial, cross-domain theme of contemporary societies that can serve as an entry point for new, interdisciplinary analyses of the

DOI: 10.4324/9781003141150-1

contemporary, combining cultural and social research. Consequently, we will go back and forth, in a sort of hermeneutic spiral, between a prominent example, its broader political and cultural contexts and the methodological and conceptual tools that we suggest are necessary for making sense of them.

In the first section, we will follow the term "elite" in political rhetoric and introduce the overall problematic. The second section begins by putting that rhetoric in the context of a specific historical situation, i.e. the political upheavals around the middle of the 2010s, more broadly. It then discusses the ways in which cultural politics in recent times have been shaped by anti-elitism and poses the question how this may have contributed to crises that are multiple and interconnected. The third section returns to a peculiar "moment" of anti-elitism between 2015 and 2018, giving an overview of journalistic and academic attempts at explaining the scepticism and enmity towards elites, primarily in the US and Europe – also highlighting the different meanings attached to the term in that discourse. In the fourth section, we introduce theoretical background assumptions that are particularly important regarding our approach to studying anti-elite articulations, focusing on the epistemological status of diagnostic narratives of different types and the notion of conjunctural analysis and its purchase. The fifth section asks what happened to anti-elite articulations and what their role might be *after* this historical "moment". Instead of summarising the chapters of this volume at the end of this text, we highlight throughout this introduction how the chapters expand on the book's overall themes and topics.

Anti-elitism and its moment

A book on anti-elite articulations with a focus on the late 2010s and early 2020s must almost inevitably begin with Donald Trump, 45th President of the United States of America. Trump's was a very public, epoch-shifting – or at least so it seemed – discourse about and against "the elites", the political and cultural "establishment", imbued with the ambiguities of calculated vilification, open resentment, reasonable critique and a palpable desire for the status, recognition and accoutrements of the chosen few. Trump's anti-elite rhetoric seems to have caught the mood of hundreds of millions or even billions worldwide and stunned and shocked at least an equal number. It also popularised specific ways of speaking and thinking about "elites". A 2018 article on *Politico* documents Trump's shifting use of the term at length and spells out some of its basic tensions.

"For Donald Trump, 'elite' used to mean a modeling agency", the article begins.

> "She was with Elite," he said of Anna Nicole Smith four days after her death in 2007 in an interview with Howard Stern, the same way some might say a person had won a prestigious prize. "She had the best body. She had the best face. She had the best hair I've ever seen."
>
> (Kruse, 2018)

As a celebrity businessman, *Politico* author Michael Kruse writes, "Trump used the world 'elite' the way the agency did, as a bit of marketing boilerplate more or less interchangeable with 'classy' or 'luxury.'" His own properties and developments were praised as "elite"; "applied to people, it was an unvarnished compliment: Eli Manning was an 'elite' quarterback."

Then came Trump's nomination and election campaigns in 2015 and 2016, where – following the global populist handbook, sophisticated electoral research and, apparently, his intuitive social analysis – he attacked "the elites", the "media elite", the "political elite", "the establishment" and so on, promising to drain the Washington "swamp", "lock up" Hilary Clinton and, equally importantly, take all these self-righteous progressives, liberal celebrities, artists and professors to whom the term was applied (by people like Trump, to a large extent) down a notch or two, promising a sort of cultural revenge in the name of ordinary, common people.[2] He railed against the elites on Twitter and in campaign rally after campaign rally. This was not only a matter of discursive content. As cultural anthropologists Kira Hall, Donna Goldstein and Matthew Ingram pointed out, it was also a matter of linguistic style and bodily performance, for example, his contortions and gestures as he ridiculed the stiff bodies of establishment politicians – or, infamously, the physical impairments of a reporter (Hall, Goldstein, and Ingram, 2016). In mocking those whom he labelled the elite, their pretensions and their corporeal inadequacies, Trump exemplified a cultural strategy that Pierre Ostiguy (2017) calls the "flaunting of the low": pleasurably exhibiting the seemingly unconstrained, "base", "mean", "vulgar", prejudiced behaviour – in content and form – that educators and modernisers of different kinds say we should overcome, like a clichéd rebellion of id against super-ego. However offensive it all was, however much it was permeated with racist and sexist messages, Trump also cleverly identified the hypocrisies, contradictions and weaknesses that characterised the self-image and the socio-political position of many liberals in the US and elsewhere – and of progressive neo-liberal formations more broadly, or, at least, some crucial tendencies within them (Fraser, 2017; see also Beyer, Wietschorke and others in this volume). Addressing them as "elites" was a crucial element of this strategy.

After the president's inauguration and its aftermath, however, came another phase in Trumpian rhetoric – one often overlooked by observers – in which a more ambiguous usage took hold. Since about 2017, the US president had been

> reclaiming the word "elite" with an almost vengeful pride. Having vanquished his opponents at the polls, having slammed the "elites" as corrupt, incompetent and out of touch, Trump now has bestowed upon himself, as well as his most fervent supporters, the mantle of "elite" as if it were a spoil of war.
>
> (Kruse, 2018)

The president often – in a first step – introduced "the elite" in a satirical and polemical tone in his speeches in 2017 and 2018: the so-called elites, enemies of the people and so forth. Then, however, he went on to make his own claims on the term:

> "Why are they elite?" he said in Minnesota. "I have a much better apartment than they do. I'm smarter than they are. I'm richer than they are. I became president, and they didn't. And I'm representing the greatest, smartest, most loyal, best people on earth – the deplorables."
>
> (ibid.)

Somewhat stunned, the *Politico* writer summarised the shift thus: "He and his voters are now the elite, the new elite, 'the super-elite,' Trump said in South Carolina." "Can he really run as the elite instead of against the elite?", the article asked.

With the benefit of hindsight, one could say that he tried, but failed. The journalist's incredulousness about the seemingly contradictory way of relating to elites, however, was closely intertwined with a wider issue: The question as to how Trump, a billionaire and serial fraudster, could present himself as a true man of the people and a champion of the working class – and, put more simply, how those who in recent decades had not been the beneficiaries of ever more capitalism could fall for it.[3] (It seemed almost natural to many observers, on the other hand, that most Republicans from the upper middle and upper classes would support him.) Did they not see that he himself – like other right-wing populists globally – was part of the ruling class, that he had seriously ripped off workers, other businesses and ordinary people, for example, at his so-called university – and that his policies, such as tax reforms planned and administered by Wall Street insiders and industry lobbyists, would benefit the rich?[4] Was not this anti-elite rhetoric so full of contradictions that it should defeat anyone's ability to live with cognitive dissonance? Apparently, it was not – and attempts to explain this seeming paradox soon began to proliferate. In our view, the success of Trump's anti-elite rhetoric should not be read as implying that people understood insufficiently what the slippery term "really" meant. Rather, it illustrates that the ascriptions to elites, and the attitudes many people have towards them, are more ambiguous than they seem at first sight.

In a psychological register, the *Politico* author concludes that, Trump's "acrobatic use of 'elite'" represented a key to his "abiding sense of grievance, his unconcealed mix of envy and resentment of this class of person". Importantly, Kruse argues that this was not only a matter of an individual character, the story often told of the real estate heir from Queens to whom old money and society hotshots in Manhattan had given the cold shoulder. Rather, Trump's personal baggage allowed him to tap "into a deep American history of anti-elitism as a potent political tool" that – over a hundred years earlier – had produced the original Populists, then figures such as Huey P. Long and, later on, Richard Nixon and Ronald Reagan or less significant figures such as Sarah Palin, who had all prominently attacked the cultural, political and media elites as well, even if they did not necessarily call them by that name. Richard Hofstadter's (1964) "paranoid tradition" was apparently alive and well – and people such as Roy Cohn and Roger Stone, Trump's mentors and advisers, represented direct links into that past. In these older instances of populist anti-elite rhetoric in the US, there had been similar juxtapositions of fake elites and true heroes of the people – and the latter could include the deservedly rich.[5] Crucially, that tradition often also recurred to a racial pecking order that becomes ideologically legitimised by the construction of moral boundaries. Figures of elite decadence and popular decency, and of the moral and economic dangers posed by racial others, belong to the same imaginary, the same process of symbolic boundary-drawing with all its material implications (Hartigan, 1997; Hochschild, 2016). Therefore, Trump made manifest latent meanings and desires inherent in an important strand of the broader populist tradition when he announced that he not only wanted to win a fight against the elites for ordinary Americans, he and those he represented actually *were* what the others only claimed to be, the elite, and, thus, *truly deserved* riches and recognition.[6]

For understanding the current situation and its genealogy, including the ambiguities inherent in the term "elite", it is important to also remember another connection. The key protagonists of the neo-liberal political–economic turn since the 1970s and 1980s also relied on a specific form of anti-elite rhetoric and an outspoken enmity to intellectuals, the state bureaucracy and labour unions and their leaders, whom they depicted as elites. As the neo-liberals – politicians such as Ronald Reagan and Margaret Thatcher, and theorists such as Friedrich Hayek, Milton Friedman and James Buchanan – argued, these elites arrogantly claimed to know better than ordinary citizens and consumers. They stood in the way of their market-democratic self-determination. Therefore, markets had to be freed from restrictive state regulations. Furthermore, democracy had to be restricted in order to ensure that competition would not be disturbed (see, *inter alia*, Slobodian, 2018; Slobodian and Plehwe, 2020) and "true", deserved elites could prosper. It was with this anti-state, -bureaucracy, -intellectual, pro-entrepreneurial, -consumer rhetoric – and its policy substance – in mind that Stuart Hall had famously termed Thatcher et al. a new breed of right-wing authoritarian *populists*, as opposed to the older formation of authoritarian *statism*.[7]

Neo-liberal reforms and deregulation helped bring about a new class of super-rich. Culturally, they fostered a sense of consumer subjectivity, and they also brought about precarity and a strong sense of threat and loss for large parts of the population, for which they also offered specific kinds of explanations (on the connections, see, e.g. work by ethnographers in Europe, such as Kalb, 2009; Kapferer and Theodossopoulos, 2019; Narotzky, 2019). This is a crucial background for subsequent waves of anti-elitism. Since the 1990s, the basic ideological suppositions of neo-liberalism have increasingly pervaded everyday consciousness and common sense: Competition is key for progress, the profit maximisation motive should permeate all spheres of action, families and traditional "communities" are needed to buffer the social costs and those that threaten or evade these principles must be repressed by authoritarian means. These attitudes became strengthened and normalised as neo-liberal "common sense" (Hall, 2011; Hall and O'Shea, 2013). In that sense, contemporary anti-elite articulations in the Trumpian vein, which present themselves as populist rebellion against the status quo, are as much inside the neo-liberal configuration as they are outside of it – but they are also pushing it in new directions.

A break-up of hegemony in politics – and in culture?

While historical contextualisations and longer-term developments, such as the ones we briefly sketched out here, are crucial for understanding recent goings-on, there was clearly something new, something emergent to this massive wave of anti-elite sentiments and rhetoric during Trump's rise. The years 2015 to 2018 seemed like the midst of an interregnum, in Antonio Gramsci's sense of the term (Hall, 2015; Fraser, 2017; Grossberg, 2018; Massey, 2018), when "long-simmering discontent suddenly shape-shifted into a full-bore crisis" (Fraser, 2017, unp.), a crisis of authority and even, possibly, of hegemony – political, but also cultural. This was a broader anti-elite "moment" in US politics and many other places as well: Brexit was supposed to return control from European Union bureaucrats to the British people, or, at least, so the rhetoric went. The crisis of political representation in countries such as Italy, Greece or France escalated and well-established parties shrunk almost into oblivion. New movements emerged: The French *gilets jaunes* (yellow vests), for example, were an unforeseen, forceful and programmatically as well as affectively and habitually anti-elite movement (Lem, 2020; Susser, 2021) that observers in France called a sign of a broader "twilight of the elites" (Guilluy, 2015, 2019).[8] At the very least, a rearrangement of leading blocs or societal–political coalitions was taking place. Other right-wing populist leaders – many of whom are extreme rightists and neofascists – famously employed similar anti-elite rhetoric as well, whether in the opposition or in government: Salvini, Orban, Farage, Johnson, Kaczynski, Le Pen, Babis, Blocher, Strache, Wilders, Netanyahu, Erdogan, Modi, Putin, Duterte or Bolsonaro. These years also saw consistent left-wing agitation against "the one percent", "for the many, not the few", where the elites, "the one percent", "the rich", "the caste", figured as the beneficiaries and the agents of a class struggle from above. The elites were primarily defined in the politics of Syriza, Podemos, Corbyn, Sanders, Mélenchon and others in economic terms and in reference to their political power (national and international ones, such as EU leaders) and also, culturally, their detachment from the lived experience and reality of ordinary people.[9] The "populism of the centre" is a much less popular topic among political scientists than right and left populism, but it also certainly exists – even technocrats such as Emmanuel Macron (see, *inter alia*, Curini, 2019, p.1416) or Matteo Renzi have railed against state elites and left-wing intellectuals; New Labour/Third Way social democrats such as Gerhard Schröder had not been all that different in that respect.[10]

Anti-elitism, as a political strategy, style and discourse, was articulated with a wide range of political positions and goals in these turbulences. More generally speaking, anti-elitism in political rhetoric is accompanied by promises that can be defined as progressive in an optimistic sense of that term, such as the levelling of undeserved privileges and the realisation of an egalitarian, democratic spirit and collective sovereignty. There are also equally constant dangers, such as bad social analysis, a reinforcement of prejudices and the many connections between anti-elitism, certain critiques of capitalism, conspiracy theories and "coded" anti-Semitism (see, *inter alia*, Reznikova in this volume). Our starting supposition, however, is that the actual meanings and effects of this strategy are, at least on this general level, open and indeterminate in important ways. This is because they depend on the concrete articulations of which they are part – nationalist or anti-nationalist, sexist or anti-sexist, anti-Semitic or not, for example – and because they were in many ways being articulated anew and along very different lines in this specific "moment". In pragmatic terms, political actors were, therefore, well-advised to fight over them rather than to leave them to their adversaries.[11]

Anti-elitism, anti-elite articulations and cultural politics

The phrase "against the elites" and the term "anti-elitism" that we have been using require some further clarification. Being against elites, against the elites, being anti-elitist and anti-elitism have been used as synonyms so far, but they can also mean different things and their usage can perform different forms of critiques. Listening to media figures, *vox pop* interviewees, protagonists in ethnographic writing, internet commenters, populist politicians and others castigate "elites", one may think that these complaints came from a place of egalitarianism: Down with the elites – there should be no elites! And indeed, anti-elite sentiments *can* be egalitarian and anti-elite in such a strong and universal sense. The anti-authoritarian tradition of the Left has had a strong and programmatic anti-elitist bent in that sense.[12] The protagonists of 1960s left-leaning, anti-authoritarian pop culture, for example, found colourful expressions against the idea and the institutions that maintain that there are legitimate elites and that they deserve to be privileged. However, anti-elite sentiments can be highly ambiguous, critical of "these elites" or "pseudo-elites" (see Dümling in this volume), while calling for "true elites" to rise to power. They can also be anti-elit*ist*, in a slightly weaker sense, i.e. opposed primarily to condescending, exclusionary behaviour or regulations of a specific kind. In that framing, the problem with elites is understood primarily as a matter of conduct, rhetoric (elitist language and other cultural codes) and institutional policies, such as membership rules in a club or admission regulations to a school or university. Speaking of an "-ism" here connotes the speaker's critical attitude to an excess of elite-ness, not necessarily a problem with the existence and high standing of elites. Anti-elitists, in that sense, can also be supportive of elites, be they supremely competent or supremely rich, who do not behave in elitist ways – down-to-earth scientists, hands-on entrepreneurs, deserving celebrities, politicians who are "demotic" and folksy or just matter-of-fact (on celebrities, see also Luthar in this volume).

In order to be terminologically precise, we use the term "anti-elite articulations" as an umbrella for these different expressions of opposition to elites and elitism, be they anti-elite or -elitist. We also use the term to refer to phenomena of different kinds, be they attitudes, sentiments, styles, rhetoric, arguments, images or narratives. The term "articulation" stands both for bringing-to-speech and the connection (as process and result) of separate, heterogeneous elements, be they fragments of meaning or cultural practices (see Clarke in this volume). "Articulation", thus, highlights the importance of cultural forms, practices and representations (media and others), as well as the contingency and complexity of meaning. Despite these terminological considerations, however, we do not always make these distinctions here. It would be cumbersome, for example, to always speak of "anti-elite articulations" instead of "anti-elitism". Hopefully, the terminological caution expressed here will suffice and meanings will be conveyed by the arguments.

Anti-elite articulations in culture

While political dynamics of the kind we mentioned above were relatively easy to follow and to name, there has also been a culmination of *cultural* anti-elite phenomena. In order to situate political rhetoric such as Trump's and its societal resonance, we must also spell out different kinds of cultural anti-elitism. For our purposes, the term "culture" can be understood as comprising not only systems or assemblages of meaning/representation (including, but not limited to, aesthetic ones; see Gilbert, 2019a, 2019b) and affect, practices of meaning-making and "affecting" but also the practical side of relating to and constituting those systems of meaning. This takes place in the realm that we usually call everyday life. It is through culture in this wide sense that consensus with a status quo is created, reinforced, challenged, rejected and reconfigured.[13]

Anti-elite articulations in culture span a broad spectrum of forms. The term "cultural politics" serves as a placeholder for processes, relations and struggles that are relevant in that context. It refers to political implications of processes that take place outside the narrowly defined sphere of politics, and it also points to the question of how hegemony, prestige and dissent are being produced culturally and play out in support for movements and parties, electoral behaviour and so forth. It also refers to expressive forms and sensibilities in ordinary life and to less clearly defined, more qualitative textural-atmospheric implications and consequences of culture whose political effects are impossible to pin down exactly.[14] Cultural forms of anti-elitism are much more heterogeneous and also more difficult to periodise than the obviously "political" forms on which we have focused so far – but ultimately no less important. The understanding of culture that we employ here – building primarily on cultural studies and socio-cultural anthropology and an updated version of hegemony theory that connects both – is not entirely congruent with the sense of culture that is used in many debates about populism, identity and political strategy. In the latter, there is often a strict contrast between "cultural issues" and "economic issues" (see, *inter alia*, Manow, 2018; Rodrik, 2021). In contrast to many positions in the cultural versus economic causes problematic and also the related cultural versus economic strategies debate, we do not claim, for example, that questions that relate to collective identities should be situated outside of the economic sphere. Gender, for example, is as much an economic as it is a cultural category. At the same time, cultural processes and questions of identity necessarily co-constitute any sense of "economic" class consciousness. Furthermore, there is a "cultural" side to the everyday worlds of work as much as there is to so-called private life. Pop-cultural representations are cultural in our sense of the term; they are an important part of the field of cultural politics, but they are also part of cultural industries and a cultural economy.

In what sense, then, is anti-elitism *expressed* in contemporary culture, *shaping* contemporary culture and shaping the current conjuncture – including on the political level – culturally? To begin to answer these questions, this section of the introduction will highlight some exemplary phenomena and processes.

Overall, it has become somewhat of a cliché that we live in an *age of participation and an ever-wider democratisation of expertise* (Jenkins, 2006; Kelty, 2008, 2019; Maasen and Weingart, 2008; Carpentier, 2011; Barney et al., 2016; Baiocchi and Ganuza, 2017; Fuchs, 2017). This is not only a matter of knowing but of doing: People are also "against elites", it can be argued, because their everyday practices have strong and increasingly egalitarian elements. They work just fine without the presence of elites – who nonetheless, often enough, claim authority over them. A lot of information that used to be esoteric and shielded is now widely accessible and can be turned into knowledge, in many cases, regardless of professional or educational status – while ongoing restrictions of information, commercial or bureaucratic, receive justified criticism (Hall, 2008). This continuously raises questions about whose knowledge and expertise counts, how it is authorised or, put differently, why some people get transferred opportunities, recognition and pay as artists, experts, intellectuals or critics and others do not (Clarke and Newman, 2017; Newman and Clarke, 2018; Hall, 2021).

In that context, the claim that people in their online lives are caught up in "filter bubbles" and "echo chambers", where their views and dispositions are being confirmed, reinforced and radicalised, has become part of contemporary common sense (Pariser, 2011; Nguyen, 2020). Commentators mostly, and for reasons that are quite understandable, view this development as highly problematic for political discourse and societal cohesion. It is relevant, however, for much broader circles than QAnon conspiracy theorists and the like, and it is usually *experienced* in a much more positive sense. Contemporary social media, with all its flaws, is not only about becoming passively exposed to influencers of various kinds. It offers affordances for communication that are structured in an at least somewhat horizontal way, especially when compared with one-way communication that is controlled directly by broadcasters who are invested with (state) authority or large amounts of private capital. So, passing information on and adding to it, expressing one's views and publicising them and having highly specialised forums for direct and trans-local communication are now popular, not elite or niche practices, even if they are, of course, also usually shaped by profit-oriented infrastructures (Fuchs, 2017; Gerbaudo, 2018; Miller and Venkatraman, 2018) and actors with strategic goals. This (restricted) participatory and popular element exists in platforms, such as Instagram, Facebook, Twitter and YouTube, in mass messaging services, such as Telegram, and in more specialised forums and boards, such as Reddit (Massanari, 2014), where self-organisation and, in some ways, self-governance (i.e. through volunteer community moderators) are practiced by millions.

Overall, in sociological terms, there seems to be an abstract homology between these kinds of participatory forms of sociality and knowledge, the networked character of late modern, post-Fordist forms of production and governance (for a classic account of this, see Castells, 2000, 2015) and a decline of respect for elites which are defined by older forms of cultural capital and their role in older institutions. Well-worn sociological metanarratives hint at some aspects of these processes that cumulatively contribute to an anti-elite moment in the cultural sense: Narratives of individualisation and value change, the informalisation of language and customs and the decline of deference, the transformation of forms of governing towards participatory regimes and the tendency towards neo-liberal governmentality that focuses on a free, self-responsible subject. A sort of habitual anti-elitism is reinforced by *dispositifs* that address citizens as participants and as consumers who make their own choices rather than having their choices made for them by experts and paternalists of different sorts.

At the same time, *motifs* and *themes* prevalent in the entertainment world are equally relevant here, including blockbuster films such as *The Hunger Games* series (2012–2015) or *Joker* (2019), with their stories of rebellions against privileged castes or classes. They serve as indicators and popularisers of anti-elite attitudes. Such films reiterate a high-versus-low distinction that has long been a central aspect of the structural grammar of popular (and older, "folk") cultural narratives. A number of authors have spelled out their anti-elitist implications. Mark Fisher (2013) described the affective dimensions of watching *Catching Fire*, the second part in *The Hunger Games* tetralogy, as a film that offers as its set-up a world split into "neo-Roman cybergothic barbarism, with lurid cosmetics and costumery for the rich" and "hard labour for the poor". For Fisher, it offers nothing less than a counter-narrative to capitalist realism: Feelings such as "rage, horror, grim resolve" merge into a "delirious experience. More than once I thought: How can I be watching this? How can this be allowed? Will everyone want to be a revolutionary after recognizing the world and the modes elites live and rule after this?", he asks.

Elites are also explicitly or implicitly denounced in smaller, "realist", thematically focused films, such as UK-based *The Riot Club* (2014; about young, deeply classist and sexist Oxford students), in TV series like the German *Bad Banks* (2018–), in advertising campaigns that spoof the rich and pretentious and confront them with a more diverse, popular world[15] or in TV series whose plots start out from conflicts ordinary people have with corrupt, arrogant elites, such as the Spanish high-school series *Élite* (2018–) or the Korean hip-youth-against-*chaebol*-conglomerates *Itaewon Class* (2020). In all of those works and in many other pop-cultural productions as well,[15] elites are ridiculed, lampooned, cast as the problem, fought – but also desired and replaced by "worthier" successors.

From a historical viewpoint, it is striking how the anti-elite motif's popularity seems to mirror economic cycles. Times of depression have produced remarkable anti-elite films and auteurs – in the 1930s, Frank Capra with his little man trilogy; Preston Sturges[17] – and major transformative periods have spawned whole genre cycles such as trucker or strike films in the 1970s. Mapping the field of anti-elitist articulations through popular art forms such as film points to the ambiguities and emerging forces in these conjunctures. It also illustrates how figurations of gender, race and sexuality have been overlaying the low-versus-high axis: Figures of the popular and of elites are "racialised" and gendered in specific ways. Films such as *Dirty Harry* (1971) showcase a right-wing perspective on countercultural movements, minorities and liberal politicians in San Francisco as straight, *white*, male backlash "from below".[18] Strike films, on the other hand, have often represented an insurgent, multi-ethnic, feminist working-class anti-elitism (with negative images of rich elites and male bosses) in labour struggles, such as in French independent classic *Coup pour Coup* (1972) or more mainstream US films *Norma Rae* (1979) and *Nine to Five* (1980). The anti-elite motif entails a rather conservative aesthetic of sexuality in many of the most popular examples of the field, such as in *The Hunger Games* films, where the privileged are depicted as camp-y and queer-like, whereas ordinary workers are as morally straightforward as they are sexually "straight". This resonates with widespread patterns of heteronormativity in the populist imaginary and the culturally conservative implications of rhetorics of the "ordinary" and communitarian. On the other hand, there is another pattern which has become more prominent recently (i.e. in the TV series and advertisement campaigns mentioned above) where affirmative images of diversity, equality, non-normativity and creativity on the "popular" side are contrasted with bland, *white*, sexually repressed and normative "elites". Here, the "elite" merges with the upper-class "square".[19]

Film history also offers insight into the continuities of certain core themes and cleavages, such as the city/country divide which have been at the forefront of New Deal films like *Mr. Smith Goes to Washington* as much as of more recent ones as *Hillbilly Elegy* (see Wietschorke in this volume; Phelps, 1979; Rogan and Morin, 2003; Walsh, 2014; Seeßlen, 2017, p.20ff.).[20] The rural/urban and periphery/centre dichotomies have also been brought up time and again in recent times to explain Trumpism, Brexit and *inter alia*. However, to treat these questions only in electoral terms – or merely as a continuous motif in film history – would mean to underestimate their cultural and life-world qualities. Researchers have highlighted new patterns of anti-urban resentment and anti-elite critique in that sense in many countries and conflicts (for the US Midwest, see Cramer, 2016; also see Schmidt-Lauber and Roohi in this volume on Austria and India, respectively). Skogen and Krange, for example, in a Norwegian case study on the reintroduction of wolves, point to the emergence of "counterpublics" among hunters and a more general sense of rural disenfranchisement where illegal wolf hunting is perceived as "more-or less-legitimate resistance against power that not only controls wolf management, but is also seen as underlying unfair urban-rural relations and advancing the interests of social segments branded as 'elites'" (2020, p.568). In stories of rural anti-elitism like this, decline manifests itself in economic terms, shrinking processes (depopulation, the deterioration of public services) and social fragmentation. All this meets with a nature conservation discourse that is perceived as jarring and orchestrated from the centres. Spatial–economic–cultural cleavages also permeate debates around climate change, environmental and conservation politics (see Schwell in this volume) and their consequences for livelihoods and traditions within local rural populations that have little representation in many of these negotiations. Against that backdrop, recent scholarship (Mamonova and Franquesa, 2020; Pied, 2021) underlines the need to understand the forces behind right-wing movements in rural contexts and calls for sounding out progressive agrarian populisms.

The *forms* of a broader field of anti-elite sentiments understood as everyday life also require further exploration and analytical consideration. An important way of approaching this field is through everyday sentiments and affects, where scepticism, disenchantment and resentment towards elites can build up, or through the informal culture of conversation and storytelling, be it online or offline, in which elites and the self-important are "levelled". This field encompasses feelings of inferiority or "secondariness" (Hall et al., 2013, pp. 333–341; also see Hürtgen in this volume), misrecognition, being on the receiving end of paternalism and tutelage from higher-ups, being exploited, talked down to, i.e. the "hidden injuries of class" (Sennett and Cobb, 1977; Bourdieu, 1999), and other aspects of inequality as they are experienced and made sense of "from below".[21] These patterns are constituted by multiple axes of inequality and oppression (e.g. class, race/ethnicity, gender, sexuality, urban/rural), and are intersectional in that sense, but these categories are usually not kept separate on the plane of experience. Instead, they exist as "underdetermined" affect and are represented and articulated in condensed, "overdetermined" cultural figures. There is beneath and within the history of anti-elite social and political movements and rhetoric, then, a micropolitical cultural archive and a folklore of relating to these figures and the social relationships they symbolise: Through jokes and knowing glances, shared laughs, brief comments, eye-rolling, shrugs and idioms, sometimes defensive, sometimes more aggressive, that people use to keep the more powerful, distant "elites" and their representatives at bay. The objects and targets of such sentiments and expressions are not necessarily elites in a social–scientific sense of the term: They can be hierarchical superiors, small-scale authorities or street-level bureaucrats. They can also be mediated figures, including politicians and celebrities. Such figures often blend into one image or figure, or, more precisely, they *are* blended/articulated through mediated cultural work of many kinds, across the divides of "representations" and "everyday life" (see Roohi and Ege/Springer in this volume).[22] As historians and social scientists, particularly in the UK, have diagnosed a general "decline of deference" of ordinary people towards the privileged and elites of all kinds over the post-war decades (Sutcliffe-Braithwaite, 2018), these old forms of anti-elite sentiments and knowledge come to the fore.

The attitudes and relationships that people have towards institutions and official figures of authority, especially those in matters of knowledge or aesthetic judgement, also contribute to the broader cultural anti-elitist wave. This includes attitudes towards journalists, cultural critics in academia and "legacy media", high-prestige cultural producers or policy administrators whose authority results from processes of institutional legitimation (Bourdieu, 1984). Compared to the high points of their prestige in previous decades, these relationships have tended towards increasing scepticism and disinterest and towards anti-elitism in that sense – even if there are recent countertendencies as well, such as the divisive popularity of medical experts in the coronavirus pandemic.[23] The right wing's fight against the legitimacy of public service media is also implicated in this trend. They have brought emerging media systems under pressure not only in countries such as Hungary and Poland but also ones which were held in high esteem over decades, such as the BBC in the UK or the SRF in Switzerland. In all of these cases, accusations abound that proximity to the state makes these broadcasters complicit in an alleged "corrupt elite power complex" (Holtz-Bacha, 2021, p.5) or that as intermediaries they pose an impediment to the direct implementation of the "people's will" (Krämer and Holtz-Bacha, 2020).

Gatekeepers, canons and cultural institutions are being challenged from different angles and genealogies. This includes the declining role and position of professional cultural criticism and journalism, their symbolic power and authority as arbiters of good taste and their economic base.[24] On German public television, for example, *Das literarische Quartett*, where four literary critics discussed the merits of contemporary novels, had been a mainstay of cultural debate since the late 1980s. It turned Marcel Reich-Ranicki, the lead critic, into a widely known and often caricatured public intellectual. In its recent relaunch, the critics were replaced by novelists, celebrities, athletes and other pundits. Public television channel ZDF explained that "the principle of the authority [...] of master critics (*Großkritiker*)" was no longer valid "in a transformed societal situation" (Rüther, 2021). Instead, the debate should be between more relatable "passionate readers". According to journalist Tobias Rüther (2021), the presenter and some guests in the new version of the show use *more* seminar-style, technical language in talking about books than the literary critics did in the old version, pointing perhaps to insecurities and uncertainties over what counts as legitimate knowledge and authority. Whether or not this is the case, this kind of programming shift illustrates that decision makers in cultural institutions feel the need to distance themselves from what they see as elite culture. What takes its place is not necessarily less elitist but follows different principles, such as personal experience, celebrity, increasing diversity (especially in terms of gender, in this case) and a conservative sense of what are assumed to be popular taste sentiments.

One important context here is that rating and counting systems of various kinds, algorithmic decision-making and alternative, "lay" or "amateur" experts (e.g. bloggers, podcasters) amplify, confirm and shape many people's experiences and opinions. In that sense, they formally orchestrate the questioning of elite judgements and the need for them more generally. Controversies over the 2019 edition of the Sanremo Italian Song festival provide a prototypical case for these dynamics. The broader public's taste, expressed through audience voting from home (for singer Ultimo), was overturned by the votes of a jury of journalists and other experts who gave the victory to the singer and rapper Mahmood. Political actors, including right-wing Matteo Salvini and politicians from the neither-left-nor-right populist *Movimento Cinque Stelle*, immediately took this occasion to diagnose an antagonism between "the people" and their taste, on the one hand, and "the elites" in media and other cultural industries, on the other – with undertones of a paternalistic and "politically correct" choice having been made by the latter. To some extent, the rift was diagnosed on grounds of ethno-nationalist understandings of Italianness. Equally significantly, this was a procedural and technical question, especially for representatives of *Cinque Stelle*. Their questioning of the voting system and its "betrayal" of the audience's tastes was embedded in a broader argument for more direct, digital democracy. Different, intersecting forms of anti-elitism are at play in cases like this. The "emergence and legitimation of new systems for expressing judgements" (Magaudda, 2020, p.149) and their relationship to populism and anti-elitism in the realm of music and pop culture deserve closer study.

Regarding the field of art, Julian Stallabrass observed that the market's recent speculative boom went along with "the rise of a great deal of populist art – that is, an art of simple character, wide popular appeal, and an enthusiastic engagement with commercial mass culture delivered through branded artistic persona" (2012, p.42), sweeping both private collections and museums. According to this argument, the rise of street art, most prominently the artist Banksy, and similar registers, which are constituted by anti-elite gestures, exemplifies how specific forms of the popular have gained a foothold in the circles of elite culture. In aesthetic terms, Stallabrass traces this pattern back to what Fredric Jameson in the 1980s dubbed "aesthetic populism", referring to the rise of postmodernism in the previous two decades, particularly in architecture. He recounts this trajectory's primal scene:

> Robert Venturi [...] took a group of students on a field trip, not to see the wonders of Rome or the modernist towers of Chicago, but to Las Vegas to examine the architecture of the Strip. The students dubbed the course "The Great Proletarian Cultural Locomotive", and this gives a clue to one of its most important aspects, which was a defence of popular culture against the taste of the cultural elite. The casino architecture of the Strip was designed to entertain popular tastes: in this way, argued the authors, it was more democratic than modernist buildings whose makers insisted that an absence of decoration and a concentration upon unadorned form imbued them with moral rectitude. They celebrated the extraordinary mix of styles and pastiched histories to be found on the Strip.
>
> (Ibid., p.40)

The Venturi example illustrates how anti-elite critique, far from being restricted to politics, a few works of cultural production or the world of media technologies, is intertwined with central aesthetic metanarratives and cultural formations that have shaped recent decades. What makes this argument about the 1970s also particularly translatable to the present and to other fields of cultural production is that

> the "populism" of Learning from Las Vegas, as with so much in postmodernism, conflated the operations of big business with popular taste, in a familiar move to which populist sentiment is often subject. What Venturi and his collaborators asked us to accept as "almost all right" was not popular taste, but popular taste as imagined by casino owners.
>
> (Ibid., p.44)

Anti-elite gestures in this tradition of aesthetic populism represent a highly ambiguous form of the democratisation of taste and recognition. Similarly, the "vulgar" is a negative criterion often used in the classification practices of the arbiters of "good taste and good pleasure" (Phillips, 2016, p.11) in the fashion world (see Eismann in this volume) and in the realm of aesthetic judgements more widely (see Weis in this volume). But it can also function as a form of subversive, ostentatious and pleasurable excess that is employed self-reflexively, be it in the medium of style, symbolic communication or discursive legitimation – using the provocation of one's "vulgar" enjoyment to stick it to the elites and their supposed refinement.[25]

Similar dynamics are at play in the broader *culture of taste and consumption*, where many people denounce older, residual aesthetics of distinction, respectability and the established upper class as overly formal, stuffy and elitist. Newly emergent forms and aesthetic practices are often no less elitist, but differently so – for example, in the world of "fine dining", when white-linen Michelin-starred restaurants serving classic French *haute cuisine* are being replaced by hipper, equally Michelin-starred restaurants serving local fare in a more informal atmosphere at similar prices. In the terminology of cultural sociology, a habitus of constant exclusivity is being replaced by an "omnivore" attitude (Peterson and Kern, 1996). The latter also provides distinction towards the "unsophisticated" who appear stuck in their traditions and univorousness (Johnston and Baumann, 2015). Observations of "emerging cultural capital" point towards a remaking of elite culture. New objects and practices of distinction are being established, often as expressions of generational conflict and challenge (Friedman et al., 2015, pp.3–6). Such reconfigurations of distinction-providing tastes under the banner of an anti-elite aesthetic are not new, but they have become particularly relevant in recent years.

Dynamics of digitalisation permeate many of these fields. Their cumulative effects are making themselves felt more strongly than ever, but the moment of single-minded the-internet-will-make-us-free enthusiasm of earlier decades (Shirky, 2009) and the rhetoric of the democratisation of knowledge through the digital is, nevertheless, over, as the destructive dynamics of platform and surveillance capitalism have become as apparent as have their manipulative and limiting aspects (Morozov, 2012; Srnicek, 2017; Zuboff, 2019). Similar ambiguities and scepticism are at play in non-digital forms of participation: Even if citizens in many contexts now expect participation in decision-making processes, for example, in urban or rural spatial planning (Baiocchi and Ganuza, 2017; Farías, 2020; Müller, Sutter, and Wohlgemuth, 2020; Bikbov, 2021), this often remains limited to a narrow range of actual options and can take on a tokenistic quality.[26] There is usually little room in such processes for challenging ownership structures, for example. Furthermore, a model of the consumer citizen, of approaching the state as a consumer and taxpayer, often prevails over more emphatically democratic senses of what it means to be a citizen (Clarke, 2013).

The point in mentioning such debates over broader cultural diagnoses is not to attempt to provide definite answers to them. These discussions take place on a higher level of generality than our approach in this book, which seeks answers to smaller-scale questions and prioritises case studies over theory-building. The processes summarised here are not about a straightforward, secular process of cultural democratisation and improvement or a movement towards egalitarianism and emancipation in strong senses of these terms. Two dynamics are particularly important: Firstly, as all that is solid melts away, the new and emergent must, at some point, also move through the field of social forces of contemporary power structures and is shaped by it – surely not exclusively, but more often than not, decisively. Secondly, as we have begun to show, authority and elitism tend to reappear in new forms. Nevertheless, the new configurations that emerge are not fully determined by these dynamics either.

These spotlights highlight the breadth of current forms of anti-elitism. In the realm of the cultural, however, there has been no clearly identifiable anti-elite "event" on the scale of Trump's election. Cultural transformations usually take place more slowly, they are difficult to quantify and – given the polysemy and instability of meaning and affect more broadly – are not very clearly delineated. Nonetheless, molecular cultural processes and conflicts accrue and may fundamentally transform the social through slower processes of "drift" (Stewart, 2007; Grossberg, 2018, p.39). Individual and collective actors can activate, strengthen, make use of or defuse them through strategic cultural politics.

One more point is apposite here: In what we have reviewed in this section, metanarratives about socio-cultural processes intertwine with normative arguments about what is good and bad about them. The normative arguments made in critique of "elite" gatekeepers of "legitimate" tastes, cultural canons and institutions have their own heterogeneous (intellectual–political) sources: They come from (neo-)liberal and libertarian thinkers who make consumer choice the only legitimate paradigm of estimation; they come from anti-authoritarian leftists and radical democrats and from critics of identity-based privileges (of older, *white*, cis-gendered men, in particular). Just as the confluence of different processes that are referenced through these metanarratives has increased the vehemence and impact of cultural anti-elitism, so does the convergence of these radically different arguments in the realm of normative discourse. Both, however, are important. Taken together, they also lead to new conflicts over the meanings, effects and legitimacy of anti-elitist articulations. The former makes their consequences more difficult to surmise, the latter should complicate facile judgements.

Observers of anti-elitism observed

What are the consequences of such cultural processes and struggles for political matters in a more restricted sense? How do they converge, resonate and interfere? Given the difficulty of providing general answers to questions of this sort, we want to take a step back in this third section of the introduction and shift the order of observation by explicitly observing other observers. By doing so, we ask a seemingly simpler question: How were these questions and interactions discussed in early attempts at diagnosing the current series of crises?

In response to recent political crises, an explanatory discourse about the interplay of the political, economic and cultural dynamics of anti-elitism with a limited number of themes, subject positions and expectable utterances has built up. Without getting overly technical, we want to highlight nine of its strands. We do this in order to give a quick overview of the state of the debate about resonances and interactions between different forms of anti-elitism and illustrate the need for more integrated perspectives.

For this purpose, we first need to have a quick look at the time after the banking and financial collapse around 2007, arguably the beginning of recent crisis cycles in the US and Europe and certainly a cause of a later *malaise*. In the aftermath and during the Occupy and other place occupation protests, critical analysts with an intellectual background in hegemony theory (see, *inter alia*, Bader et al., 2011; Candeias, 2011; Demirović, 2011, p.65, 2013; Hall, 2011, 2015) downplayed the role of culture as a factor in the escalation of the crisis. This is the first strand we want to mention. According to these writings, there was clearly an economic crisis, for which – to the extent that they can be personalised – elites were responsible, and there was ongoing ecological collapse and a crisis of social reproduction, and, hence, a "multiple crisis". But there was, as of yet, relative stability in everyday life, politics and the state. Writing about Germany and the UK, respectively, Alex Demirović and Stuart Hall stated in 2011 that, so far, no real crisis of hegemony had occurred, even if there were "moments of a crisis of legitimation, the political crisis, and the state crisis" (Demirović, 2011, p.74). Hall, who – unlike other hegemony theorists – paid significant attention to the cultural in his analyses of a 30-year-long "neo-liberal conjuncture" that culminated in the banking and fiscal crisis, suggested that everyday consciousness and pop culture were to be seen primarily as conservative forces at this point. They stopped an actual hegemonic crisis from emerging, as they were thoroughly imbued, for example, with the ideology of consumption, profit-seeking and meritocracy. Countertendencies received relatively scant attention in these diagnoses.[27] Research on cultures of participation, democratisation, the decline of deference and so forth and conjunctural analyses of economic and political crises remained separate.

Since then, the overall crisis deepened and shape-shifted. The cross-domain ascent of anti-elitism was part of this process. Commentators in leading Western media outlets quickly picked up on anti-elite dynamics during Trump's rise and the populist wave that it was part of, offering initial interpretations, focusing mostly on the political side of this broader theme but asking about its cultural aspects and potentially root causes as well. In many cases, attempts at making sense of seemingly similar developments in other countries took this (initially US-based) discourse as a starting point. The ways in which different media – newspaper articles and op-ed pieces, blog and social media (particularly Twitter) posts, academic journal texts, electoral campaign communications and so forth – came together and involved lay people, such as Twitter users and podcasters, and "legacy" media, was new and exciting and in itself part of these cultural shifts.

One strand that emerged in this context – the second one we want to mention here – was concerned primarily with the historical contextualisation of anti-elite rhetoric. Historian Beverly Gage, for example, explained in the *New York Times* "How 'elites' became one of the nastiest epithets in American politics" (2017), tracing anti-elite rhetoric back to the Founding Fathers in the 18th century, making the general point that there is a close and positive connection between anti-elite impulses, democracy and popular sovereignty. She argued that the decisive turning point in anti-elite discourse in the US had occurred during the 1990s when "bashing 'the liberal elite' had become a favorite blood sport of the American right" – even if similar utterances could also, if less frequently, be found in the 1970s in the context of the "silent majority" discourse.

In a third, particularly influential strand of the explanatory discourse, the term "elite" is given a wide, vaguely sociological definition. Christopher Lasch's politically ambiguous, communitarianist jeremiad "The Revolt of the Elites and the Betrayal of Democracy" – reportedly a favourite book of Steven Bannon and other leading figures of the alt-right, but also many liberals and leftists (Lehmann, 2017) – had been published in 1996. It cemented the link between the term "elites" and the wider strata of the professional–managerial or "knowledge classes", "all those professions that produce and manipulate information" (Lasch, 1996, p.5) with their apparent tendencies to self-isolate among themselves in suburbs and gentrified urban neighbourhoods, fall for new forms of consumer and lifestyle distinction, and their cosmopolitan self-image. The link between this group – actually, in our view, too heterogenous a formation to really be called a "group" – and the term "elite" had been much less self-evident before that time.[28] This relatively new usage picked up on actual changes in class structures, such as the increasing importance of middle-class professions and the people who hold these positions, and the concomitant decline of working-class jobs and the recognition they used to command. It presented them in polemical, accusatory narratives of decline and gave them a very specific structure.

In this sociologising strand, the "elite" – again, conceived of much more broadly than in earlier decades – is contrasted with popular antagonists whose anti-elite sentiments are explained economically and culturally. This leads back to the question, briefly raised above, why Trump appealed to working-class voters. Trump's son Donald Jr., also a contributor to and commentator on these debates, had rhetorically solved the issue simply by presenting his father as a "working-class billionaire" (see Hall, Goldstein, and Ingram, 2016, p.71). In an influential piece of Trump explanation in the *Harvard Business Review*, legal scholar Joan C. Williams (2016) made a similar point. She argued that in cultural terms, Trump was indeed closer to many in the (particularly *white*) working class than upper-middle-class professionals would like to believe: to their aspirations, tastes, sense of a good life and of what being a successful and admirable person meant, and also to their dislike and distaste for certain social types or figures whom they perceived as condescending and undeservedly privileged. Furthermore, of course, there was a material aspect to Trump's appeal as he also promised a revival of manufacturing and higher wages through tariffs – the rise of China, other competitive economic pressures and the expectation to be protected from them surely play a role (Rosenberg and Boyle, 2019) – and less competition on the labour market from immigrants. Quoting cultural sociologist Michèle Lamont's qualitative interview-based study *The Dignity of Working Men* from the mid-1990s, Williams pointed out that many working-class people generally resented professionals whose credentials are strongly based on educational degrees, i.e. teachers, doctors, lawyers or professors, more than they resented "deservedly" rich businesspeople (Williams, 2016; see also Lamont, Park, and

Ayala-Hurtado, 2017).[29] Exploiting this constellation, one basic technique of recent right-wing discourse has been a terminological slippage where the vaguely defined group of upper-middle-class, college-educated and credentialed professionals (or, in a slightly different terminology, the knowledge class or professional–managerial class, see Ehrenreich and Ehrenreich, 2012; Graeber, 2014), and particularly those among them who identify with a politically "progressive" worldview, merges with the elite in the sense of the "power elite", well-connected billionaires, the Davos set.[30] This slippage reliably puzzles and enrages progressives and the Left. It is both a result of successful political–ideological work of the organised right *and* based on a relatively long-standing "structural" enmity that many people – across different social classes – sense towards people who may not be part of the ruling class or elite in a stricter sense of the term but who, nonetheless, reap some of the benefits of the current economic order, administer it and also wield forms of cultural domination that they tend to underestimate and deny, as artist Andrea Fraser (2018) put it in one of the relatively few texts written in a mode of leftist self-critique.[31] A few years earlier, anthropologist David Graeber had gone so far as to claim that "members of the professional-managerial classes themselves – who typically inhabit the top fifth of the income scale" were "the traditional enemies of the working classes" (Graeber, 2014). Quoting older work by political activist and author Michael Albert, Graeber claimed that

> actual members of the working classes have no immediate hatred for capitalists because they never meet them; in most circumstances, the immediate face of oppression comes in the form of managers, supervisors, bureaucrats, and educated professionals of one sort or another

and that "members of the working class (or, in America and Europe at least, the *white* working class) have become increasingly prone to identify, out of sheer rejection of the values of the professionals and administrators, with the populist right" (Graeber, 2014, p.77).[32] Such a strong causal connection ("out of sheer rejection"), or, at least, its immediacy, is debatable – there is little hard evidence for it. Claims like these also lack differentiation – it is unlikely, for example, that managers and supervisors in different industries by default embody progressive–liberal values. Furthermore, the borders of the term (are teachers part of the professional–managerial class?) are difficult to define as well. However, acknowledging these complications should not lead to a facile denial of the overall problem of cultural and economic domination through different fractions of the middle and upper classes and their association with progressive and left politics, which remains virulent either way.

This, then, was a third major discursive strand. A fourth emerged through the work of more positivist-inclined political scientists with quantitative methodologies who began to measure the geographic spread of what they called "anti-elite parties" (Marx and Nguyen, 2018), "anti-elite rhetoric" (Curini, 2019) or the "anti-elite" and "anti-European vote" (Ferrante and Pontarollo, 2020) and correlated them with the usual variables of psephology.

In doing so, they built on authors such as Cas Mudde or Jan Werner Müller and an older tradition of populism research in which anti-elitism is a defining feature of that political style and related movements and ideologies (Mudde, 2004). This produced some interesting insights, but – at least as seen from the perspective of our undertaking – it also remained limited to a fairly narrow sense of (electoral) politics. More problematically, it also reified anti-elitism into an evident-seeming concept or variable that supposedly stood for something clearly definable out there in the world, and often built on hypotheses that were derived from and embedded in debatable diagnostic narratives.[33] Just as the question what kind of "thing" populism really is – an ideology, a style, a strategy, a political logic – remains hotly debated in populism studies, anti-elitism was conceptualised in different ways in studies about such patterns as well: Curini, for example, terms it a "non-policy vote-winning strategy" (2019) that attracts voters of otherwise very different persuasions. Others conceptualise it as a specific style or as part of a ("weak") ideology that can articulate with more specific left or right ideologies. However, the different implications of these definitions are hardly spelled out. "Elite" quickly becomes a quasi-common sense, seemingly self-evident term in these contexts. Prominent political scientists and "cultural cleavage" theorists (Koopmans and Zürn, 2019; Merkel and Zürn, 2019) even designed a survey study so as to directly compare "elite" and "mass" opinion.[34] In doing so, they tend to make the cultural cleavage they seek to prove – in their case: communitarian masses, cosmopolitan elites – seem self-evident by design, presupposing these categories and examining one data set for each of the two, "elite" and "mass", as if these were unproblematic sociological categories.

Academic and feuilletonistic observations of the anti-elite moment also had other central themes. We want to briefly shift the focus to German-speaking countries, our own primary context, where this was initially, to a large extent, an imported and recontextualised discussion (e.g. see the contributions in Geiselberger, 2017). Here, one prominent strand of the debate in reaction to worldwide anti-elite rhetoric in politics – the fifth in our list – was primarily defensive: Liberal-conservative authors, such as political scientist and geopolitical strategist Herfried Münkler (2018) or philosophy journalist Wolfram Eilenberger (2018), quickly stepped in to defend "the elites", the irreducible complexity of functionally differentiated society, the necessity of specialised expertise, the benefits brought to society by great achievers and so forth, in newspapers and magazines such as *Neue Zürcher Zeitung* and *Die Zeit*, in the popular pose of the contrarian but serious realist: against too much idealism, too much democracy, too much egalitarianism and equality. Mainstream Social Democrat Sigmar Gabriel – a centrist, corporatist, at times also populist – warned that anti-elitism from right *and* left would undermine democracy (Gabriel, 2018). *Philosophie Magazin* defensively asked "Do we need elites?" (*Brauchen wir Eliten?*) and collected suggestions on how "the legitimation crisis" of elites could be overcome – the elites should do better, then dangerous anti-elitists would go away.[35]

In Germany, too, there were also a few authors and political strategists who – a sixth strand – suggested and also tried to direct anti-elite energies in a different, left-populist direction. Sanders, Mélenchon, Corbyn and Podemos had at least been able to gather and gain some of the ground usually occupied by the liberal centre. In Germany, however, the "*Aufstehen*" (rise up) campaign made few inroads, despite its very obvious attempt to benefit from anti-elite attitudes. In the aftermath of the 2015 "summer of migration", it became positioned particularly strongly against the pro-migration fraction within society and particularly within the Left party, which it attacked in the name of the welfare state and the national working class; some of its spokespeople, such as Bernd Stegemann and Sahra Wagenknecht, increasingly spent their time criticising the Merkel government for its perceived pro-immigration stance, bashing "identity politics", "cancel culture" and what they took to be the loony left, primarily for right-wing audiences in newspapers like *Die Welt* (Ege and Gallas, 2019; Slobodian and Callison, 2019). Political activists of very different backgrounds – primarily anti-racist, pro-migrant, anti-sexist – were labelled "elite". In the post-2015 backlash, right-wing activists had worked to merge the pro-migration position with the figure of an out-of-touch, privileged elite (in the sense discussed in the previous paragraphs). Many liberals and economically left-wing social democrats, both populist and more traditionally corporatist, followed suit. This resultant post-left political formation was a seventh strand, which branched off from the sixth (left populist). It is closely connected to the sociological explanation of anti-elite populism, the third strand mentioned above.

It was the radical right that had led the way. Alexander Gauland, a figurehead for the Alternative für Deutschland (AfD) and long a card-carrying member of the conservative political and media establishment, in an opinion piece in the *Frankfurter Allgemeine Zeitung* titled "Why it must be populism" (Gauland, 2018) appropriated the language of the Lasch-inspired anti-elite intellectual right-wing discourse from the US (or the *anywhere-somewhere* pop sociology from the UK, see Goodhart, 2017) and the types of arguments against anti-discrimination policy that became known in the US and elsewhere as the "anti-woke" position. He presented the AfD's overall project and strategy as a defence of both the traditional bourgeoisie and the lower middle and working class against "a new urban elite", a "globalist class" or "globalist elite", whose members lived in an aloof society of their own ("*abgehobene Parallelgesellschaft*") and looked down upon those with a strong sense of home (*Heimat*), locality and regional and national identity. They felt at home in London or Singapore as much as in Berlin, Gauland – or someone on his staff – wrote.

Critics pointed out that while the language of the "urban elite" and the "global elite" seemed borrowed from an international discourse, the gist of Gauland's anti-urban, anti-cosmopolitan argument was reminiscent of something very close to home, i.e. Hitler and Goebbels speeches from the Nazi era.[36] In an address to workers at a Siemens plant in 1933, for example, Hitler – not using the word "elite", but evoking its semantics – had hailed the ordinary German working people bound to their soil, the factory, the *Heimat* and the nation. These people needed a strong state for protection from global economic forces, he argued, whereas a "rootless international clique" was creating strife among the peoples of the world and had no need for a nation. His tropes and even the cities he mentions overlap with Gauland's:

> These are the people who are at home everywhere and nowhere but live in Berlin today, may well be in Brussels tomorrow, in Paris the day after, and then again in Prague or Vienna and London, and who feel at home everywhere.
>
> (translation M.E.)[37]

At this point, people in the audience yelled "Jews!", according to a transcript.[38] As political scientist and blogger Floris Biskamp (2019) noted, Gauland probably had not actually consulted Hitler's speech there, and, as always, the "argument *ad hitlerem*" (ibid.) had its limits. The tropes, however, had been and continued to be popular among conservative and reactionary writers much more broadly; they diffused into broader discourses, were repeated by the reactionary wing of would-be left populists and continued to reiterate anti-Semitic codes. Even if they were at best superficial as social science, they needed to be taken seriously as a political strategy. In that context, many German public intellectuals and politicians felt the need to line up in one of two camps, either progressive–liberal–cosmopolitan (and, rhetorically, anti-anti-elite) or national–communitarian (rhetorically anti-elite), accepting the way the field had been construed by Gauland and the like.[39]

At the same time, some analyses also provided narrower definitions of the "elite" or of different elite and elite-like groups and came to different conclusions. In doing so, they shifted the focus to other social domains than politics proper and the economy, especially to the realm of knowledge and its social and technological organisation. Some particularly insightful analyses – which for the sake of simplification we subsume as an eighth strand here – could be found in the UK. Social and political theorist Will Davies (2018) provided a diagnosis that also took into account "molecular" transformations in culture, technology and knowledge. Firstly, he argued that trust in politicians had declined particularly rapidly in the early 2010s, after the 2007–2009 crises and in the midst of their austerity aftermath and the places and Occupy movements. Sketching out an answer to the rhetorical question "why we stopped trusting the elites", particularly in the UK, Davies reminds us of the rational core of that scepticism as it was expressed in recent elite failures and misdeeds, particularly as they were uncovered through "leaks" of data to the press (see Bramall in this volume): from politicians' expense scandals, the discovery of long ongoing sexual abuse by celebrities, such as Jimmy Savile, corporate reporting scandals, LIBOR rate fixing, Volkswagen's emissions fraud or the WikiLeaks complex. There was famously little in terms of punishment of the perpetrators within the financial industry or its regulators and enablers.[40]

Popular distrust in elites, Davies points out, is closely related to a distrust in media reporting and conventional news about them. Rather than people believing in their credibility, "truth is now assumed to reside in hidden archives of data, rather than in publicly available facts" (Davies, 2018, unp.). As "the elites" are the only ones who are thought to be capable of hiding these reservoirs of truth, "suspicions of this nature – that the truth is being deliberately hidden by an alliance of 'elites' – are no longer the preserve of conspiracy theorists, but becoming increasingly common" (ibid.). Here, we return to a narrative of shifting regimes of knowledge and authority. The decline of trust in politics and politicians corresponds, thus, to a new regime of truth in Foucault's sense of the term: The conditions for believing that something can be true have changed. Davies's sketch of an explanation of anti-elitism's recent rise sees the latter as not primarily motivated in economic, cultural or political terms, but in relation to this regime of truth, media technologies and broader "diagrams" of knowledge and power that cross those domains. In that respect, he argues that anti-elite tendencies are part of long-running, epochal transformations in the organisation of knowledge – rather than limited to more contingent, superficial developments or, again, to political–economic transformations alone. This was also connected to a different make-up and structure of the people who had inherited the role of the "power elite", as C. Wright Mills had defined it in the 1950s (see Gerbaudo in this volume).

A ninth strand we want to mention here concerns this sense of shifting elites in a much narrower sense – and brings forth the argument that these shifts precipitated new forms of anti-elitism as well. Davies argued (2016) that "financial intermediaries" in banking and related industries, an important segment of the new super-rich, represented a novel type of elite: In comparison to their predecessors, they were much less preoccupied with their own cultural authority and with the normative legitimation of the social order in public discourse more broadly. These virtuosos of coding and deciphering data in a deregulated, digitally financialised environment "lose their extraordinary public status, and gain extraordinary profitability instead".[41] This is a very different "group" than corporate barons, but also than the broader professional–managerial class, much less recognisable through a sociology of lifestyles. In highlighting this, Davies's account stands for a broader literature that stresses the abstract and invisible character of contemporary economic elites. Building on the work of Italian post-operaist theorist Maurizio Lazzarato and Gilles Deleuze's classic opposition between disciplinary modern societies and a new society of control, Davies argued that today's most powerful elites are "post-juridicial" in that the systems upon which their power is built – financial markets and especially price-setting mechanisms arrived at through vast computer networks – in important ways operate *outside* of the realm of juridical norms, disciplinary apparatuses and even conscious reflection: "This elite inhabits and interprets an encoded semiotic system which derives from machines, rather than from political or juridical discourse" (ibid.). At its core, the power of these new elites is, therefore, neither disciplinary, nor about shaping subjectivities, nor about ideological or hegemonic consensus. Instead, it is "machinic", automatic and distributed.[42] Importantly, according to Davies, the critical presentation of new facts and scandalisations of elite misdeeds in the media cannot truly challenge this elite formation and its resources.

Conjunctural diagnoses and anti-elitism as an entry point

These strands have illustrated the ways in which the recent wave of anti-elitism has been made sense of in journalistic, essayistic and academic writing, and they have given a first overview of some particularly relevant positions and patterns. Through them, we have identified an initial range of definitions, interpretations and contextualisations of anti-elitism and anti-elite articulations. They are "diagnostic stories" (Grossberg, 2018, p.28): Narratives that are supposed to describe a malady and its aetiology. Such diagnostic stories attempt to make sense of a historical moment. They ask which kinds of symptoms stand for which kinds of conditions, what is fading away and what is emerging, what societal forces are gaining influence and why. Crucially, diagnostic stories such as these are inevitably also part of and contribute to the shape of that historical moment itself: They express the worldview and the ideological strategies of particularly "positioned" intellectuals, the ways in which they organise and produce knowledge and the groups they represent or want to represent or bring about. It has been said that ethnographies are always and necessarily "partial truths" (Clifford, 1986). The same is true for diagnostic and conjunctural narratives. The point in highlighting this is not to dwell on epistemic scepticism or to want to limit legitimate intellectual to the deconstruction of such narratives, a rather tired gesture at this point. Instead, it is a call to pay close attention to the performativity of such representations within broader hegemonic struggles and an overall war of position. Diagnostic stories and the figures that populate them become part of the common sense through which people make sense of their own and others' place in the world (Sutter, 2016). They contribute to shaping identities and selves, stereotypes of others ("the left-behind", "cosmopolitans"), affects (resentment, anger, concern, care), self-reflections ("am I really an elite? How can I be/not be one?" "they really think I'm deplorable?"), forms of mobilisation and other political strategies. A variety of agents use diagnostic stories to shape a conjuncture and its future and shift the balance of political forces. Morally loaded terms such as "betrayal", "abandonment" or "ignorance" (of ordinary people or the working class from the side of the elites) that find their way into such diagnostic stories are striking examples; the dividing up of the world into the old-fashioned and to-be-overcome and the "more advanced" and truly contemporary is another. Again, the point is not that this is necessarily wrong, the point is that it matters. *How* it matters remains to be spelled out in light of more specific situations and research questions.

Conjunctural analysis requires a concrete entry point into its object of analysis. Following this approach, we take anti-elitism as our entry point for a collection of independent but thematically interconnected chapters. These chapters consider how a wide range of anti-elite phenomena is connected to a range of contexts characteristic for the current conjuncture and its vectors of change. Thereby, the authors of the chapters suggest new readings and new diagnostic stories that, hopefully, are helpful for making sense of this conjuncture and gaining a better sense of how to act in it politically.[43] As explained above, anti-elitism is not strictly defined as one thing or another but heuristically understood as a conjunctural "theme" or motif. For that reason, it also cannot be sufficiently claimed as an object by any one disciplinary perspective, be it political science, cultural studies, sociology, media studies, anthropology or history. The challenge – both for this book and future research – is to explore these articulations, in a phenomenological sense, and to learn more about the ways in which these articulations interact, resonate and interfere, without, on the other hand, producing overly grand (and thereby analytically worthless) declarations about everything being connected to everything else.[44] The editors and many of the authors of the chapters come to these topics from the interdisciplinary cultural studies and conjunctural analysis tradition, be it more from an ethnographic or from a media studies angle. They therefore share, at least to some extent, a common analytical framework across their disciplines. In Clarke's contribution to this volume, some implications of this overall approach are spelled out in more detail (also see Hall, 2011; Clarke, 2014; Grossberg, 2018, 2019; Massey, 2018; Ege and Gallas, 2019; Gilbert, 2019b).[45] Conjunctural analysis requires empirical work, be it ethnographic-qualitative, historical, discourse-centred or quantitative. It requires further work to assemble them for broader analyses. This common conjunctural interest – an interest, briefly put, in the ways in which political, cultural, economic and other forces interact in specific historical situations – also allows distinct theoretical approaches in other respects. Bramall and Gerbaudo, in their chapters, rely particularly on Laclau and Mouffe's discourse theory, the former more strongly within a conjunctural framework, the latter in combination with a classic sociology of elites. The chapters by Hürtgen, Reznikova, Schwell, Weis, Eismann, Luthar, Schmidt-Lauber and Dümling bring other theoretical approaches into the mix, such as critical psychology and Marxist labour sociology, Frankfurt School critical theory, securitisation studies, feminist media studies, sociological systems theory and narratology. This is not to argue that these are completely distinct or even incompatible with conjunctural analysis but to stress that the book also implicitly includes a debate about appropriate concepts and theoretical approaches.

Can the accounts of anti-elitism presented in this book, then, be of a higher order of observation than the diagnostic stories we have highlighted above? Ultimately, we do not claim that these analyses are categorically on a different analytical plane than the texts they take as their data. We do not present a unified theory of anti-elite articulations – in our view, no such theory exists, and a combination of domain-specific approaches is more useful than an attempt at a grand synthesis.

Similarly, in his reflections on the Trump moment and the need for a new conjunctural analysis of the present, Lawrence Grossberg, a pioneering author in this line of thought, stated that

> Cultural Studies is made for moments where we don't know what's going on, and we don't yet know what theories, concepts and methods may enable us to find useful answers, or even to specify the questions. Profound changes with high stakes are taking place, and we cannot fuse the many struggles, contradictions and crises together into a neat, predefined totality or narrative.
>
> (Grossberg, 2018, p.35)

We also cannot fuse all the diagnostic narratives into a single, overarching one – for good reasons:

> All such stories define moments of unity (identities) and relations of difference: white vs. people of color, rural vs. urban, parochial vs. cosmopolitan, educated vs. ignorant, self-conscious vs. duped, open-minded vs. close-minded, good people vs. racists, reasonable vs. fanatical people, reason vs. emotion, etc. Such identities and relations are not illusory; they are real but contingent. Reality is an organized multiplicity (chaos) but any particular organization is neither necessary nor guaranteed. Conjunctural stories are expressions of and responses to the lived realities, struggles and crises of people's lives.
>
> (Grossberg, 2018, p.31)

Grossberg, thereby, highlights the grounded nature of all conjunctural diagnoses. Conjunctural analyses can only depict and narrate conjunctures in light of specific interests (i.e. concrete research questions and *Erkenntnisinteressen*: epistemic interest) and thought objects. The all-seeing position that construes a conjunctural totality will remain imaginary and inaccessible. This is the reason why, in theoretical terms, it is so difficult to *categorically* distinguish one conjunctural account from another – even if, of course, by all sorts of gradual measures, they can be better or worse, more or less supported by evidence, and worthy of defence or critique in light of their normative presuppositions and reasonings. In that sense, our accounts are on the same level as the ones we write about – even if they employ different strategies to gain analytical distance, will hopefully provide new insights, and the conclusions will be supported by transparent methods of data gathering and interpretation. Conjunctural analysis also has a practical, interventionist bent: the "commitment to politicising the conjuncture in the first instance is defined by cultural studies' project itself: to offer better knowledges, better understandings or narratives of the conjuncture in order to provide resources for changing the world", as Grossberg puts it (2018, p.45; see also Gilbert, 2019b). "Cultural studies completes its conjunctural analysis by entering into the struggle over whether and how to construct an organic crisis (and thus a conjunctural unity)" (Grossberg, 2018, p.54).

This also raises questions about geography and scale. It is no coincidence that we have chosen a geographically somewhat rambling approach in this introduction, moving from Trump and his analysts in the US to German newspaper articles to British reflections to the goings-on in many other countries. To put it slightly differently: If anti-elitism as an entry point can lead into analyses of specific situations in many different countries separately, then this fact alone *also* suggests that the phenomenon, the theme, points towards a conjuncture on a larger, "more global" scale and towards connected and common processes and forces. We mean this in two senses: Firstly, in the sense of existing connections between units that we otherwise consider separately (i.e. parallel processes and convergences; the collaboration between concrete agents, such as political circles and movements; the transnational – if strongly hierarchy-based – reach of technology and media, such as movies or Tweets or academic papers; the transfer and recontextualisations of, for example, political strategies and technologies, and arguments and narratives in style and in substance). Secondly, in the sense that larger blocs of countries can be seen as forming an interdependent conjuncture where there are relations of forces between a range of actors and forces, and developments in one area (e.g. China's increasing economic power, the long-standing primarily German hold over European fiscal policy, the Arab spring, movements of migration or even Korean pop-cultural influence) lead to anti-elite reactions in another. That being said, this book cannot do justice to all of these connections. The introduction has taken the US as a starting point and an over-proportional number of chapters focus on the UK and German-speaking countries. Some of the book's chapters will bypass the Anglo- and, to some extent, German-/Austrian-centric approach, inner-European differences will also be highlighted, but there are certainly large lacunae. While these considerations betray some serious limitations to which we must admit, they also connect – we hope productively – to the difficult methodological question regarding what kind of reach a particular conjuncture is assumed to have.

Post-2016: a new conjuncture?

At the end of this introduction, we return to the course of chronology. Around 2019, the anti-elite wave was starting to ebb in many countries. In the US, Trump's reversals, and his defeat in the election in 2020, could be read as signs of this. There, the anti-elite wave apparently crested somewhere between 2015 and 2018. Has that "moment" ended? If so, why? And is this a new conjuncture in a broader sense as well?

By 2020, it seemed as if the old regime was back, even if it promised real political changes, not only from Trump's time, but from the previous version of neo-liberalism as well (on the US, the UK and the Eurozone, see Watkins,

2021). There seemed to be no new wave of anti-elite "content" in popular culture overall either, even if, for example, films such as *Hillbilly Elegy* popularised the by now well-known diagnostic narratives about *white* working-class conservatism, and superhero movies continued to present political elites as clueless to the extreme. As a topic, however, anti-elitism had apparently lost the lustre of the new, even if there was still a baseline of anti-elitism and the overall semantics remain in place, ready to be used for new purposes. Progressives and the Left in many countries focused their political energies more strongly on anti-racist and -sexist politics, and on averting further climate change, and attacked the police, sexists of all classes and milieus, and an overall "society of externalization" (Lessenich, 2016) in which millions partake. These political struggles are critical of power structures of different kinds, but they are not primarily directed against "the elites" – by contrast, for example, to the left-populist election campaigns and mobilisations before.

Nevertheless, anti-elitism went into overdrive again in the last two years, in the late phase of the 2020 US election and among denialists about the coronavirus pandemic. What had smouldered subterraneously – the "Pizzagate" conspiracy theory, for example – turned into a firestorm. The QAnon complex, according to which "elite" cliques are secretly feasting on adrenochrome harvested from tortured children, received ever more popularity. With the coronavirus pandemic came newly visible medical elites and a backlash against them, but also against the Chinese, the pharmaceutical industry, philanthropical capitalists and the media. As we have shown in this introduction, between the mid-2010s (and, arguably, earlier from the 1990s) and around 2019, anti-elitism had in many cases turned primarily sociological: relatively large parts of society – the professional–managerial class, the knowledge class, whichever word one uses to signal them – were "the elites", according to widespread rhetoric and the discourse of which it forms part.[46] Now, a few years later, among anti-vaxxers and Trump "dead-enders", "the elite" again primarily appeared as a small, sinister clique pulling strings. It probably helped that there were prominent and very real cases, such as the Jeffrey Epstein saga, where "elites" were again shown to be evil and its members connected. However, this was not necessary for anti-elitism to turn full-throttle paranoid during the pandemic and in the "Covid conjuncture" (Means and Slater, 2021; Morley, 2021), where the "oligarchic plunder of public wealth" during the pandemic (Means and Slater, 2021, p.517) was much less debated than alleged plots by secret powers. Anti-Semitic dog whistles were in many cases replaced by straightforward anti-Semitic tirades.[47] These movements were closely intertwined with the radical right. The latter has its own complex and contradictory relationship to anti-elitism, given its belief in "true", natural elites (see Dümling and Gebhardt in this volume). But these movements also offered something else, a do-it-yourself epistemic tool-kit where those who were open to it could discover "the truth", to a large extent in messenger service group chats. Truth and – particularly among the anti-vaxxers – "love" would conquer alienation, abstraction and disease. In that sense, an emphatically anti-elit*ist* theme – primarily in a cultural and epistemic sense – shapes and powers these amorphous groups and formations just as much as older radical right ideology, with which it to some extent merges.

It seems, then, that anti-elite articulations are here to stay, and they will continue to take on new forms. Many structures remained remarkably stable in the economic and political order. Levelling and reconfiguration processes in the logic of digitalisation under capitalism/neo-liberalism are moving ahead continuously. In the terminology of hegemony theory, even if there had been something like a hegemonic crisis, at least in the US, and a "settlement" was coming undone, there was certainly no revolution, no "ruptural unity" (Althusser), even if some elements of capitalist globalisation are being recalibrated and regulations of international trade on a national level have again gained greater legitimacy, especially in relatively powerful countries (Slobodian, 2021; Watkins, 2021). Nevertheless, it seems safe to say that this was more than just a surface movement. On the political right, the anti-elite pattern has, in many cases, become indissoluble from the overall war of position. There was no return to the normalcy of centrist-leaning politics and a relatively broad consensus, be it on trade policy or the conditions for truth, but a deepening cleavage and polarisation, for which anti-elite rhetoric remains instrumental.

In our view, it is primarily the political Left that does not know what to do with the wide spectrum of anti-elite articulations at this point. During the left-populist wave, its potential power was acknowledged and made use of, but its dangers became increasingly manifest as well, as some would-be left populists, not least in Germany, moved ever more strongly towards resentment politics against so-called woke elites and nostalgia for earlier stages of capitalism. Furthermore, the strategy usually worked much better on the political right, as many aspiring left populists found out. Rhetoric – this rhetoric included – could not compensate for a weakness in organising and media access, for example, and many potential Left supporters and voters were turned off by populist exercises. Equally importantly, this is not only a tactical matter. The political analyses of many current Left movements have no real use for an explicit anti-elite mode. To some extent, this is because of the aforementioned dangers and the equally sound reason that Left movements have better and more complex analytical tools and analyses of the situation at hand. However, the Left seems partially hesitant to make use of anti-elite articulations, despite their power, for two reasons: Out of a fear of popular anger and because it explicitly or implicitly sees the formation of a new bloc that includes liberal-leaning centrist upper-middle-class milieus, and revolves around them, as the inevitable way forward. A progressive-egalitarian anti-elitism that embraces difference, a "cosmopolitism from below" and radical economic and ecological demands, seems out of reach. This might be even more dangerous, however, because it leaves the anti-elitist tool-kit, and also the promise of an egalitarianism that starts here and now, for others to use.

Notes

1 We will return to the concept "conjuncture" and conjunctural analysis as an approach below.

2 On the populist strategy, see, *inter alia*, Worsley (1969), Laclau (2005) and Kapferer and Theodossopoulos (2019).

3 We will not go deeper into voter analysis here. Trump's electorate included large numbers of well-off Republicans, but his win was also made possible by (primarily *white*) Rustbelt working-class voters, many of whom had voted for Obama before (see Grossberg, 2018; Karp, 2020).

4 The role played by industries or factions of capital and their representatives who were discontent with some elements of neo-liberalism will probably play a larger role in future political–economic analyses. Quinn Slobodian stresses that "the contemporary challenge to neoliberal globalization [...] is not simply a backlash from below; it is also a back-lash from above" (Slobodian, 2021, p.5), using the steel industry as his main case.

5 Because of these different meanings of the term, it seemed beside the point when, for example, Trump-sceptic conservative pundits tried to poke fun at his new elites that seemed worse than the old ones (Brooks, 2017), or when Republican primary candidate and US senator Ted Cruz called Trump "an elite", just like Hilary Clinton (Corasaniti, 2016). People knew Trump was rich; that was part of his appeal.

6 Leo Lowenthal (2015) documented similar usages in his classic analyses of the rhetoric of quasi-fascist "agitators", such as Charles Edward "Father" Coughlin in the US during the 1940s.

7 See Hall (1979, p.15, 1985).

8 US journalist Chris Hayes had published a book with the same title in 2012.

9 These recent anti-elite populist positions on the Left resonate, to some extent, with a longer tradition of left-wing politics and the democratic–majoritarian impulse ("for all", the "popular"). The class struggle, after all, almost inevitably targets the economic elite. The terms "ruling class" and "bourgeoisie" are, however, conceptually quite distinct. "Elite" has not been a prominent term in Marxism and many other left-wing theories. On the other hand, at least in many Western European countries, Left-wing anti-elite populism's tendency to speak in the name of "the people" rather than in the name of a class, social movement or other identity *and* naming "the elite" as the main antagonist is a relatively recent phenomenon. It is not only instinctual but also, in many cases, calculated and strategic (Stavrakakis, 2014; Stavrakakis and Katsambekis, 2014; Errejon and Mouffe, 2016). However, when, by 2020, Bernie Sanders had failed to secure the nomination as presidential candidate to Joseph Biden, Jeremy Corbyn was ousted by centrist Keir Starmer and movements such as Podemos lost their lustre, while others such as "*Aufstehen*" (rise up) in Germany failed to gain momentum, the "Left populist moment" seems to have passed.

10 In the French context, Macron's announcement in 2021 that he will close down – or significantly reform and rename – the École nationale d'administration, the school for elite civil servants (from which he also graduated) and politicians, seems like a public gesture towards meeting anti-elite demands.

11 The distinction between popular–democratic and authoritarian–populist approaches has been one way of articulating this (Hall, 1980; Grossberg, 2018, pp.5–6). See Reznikova in this volume for a critique.

12 More authoritarian traditions of the Left, be they Leninist or social democratic, proclaim universal-egalitarian values but see a strong necessity of (primarily party) cadres (or, in other strands, experts) and, thereby, implicitly a counter-elite.

13 This view of culture and the cultural is based primarily on Stuart Hall's "conjunctural" version of Gramscian hegemony theory (see below) and subsequent

positions in cultural studies and social and cultural anthropology that have attempted to strengthen the connections to related theoretical strands.

14 This implies that all politics is also in some way cultural. At the same time, to highlight cultural politics is not to be in denial of political economy (or its critique) or to pretend the tools of cultural analysis would suffice for understanding entire conjunctures. These kinds of reductionism (be they economistic or culturalist/ideologist) are counterproductive. The methodological point here is not to overstretch the concept, but to elaborate non-reductionistic analyses that, ideally, also reflect the historicity of their own terms. As culture is always in danger of being reified and used instrumentally for the construction of collective identities and differences, particularly in ethnic terms, it can make sense to use the concept cautiously, for example, by primarily employing it as an adjective or as the designation of something like a level or dimension, "the cultural", in order to stress the processual, non-homogeneous, diffuse nature of cultural processes. Any designation of culture as a specific domain is of a heuristic nature, as the economy and politics, for example, are always and necessarily also "cultural" (i.e. dependent upon meanings and everyday practices) and vice versa, and the distinction between these domains remains a historical and political struggle. They are, nevertheless, heuristically indispensable.

15 German-market advertisements for French-Romanian car maker Dacia were a case in point: They positioned lively, diverse families driving relatively inexpensive Dacias against a lifeless country club elite in need of conspicuous consumption.

16 We leave out popular music here – this would be too large a field (for instance, country music, punk and rap having developed quite distinct anti-elite vocabularies). While there are certainly examples of direct anti-elite texts and imagery (see, *inter alia*, the beginning of Bramall's chapter), our sense is that (a) anti-elite sentiments find different, less obvious forms of expression in contemporary popular music and (b) critiques of power and inequalities that are not primarily articulated as anti-elite struggles (for example, the "Black Lives Matter" movement) have become more relevant than anti-elite narratives in public statements by pop musicians.

17 Whereas Capra, his conservative personal politics notwithstanding, arguably produced very New Deal-friendly, progressively inclusionary films in those years, contemporary German films such as *Paracelsus* (G.W. Pabst, 1943) showed the much more dangerous, anti-Semitic side of anti-elitist affects at that time.

18 We put the term "white" in italics as a signal that it is not a self-explanatory designation of skin colour or biological "race", but a complex, historically somewhat variable (but nonetheless powerful) social construction.

19 This fits well with the pattern of austerity and gender representations outlined in Negra and Tasker (2014) and Davies and O'Callaghan (2017).

20 A similar story could be told about many other media and pop-cultural registers, such as the music video, where it was the electronic musician John Maus, now infamous for his presence at the January 2021 MAGA rally, who produced a widely received single in 2017 that carried the lyrics "I see the combine coming, I see the combine coming, It's gonna dust us all to nothing".

21 This, of course, cannot be wholly disentangled from economic conflicts and class struggles in a narrower sense – but it makes little sense to subsume the latter under the cultural, and the reverse strategy would be reductionist as well.

22 See the classic writings by John Fiske (2011), building on the works of Mikhail Bakhtin, Peter Stallybrass and Allon White and many others; Laura Kipnis's exemplary analysis on gender, sexuality and anti-elitist aesthetics (1992). Alex Niven (2012) has attempted an essayistic synthesis of cultural and political anti-elitism from the populist Left based on similar diagnoses. Upon closer inspection, of course, many seemingly anti-elite representations in the pop-cultural sphere are – like Trump's rhetoric – ambivalent: while the gaze upon "the

elites" may be critical and disarming, it turns out as envious and desiring as well. The rebel becomes the new tyrant, and the "moral of the story" is that power corrupts and there will always be rulers and ruled.

23 On the public's declining confidence in educational, media and medical institutions in the US, see Funk and Kennedy (2020). Trust in science and scientists, on the other hand, has remained constant since the 1970s – but is strongly polarised and politicised.

24 On declining trust, see snapshots such as Brenan (2020).

25 Maak (2016) recounts the mostly American phenomenon of "rolling coal": diesel trucks that are manipulated into emitting larger than normal amounts of thick black smoke in order to anger, for example, drivers of hybrid or electric cars – a conspicuous form of anti-moral anti-environmentalism.

26 Calls for a participatory paradigm also strongly shape academic research. There are obviously many ambiguities and opportunities in all of these domains. For a defence of the participation paradigm, see, *inter alia*, Carpentier, Duarte Melo and Ribeiro (2019).

27 This was different in the literature on commons/commoning, protests and new forms of solidarity, which usually had a much more optimistic bent but had less to say about the actual shape of the crisis.

28 See Lemann (1996): "Populists used to hate the rich, but now they hate the elite. This shift has made possible the migration of populism from the Democratic to the Republican Party." The debate also connects with older right-wing versions of the sociology of intellectuals and "new class" discourse, such as German sociologist Helmut Schelsky's anti-New Left book *Die Arbeit tun die anderen: Klassenkampf und Priesterherrschaft der Intellektuellen* (And the others do the work. The intellectuals' class rule and rule of the priests) (1975). The connections between the new right's anti-elite discourse and the intellectual history of the sociology of the middle class and the intellectuals are yet to be written.

29 The view of professionals as being clearly on the side of "them" rather than the working-class "us" is already mentioned by Richard Hoggart (1957). Williams, however, leaves out many ambiguities that the authors she cites highlight in an update on their previous research: "On one hand, working-class men in the 1990s often expressed respect for economic success, and when queried about possible heroes, a number mentioned Donald Trump due to their belief that "becoming rich" is proof of intelligence. At the same time, Lamont "[…] found that 75 percent of her respondents were critical of the morality of 'people above', who are perceived as too self-centred and ambitious, lacking in sincerity, and not concerned enough 'with people'" (Lamont, Park, and Ayala-Hurtado, 2017, p.162).

30 The latter could be defined in a more precise sense, as in the sociology and journalism of elites (Rothkopf, 2008), or in the style of conspiracy theories.

31 Andrea Fraser calls for a "reflexive resistance" where cultural producers on the political Left "recognize cultural capital [including and particularly their own, M.E. and J.S.], not only as a socially effective form of power but also as a form of domination, not only substantively, in its particular forms, but also structurally and relationally, in its distributions and through the social differences and hierarchies that it articulates and performs" (2018).

32 Graeber's point is part of a traditional anarchist critique of bureaucratic socialism that threatens to empower professionals and intellectuals rather than workers.

33 The list of anti-elite parties is debatable. The arguments are certainly of great interest: "the incentive to adopt a strong anti-elite stance grows as the ideological space separating one party from the other(s) shrinks", argues Curini (2019, p.1416). Marx and Nguyen show that "anti-elite rhetoric tends to reduce the gap between the poor and the rich" (Marx and Nguyen, 2018, p.935) in voting patterns.

34 In this research, "elites" basically refers to what is usually called functional elites; in concrete terms, a survey among "more than 1,600 occupants of leading positions across twelve societal sectors (politics, administration, justice, military and police, labour unions, finance and economy, other lobbyism, research, religion, culture, media and other civil society)" (Koopmans and Zürn, 2019, p.25). It is unclear whether this sample stands for the broader social groups that anti-elite discourse addresses and helps constitute. This, however, seems to be at least their implicit message to policymakers, political strategists and social analysts.

35 An American (and sarcastic-humoristic) version of this was articulated by P.J. O'Rourke in the (by now old-school) neo-conservative magazine *Weekly Standard* (2017).

36 https://twitter.com/znuznu/status/1048912907612934144

37 See Benz (2018). It then turned out that the text was also very close to some passages from an article in *Der Tagesspiegel* from two years before, during the apparent apex of the anti-elite conjuncture in the US. In that article, progressive author Michael Seemann had summarised the right-wing critique of elites and reflected upon its justification and the necessity for urban progressives to be self-critical (Seemann, 2016).

38 Speech held on November 10, 1933. Transcript by the film archive at Bundesarchiv Koblenz, available at http://www.filmarchives-online.eu/viewDetailForm?FilmworkID=aaa546b529f11070db80581 1df326094.

39 See Beyer and Wietschorke in this volume. A particularly important figure here was renowned political scientist Wolfgang Merkel, a director at *Wissenschaftszentrum Berlin für Sozialforschung*, who highlighted the communitarians-against-cosmopolitans cultural cleavage (Merkel, 2017; Merkel and Zürn, 2019).

40 Empirical sociologists of corporate elites, such as Michael Hartmann (2018), also came to the conclusion that corporate elites are difficult to visualise and represent (aside from images of celebrity wealth, etc.), as will become clearer in a few chapters of this book as well (see, *inter alia*, Bramall). A brilliant filmic exploration of these questions can be found, for example, in Gerhard Friedl's *Hat Wolff von Amerongen Konkursdelikte begangen?* (2004), a reflection on the invisibility of capital ownership in 20th-century Germany.

41 Davies highlights the contrast to "traditional professions (such as doctor, teacher, lawyer)", often termed "elites" in political discourse: They "retain their epistemological jurisdictions, but are no longer amongst the beneficiaries of capitalist expansion […]." (Davies, 2016, p.238).

42 In this regard, Davies's argument is reminiscent of Scott Lash's claims about a post-hegemonic phase of political and economic domination (2007). We remain sceptical about the epochal reach of such diagnoses.

43 Broadly speaking, in our view, there are two primary directions in conjunctural analysis. In the first paradigm, which is, on the face of it, the "critical" and "political–economic" one, it is particularly relevant to figure out strategies of powerful agents, be they within or outside the state – and, in this case, how anti-elite articulations connect with them. In the second, more "culturalist" one, cultural dynamics and the conjunctural nature of meanings and practices play a larger role. Matters of strategy from the centre of power (and also from self-reflexive counterhegemonic oppositional forces) take a back seat, to some extent, because they are assumed to have a more limited reach. These paradigms need not necessarily be in contradiction, but it is helpful to be able to distinguish them so as not to raise false expectations. Both are present in the book.

44 Similar stories can be told about other countries and parts of the world in recent years and decades as well: Islamists and Ottoman revivalists denouncing Kemalist, secular elites in Turkey; self-described ethno-nationalist illiberals denouncing liberal reformers, the European Union, communists and George Soros in Hungary or Poland; Bolsonaro and his ilk fighting leftists, intellectuals

and activists in Brazil; the Israeli right wing garnering much of the Sephardic and Mizrahi vote and that of recent Russian immigrants against the Ashkenazi establishment. Different kinds of Left anti-elitism dominated Latin American politics for much of the 2000s and proved highly influential on other continents as well.

45 Conjunctural analysis as a methodology or an approach flourished particularly in the late 1970s and 1980s in analyses of early neoliberalism (Hall, 1985; Hall et al., 2013). For this tradition, the study *Policing the Crisis. Mugging the State, and Law and Order* (Hall et al., 2013), a conjunctural analysis of 1970s Britain and its crises first published in 1978, remains an important resource and inspiration. After the 2008/2009 crash and the subsequent upheaval, a number of authors – some of whom had been working within this approach in the meantime – revived the concept and called for new conjunctural analyses, which was perhaps also indicative of a search for different forms of collaboration and cumulative work in cultural studies and critical political economy in the neo-liberal academy.

46 This, admittedly, has recently intensified in "anti-woke" polemics, primarily from the side of contrarian-conformist neo-communitarians.

47 German vegan chef turned conspiracy theorist and right-wing extremist Attila Hildmann (Callison and Slobodian, 2021) was a case in point.

Bibliography

Bader, P., Becker, F., Demirović, A. and Dück, J., 2011. Die multiple Krise – Krisendynamiken im neo-liberalen Kapitalismus. In: A. Demirović, J. Dück, F. Becker and P. Bader, eds. *Vielfachkrise im finanzmarktdominierten Kapitalismus*. Hamburg: VSA, pp. 11–28.

Baiocchi, G. and Ganuza, E., 2017. *Popular Democracy: The Paradox of Participation*. Stanford, CA: Stanford University Press.

Barney, D., Coleman, G., Ross, C., Sterne, J. and Tembeck, T., 2016. The participatory condition: an introduction. In: D. Barney, G. Coleman, C. Ross, J. Sterne and T. Tembeck, eds. *The Participatory Condition in the Digital Age*. Minneapolis: University of Minnesota Press (Electronic mediations, 51), pp. vii–xxxix.

Benz, W., 2018. Wie Gauland sich an Hitlers Rede anschmiegt. *Der Tagesspiegel* [online] 10 October. Available at: https://www.tagesspiegel.de/wissen/analyse-des-historikers-wolfgang-benz-wie-gauland-sich-an-hitlers-rede-anschmiegt/23166272.html [Accessed 13 April 2021].

Bikbov, A., 2021. Keep the city clean: the ambivalent ethics of ownership in urban routine and non-violent protest in Moscow. In: M. Ege and J. Moser, eds. *Urban Ethics. Conflicts about the "Good" and "Proper" Life in Cities*. London; New York: Routledge, pp. 243–260.

Biskamp, F., 2019. Rechter Ideologe und schlechter Soziologe. Alexander Gaulands Rede über Populismus und Demokratie gelesen als Theorie, Ideologie und politische Herausforderung. Available at: https://florisbiskamp.com/2019/02/11/rechter-ideologe-und-schlechter-soziologe-alexander-gaulands-rede-ueber-populismus-und-demokratie-gelesen-als-theorie-ideologie-und-politische-herausforderung/#more-517 [Accessed 13 April 2021].

Bourdieu, P., 1984. *Distinction: A Social Critique of the Judgement of Taste*. London; New York: Routledge.

Bourdieu, P., 1999. *The Weight of the World: Social Suffering in Contemporary Society*. Translated by A. Accardo. Stanford, CA: Stanford University Press.

Brenan, M., 2020. Americans remain distrustful of mass media. *Gallup* [online] 30 September. Available at: https://news.gallup.com/poll/321116/americans-remain-distrustful-mass-media.aspx [Accessed 18 April 2021].
Brooks, D., 2017. The Trump elite. Like the old elite, but worse! *The New York Times* [online] 24 March. Available at: https://www.nytimes.com/2017/03/24/opinion/the-trump-elite-like-the-old-elite-but-worse.html [Accessed 24 March 2017].
Caiani, M. and Padoan, E., 2020. Populism and the (Italian) crisis. the voters and context. *Politics*, online first, pp. 1–17.
Callison, W. and Slobodian, Q., 2021. Coronapolitics from the Reichstag to the Capitol. *Boston Review*[online] 12 January. Available at: http://bostonreview.net/politics/william-callison-quinn-slobodian-coronapolitics-reichstag-capitol [Accessed 11 April 2021].
Candeias, M., 2011. Interregnum – Molekulare Verdichtung und organische Krise. In: A. Demirović, J. Dück, F. Becker and P. Bader, eds. *Vielfachkrise im finanzmarktdominierten Kapitalismus*. Hamburg: VSA, pp. 45–62.
Carpentier, N., 2011. *Media and Participation: A Site of Ideological-democratic Struggle*. Bristol: Intellect.
Carpentier, N., Duarte Melo, A. and Ribeiro, F., 2019. Resgatar a participação: para uma crítica sobre o lado oculto do conceito. *Comunicação e Sociedade*, 36, pp. 17–35.
Castells, M., 2000. *The Rise of the Network Society*. 2nd ed. Oxford; Malden, MA: Blackwell Publishers.
Castells, M., 2015. *Networks of Outrage and Hope: Social Movements in the Internet Age*. 2nd edn. Cambridge: Polity Press.
Clarke, J., 2013. In search of ordinary people: the problematic politics of popular participation. *Communication, Culture & Critique*, 6 (2), pp. 208–226.
Clarke, J., 2014. Conjunctures, crises, and cultures. *Focaal*, 2014 (70), pp. 113–122.
Clarke, J. and Newman, J., 2017. "People in this country have had enough of experts": Brexit and the paradoxes of populism. *Critical Policy Studies*, 11 (1), pp. 1–16.
Clifford, J., 1986. Introduction: partial truths. In: J. Clifford and G.E. Marcus, eds. *Writing Culture*. Berkeley, CA: University of California Press, pp. 1–26.
Corasaniti, N., 2016. Ted Cruz ad says Donald Trump is an elite, just like Hillary Clinton. *The New York Times* [online] 21 April. Available at: https://www.nytimes.com/2016/04/22/us/politics/ted-cruz-ad.html [Accessed 10 April 2021].
Cramer, K., 2016. *The Politics of Resentment. Rural Consciousness in Wisconsin and the Rise of Scott Walker*. Chicago: University of Chicago Press.
Curini, L., 2019. The spatial determinants of the prevalence of anti-elite rhetoric across parties. *West European Politics*, 43 (7), pp. 1415–1435.
Davies, H. and O'Callaghan, C., 2017. Introduction: boom and bust? Gender and austerity in popular culture. In: H. Davies and C. O'Callaghan, eds. *Gender and Austerity in Popular Culture: Femininity, Masculinity & Recession in Film & Television*. London; New York: I.B. Tauris, pp. 1–19.
Davies, W., 2016. Elite power under advanced neoliberalism. *Theory Culture and Society*, 34 (5–6), pp. 227–250. Available at: http://research.gold.ac.uk/18744/ [Accessed 13 March 2017].
Davies, W., 2018. Why we stopped trusting elites. *The Guardian* [online] 29 November. Available at: https://www.theguardian.com/news/2018/nov/29/why-we-stopped-trusting-elites-the-new-populism [Accessed 2 December 2018].
Demirović, A., 2011. Ökonomische Krise – Krise der Politik? In: A. Demirović, J. Dück, F. Becker and P. Bader, eds. *Vielfachkrise im finanzmarktdominierten Kapitalismus*. Hamburg: VSA, pp. 63–78.

Demirović, A., 2013. Multiple Krise, autoritäre Demokratie und radikaldemokratische Erneuerung. *PROKLA. Zeitschrift für kritische Sozialwissenschaft*, 43 (171), pp. 193–215. doi: 10.32387/prokla.v43i171.266.

Ege, M., 2011. Carrot-cut jeans: an ethnographic account of assertiveness, embarrassment and ambiguity in the figuration of working-class male youth identities in Berlin. In: D. Miller and S. Woodward, eds. *Global Denim*. Oxford: Berg, pp. 159–180.

Ege, M., 2013. *"Ein Proll mit Klasse": Mode, Popkultur und soziale Ungleichheiten unter jungen Männern in Berlin*. Frankfurt: Campus Verlag.

Ege, M. and Gallas, A., 2019. The exhaustion of Merkelism: a conjunctural analysis. *New Formations*, 96 (96), pp. 89–131.

Ehrenreich, B. and Ehrenreich, J., 2012. Death of a yuppie dream. The rise and fall of the professional-managerial class. Rosa Luxemburg Foundation New York Office [online] Available at: https://www.rosalux.de/fileadmin/rls_uploads/pdfs/sonst_publikationen/ehrenreich_death_of_a_yuppie_dream90.pdf [Accessed 31 May 2021].

Eilenberger, W., 2018. Elite: Asozial, autonom, autark! *ZEIT Online*, 24 September. Available at: https://www.zeit.de/kultur/2018-09/elite-exzellenz-gesellschaftliche-bedeutung [Accessed 8 April 2021].

Errejon, I. and Mouffe, C., 2016. Constructing a new politics. *Soundings*, (62), pp. 43–56.

Farías, I., 2020. Für eine Anthropologie des Urbanismus. Ethnographisch Städte bauen. *Zeitschrift für Volkskunde*, 2020 (2), pp. 171–192.

Fenster, M., 1988. Country music video. *Popular Music*, 7 (3), pp. 285–302.

Ferrante, C. and Pontarollo, N., 2020. Regional voting dynamics in Europe: the rise of anti-elite and anti-European parties. *Environment and Planning A: Economy and Space*, 52 (6), pp. 1019–1022.

Fisher, M., 2013. Remember who the enemy is. *k-punk* [online], 25 November. Available at: https://k-punk.org/remember-who-the-enemy-is/ [Accessed 21 May 2021].

Fiske, J., 2011. *Reading the Popular*. 2nd ed. London; New York: Routledge.

Fraser, A., 2018. Artist writes no. 2 toward a reflexive resistance, *X-TRA*, 20 (2). Available at: https://www.x-traonline.org/article/artist-writes-no-2-toward-a-reflexive-resistance/ [Accessed 10 June 2019].

Fraser, N., 2017. From progressive neo-liberalism to Trump – and beyond. *American Affairs Journal*, 1 (4). Available at: https://americanaffairsjournal.org/2017/11/progressive-neoliberalism-trump-beyond/ [Accessed 9 April 2021].

Friedman, S., Savage, M., Hanquinet, L. and Miles, A., 2015. Cultural sociology and new forms of distinction. *Poetics*, 53, pp. 1–8. doi: 10.1016/j.poetic.2015.10.002.

Fuchs, C., 2017. *Social Media: A Critical Introduction*. London: SAGE Publications Ltd. doi: 10.4135/9781446270066.

Funk, C. and Kennedy, B., 2020. Public confidence in scientists has remained stable for decades. *Pew Research Center Fact Tank* [online] 27 August. Available at: https://www.pewresearch.org/fact-tank/2020/08/27/public-confidence-in-scientists-has-remained-stable-for-decades/ [Accessed 16 May 2021].

Gabriel, S., 2018. Gaulands Plädoyer ist im Kern antidemokratisch. *Der Tagesspiegel* [online] 10 October. Available at: https://www.tagesspiegel.de/politik/populismus-beitrag-in-der-faz-gaulands-plaedoyer-ist-im-kern-antidemokratisch/23166172.html [Accessed 10 April 2021].

Gage, B., 2017. How "Elites" became one of the nastiest epithets in American politics. *New York Times Magazine*. Available at: https://www.nytimes.com/2017/01/03/magazine/how-elites-became-one-of-the-nastiest-epithets-in-american-politics.html [Accessed 10 April 2021].

Gauland, A., 2018. Fremde Federn: Warum muss es Populismus sein? *Frankfurter Allgemeine* [online] 6 October. Available at: https://www.faz.net/1.5823206 [Accessed 10 April 2021].

Geiselberger, H., ed., 2017. *Die grosse Regression: eine internationale Debatte über die geistige Situation der Zeit*. 2nd ed. Berlin: Suhrkamp.

Gerbaudo, P., 2018. Social media and populism: an elective affinity? *Media, Culture & Society*, 40 (5), pp. 745–753.

Gilbert, J., 2019a. Das Kulturelle in politischen Konjunkturen. *Zeitschrift für Kulturwissenschaften*, (2), pp. 104–114, 128–136.

Gilbert, J., 2019b. This conjuncture: for Stuart Hall. *New Formations*, 96 (96), pp. 5–37.

Goodhart, D., 2017. *The Road to Somewhere: The Populist Revolt and the Future of Politics*. London: Hurst & Company.

Graeber, D., 2014. Anthropology and the rise of the professional-managerial class. *HAU: Journal of Ethnographic Theory*, 4 (3), pp. 73–88.

Grossberg, L., 2018. *Under the Cover of Chaos: Trump and the Battle for the American Right*. London: Pluto Press.

Grossberg, L., 2019. Cultural studies in search of a method, or looking for conjunctural analysis. *New Formations*, 96 (96), pp. 38–68.

Guilluy, C., 2015. *La France périphérique: comment on a sacrifié les classes populaires*. Paris: Flammarion.

Guilluy, C., 2019. *Twilight of the Elites: Prosperity, the Periphery, and the Future of France*. Translated by M.B. DeBevoise. New Haven: Yale University Press.

Hall, G., 2008. *Digitize this Book! The Politics of New Media, or Why We Need Open Access Now*. Minneapolis: University of Minnesota Press (Electronic mediations, 24).

Hall, G., 2021. *A Stubborn Fury: How Writing Works in Elitist Britain*. London: Open Humanites Press. Available at: http://www.openhumanitiespress.org/books/titles/a-stubborn-fury/ [Accessed 18 April 2021].

Hall, K., Goldstein, D.M. and Ingram, M.B., 2016. The hands of Donald Trump: entertainment, gesture, spectacle. *HAU: Journal of Ethnographic Theory*, 6 (2), pp. 71–100.

Hall, S., 1979. The great moving right show. *Marxism Today*, (January), pp. 14–20. Available at: https://f.hypotheses.org/wp-content/blogs.dir/744/files/2012/03/Great-Moving-Right-ShowHALL.pdf [Accessed 31 May 2021].

Hall, S., 1980. Popular democratic vs. authoritarian populism: two ways of taking democracy seriously. In: A. Hunt, ed. *Marxism and Democracy*. London: Lawrence and Wishart, pp. 157–185.

Hall, S., 1985. Authoritarian populism: a reply. *New Left Review*, (151), pp. 115–124.

Hall, S., 2011. The neo-liberal revolution. *Cultural Studies*, 25 (6), pp. 705–728.

Hall, S., ed., 2015. *After Neoliberalism? The Kilburn Manifesto*. London: Lawrence and Wishart (Soundings).

Hall, S., Critcher, C., Jefferson, T., Clarke, J. and Roberts, B., 2013. *Policing the Crisis: Mugging, the State and Law and Order*. London: Macmillan International Higher Education.

Hall, S. and O'Shea, A., 2013. Common-sense neoliberalism. *Soundings*, 55, pp. 9–25.

Hartigan, J., 1997. Unpopular culture: the case of "white trash". *Cultural Studies*, 11 (2), pp. 316–343.
Hartmann, M., 2018. *Die Abgehobenen: wie die Eliten die Demokratie gefährden*. Frankfurt; New York: Campus Verlag.
Hayes, C., 2012. *Twilight of the Elites: America after Meritocracy*. 1st ed. New York: Crown Publishers.
Hochschild, A.R., 2016. *Strangers in Their Own Land: Anger and Mourning on the American Right*. New York: New Press.
Hofstadter, R., 1964. The paranoid style in American politics. *Harper's Magazine*, (November), pp. 77–86.
Hoggart, R., 1957. *The Uses of Literacy. Aspects of Working-class Life with Special References to Publications and Entertainment*. London: Chatto and Windus.
Holtz-Bacha, C., 2021. The kiss of death. Public service media under right-wing populist attack. *European Journal of Communication* 44 (320), pp. 221–237.
Jenkins, H., 2006. *Fans, Bloggers, and Gamers: Exploring Participatory Culture*. New York: New York University Press.
Johnston, J. and Baumann, S., 2015. *Foodies: Democracy and Distinction in the Gourmet Foodscape*. 2nd ed. New York: Routledge (Cultural spaces).
Kalb, D., 2009. Conversations with a Polish populist: tracing hidden histories of globalization, class, and dispossession in postsocialism (and beyond). *American Ethnologist*, 36 (2), pp. 207–223.
Kapferer, B. and Theodossopoulos, D., 2019. Introduction: populism and its paradox. In: B. Kapferer and D. Theodossopoulos, eds. *Democracy's Paradox. Populism and Its Contemporary Crisis*. New York; Oxford: Berghahn (Critical Interventions. A Forum for Social Analysis), pp.1–34.
Karp, M., 2020. Bernie Sanders's five-year war. *Jacobin* [online] 28 August. Available at: https://jacobinmag.com/2020/08/bernie-sanders-five-year-war [Accessed 8 September 2020].
Kelty, C.M., 2008. *Two Bits: The Cultural Significance of Free Software*. Durham, NC: Duke University Press (Experimental futures).
Kelty, C.M., 2019. *The Participant: A Century of Participation in Four Stories*. Chicago; London: The University of Chicago Press.
Kipnis, L., 1992. (Male) Desire and (female) disgust: reading hustler. In: L. Grossberg, C. Nelson and P. Treichler, eds. *Cultural Studies*. London; New York: Routledge, pp. 373–391.
Koopmans, R. and Zürn, M., 2019. Cosmopolitanism and communitarianism – How globalization is reshaping politics in the twenty-first century. In: P. de Wilde, R. Koopmans, W. Merkel, O. Strijbis and M. Zürn, eds. *The Struggle Over Borders: Cosmopolitanism and Communitarianism*. Cambridge: Cambridge University Press, pp.1–34.
Kruse, M., 2018. *Trump Reclaims the Word "Elite" with Vengeful Pride*, *POLITICO Magazine* [online] November/December. Available at: https://politi.co/2EVyEsG [Accessed 15 March 2021].
Laclau, E., 2005. *On Populist Reason*. London; New York: Verso.
Lamont, M., Park, B.Y. and Ayala-Hurtado, E., 2017. Trump's electoral speeches and his appeal to the American white working class. *British Journal of Sociology*, 68, pp. 153–180.
Lasch, C., 1996. *The Revolt of the Elites and the Betrayal of Democracy*. First published as a Norton paperback. New York: W.W. Norton & Company.
Lash, S., 2007. Power after hegemony: cultural studies in mutation? *Theory, Culture & Society*, 24 (3), pp. 55–78.

Lehmann, C., 2017. The betrayal of democracy. *The Baffler* [online] 13 March. Available at: https://thebaffler.com/latest/revolt-lasch-bannon-lehmann [Accessed 11 April 2021].

Lem, W., 2020. Notes on militant populism in contemporary France: contextualizing the gilets jaunes. *Dialectical Anthropology*, 44 (4), pp. 397–413.

Lemann, N., 1996. A cartoon elite. *The Atlantic* [online] November. Available at: https://www.theatlantic.com/magazine/archive/1996/11/a-cartoon-elite/376719/ [Accessed 10 June 2019].

Lessenich, S., 2016. *Neben uns die Sintflut: die Externalisierungsgesellschaft und ihr Preis*. München: Hanser Berlin, Carl Hanser Verlag.

Lowenthal, L., 2015. *False Prophets. Studies on Authoritarianism*. London; New York: Routledge.

Maak, N., 2016. Ganz unten. *Frankfurter Allgemeine Sonntagszeitung*, 20 (11), pp. 45–46.

Maasen, S. and Weingart, P., eds., 2008. *Democratization of Expertise? Exploring Novel Forms of Scientific Advice in Political Decision-making*. Dordrecht; London: Springer.

Magaudda, P., 2020. Populism, music and the media. The Sanremo Festival and the circulation of populist discourses. *Partecipazione e Conflitto*, 13 (1), pp. 132–153.

Mamonova, N.; Franquesa, J. 2020. Populism, Neoliberalism and Agrarian Movements in Europe. Understanding rural support for right-wing politics and looking for progressive solutions. *Sociologia Ruralis*, 60 (4), pp. 710–731.

Manow, P., 2018. *Die politische Ökonomie des Populismus*. 1st ed. Berlin: Suhrkamp.

Marx, P. and Nguyen, C., 2018. Anti-elite parties and political inequality: how challenges to the political mainstream reduce income gaps in internal efficacy. *European Journal of Political Research*, 57 (4), pp. 919–940.

Massanari, A.L., 2014. *Participatory Culture, Community, and Play Learning from Reddit*. New York: Peter Lang Inc., International Academic Publishers. Available at: https://www.peterlang.com/view/product/30949?format=EPDF (Accessed 18 April 2021).

Massey, D., 2018. The Soundings Conjuncture Projects: the challenge right now. In: J. Henriques and D. Morley, eds. *Stuart Hall: Conversations, Projects, and Legacies*. London: Goldsmiths Press, pp. 64–71.

Means, A.J.; Slater, G.B., 2021. Collective disorientation in the pandemic conjuncture. *Cultural Studies* 35 (2–3), pp. 514–522.

Merkel, W., 2017. Kosmopolitismus versus Kommunitarismus: Ein neuer Konflikt in der Demokratie. In: P. Harfst, I. Kubbe and T. Poguntke, eds. *Parties, Governments and Elites*. Wiesbaden: Springer, pp. 9–23.

Merkel, W. and Zürn, M., 2019. Conclusion: the defects of sosmopolitan and communitarian democracy. In: P. de Wilde, R. Koopmans, W. Merkel, O. Strijbis and M. Zürn, eds. *The Struggle over Borders: Cosmopolitanism and Communitarianism*. Cambridge: Cambridge University Press, pp. 207–237.

Miller, D. and Venkatraman, pp., 2018. Facebook interactions: an ethnographic perspective. *Social Media + Society*, 4 (3).

Morley, D., 2021. In a viral conjuncture. *Cultural Politics*, 17 (1), pp. 17–27.

Morozov, E., 2012. *The Net Delusion: How Not to Liberate the World*. London: Penguin Books.

Mudde, C., 2004. The populist zeitgeist. *Government and Opposition*, 39 (4), pp. 541–563.

Müller, O., Sutter, O. and Wohlgemuth, S., 2020. Learning to LEADER. Ritualised performances of "participation" in local arenas of participatory rural governance. *Sociologia Ruralis*, 60(1), pp. 222–242.

Münkler, H., 2018. Elitenkritik ist zum Volkssport geworden. Dabei erweist sie sich als wohlfeile Spiegelfechterei. Die Konsequenzen trägt – das Volk | NZZ. *Neue Zürcher Zeitung* [online] 14 June. Available at: https://www.nzz.ch/feuilleton/elitekritik-ist-zum-volkssport-geworden-dabei-erweist-sie-sich-als-wohlfeile-spiegelfechterei-die-konsequenzen-traegt-das-volk-ld. 1394211 [Accessed 14 June 2018].

Narotzky, S., 2019. Populism's claims: the struggle between privilege and equality. In: B. Kapferer and D. Theodossopoulos, eds. *Democracy's Paradox: Populism and the Contemporary Crisis*. New York; Oxford: Berghahn (Critical interventions), pp. 97–121.

Negra, D. and Tasker, Y., 2014. Introduction: gender and recessionary culture. In: D. Negra and Y. Tasker, eds. *Gendering the Recession: Media and Culture in an Age of Austerity*. Durham, NC: Duke University Press, pp.1–30.

Newman, J. and Clarke, J., 2018. The instabilities of expertise: remaking knowledge, power and politics in unsettled times. *Innovation: The European Journal of Social Science Research*, 31 (1), pp. 40–54.

Nguyen, C.T., 2020. Echo chambers and epistemic bubbles. *Episteme*, 17 (2), pp. 141–161.

Niven, A., 2012. *Folk Opposition*. Winchester; Washington: Zero Books.

O'Rourke, P.J., 2017. The revolt against the elites. *The Weekly Standard* [online] 3 February. Available at: https://www.weeklystandard.com/pj-orourke/the-revolt-against-the-elites [Accessed 10 June 2019].

Ostiguy, P., 2017. Populism: a socio-cultural approach. In: C.R. Kaltwasser, P. Taggart, P. Ochoa Espejo and P. Ostoguy, eds. *The Oxford Handbook of Populism*. Oxford: Oxford University Press, pp. 73–98.

Pariser, E., 2011. *The Filter Bubble: What the Internet Is Hiding from You*. New York: Penguin Press.

Peterson, R.A. and Kern, R.M., 1996. Changing highbrow taste: from snob to omnivore. *American Sociological Review*, 61 (5), pp. 900–907.

Phelps, G., 1979. The "Populist" Films of Frank Capra. *Journal of American Studies* 13 (3), pp. 377–392.

Phillips, A., 2016. The Vulgar. In: S. McCarthy and J. Alison, eds. *The Vulgar: Fashion Redefined. Exhibition "The Vulgar: Fashion Redefined"*. London: Koenig Books, pp. 10–15.

Pied, C., 2021. Negotiating the Northwoods: anti-establishment rural politics in the Northeastern United States. *Journal of Rural Studies* 82 (2), S. 294–302.

Rodrik, D., 2021. Why does globalization fuel populism? Economics, culture, and the rise of right-wing populism. *Annual Review of Economics*, 13 (1), pp. 6.1–6.38.

Rogin, M.; Moran, K. 2003. Mr. Capra goes to Washington. *Representations*, 84 (1), pp. 213–248.

Rosenberg, J. and Boyle, C., 2019. Understanding 2016: China, Brexit and Trump in the history of uneven and combined development. *Journal of Historical Sociology*, 32 (1), pp. e32–e58.

Rothkopf, D., 2008. *Superclass: The Global Power Elite and the World They Are Making*. New York: Farrar, Straus and Giroux.

Rüther, T., 2021. Lesekreis Lockdown-Kritik. *Frankfurter Allgemeine Sonntagszeitung* [online] 17 May. Available at: https://www.faz.net/aktuell/feuilleton/buecher/themen/was-ist-aus-dem-literarischen-quartett-im-zdf-geworden-17341127.html [Accessed 31 May 2021].

Schelsky, H., 1975. *Die Arbeit tun die anderen: Klassenkampf und Priesterherrschaft der Intellektuellen*. 2nd ed. Opladen: Westdeutscher Verlag.

Seemann, M., 2016. Eine andere Welt ist möglich – aber als Drohung. *Der Tagesspiegel* [online] 25 October. Available at: https://www.tagesspiegel.de/politik/die-globale-klasse-eine-andere-welt-ist-moeglich-aber-als-drohung/14737914.html [Accessed 13 April 2021].

Seeßlen, G., 2017. *Trump! Populismus als Politik*. Berlin: Bertz + Fischer.

Sennett, R. and Cobb, J., 1977. *The Hidden Injuries of Class*. Cambridge: Cambridge University Press.

Shirky, C., 2009. *Here Comes Everybody: The Power of Organizing Without Organizations: [with an updated epilogue]*. New York; Toronto; London: Penguin Books.

Skogen, K. and Krange, O., 2020. The political dimensions of illegal wolf hunting: anti-elitism, lack of trust in institutions and acceptance of iilegal wolf killing among Norwegian hunters. *Sociologia Ruralis*, 60 (3), pp. 551–573.

Slobodian, Q., 2018. *Globalists: The End of Empire and the Birth of Neoliberalism*. Cambridge, MA: Harvard University Press.

Slobodian, Q., 2021. The backlash against neoliberal globalization from above: elite origins of the crisis of the new constitutionalism. *Theory, Culture & Society*, pp.1–19.

Slobodian, Q. and Callison, W., 2019. Pop-up populism: the failure of left-wing nationalism in Germany. *Dissent Magazine* [online] Summer. Available at: https://www.dissentmagazine.org/article/pop-up-populism-the-failure-of-left-wing-nationalism-in-germany [Accessed 11 April 2021].

Slobodian, Q. and Plehwe, D., 2020. Introduction. In: D. Plehwe, Q. Slobodian, and P. Mirowski, eds. *Nine Lives of Neoliberalism*. Brooklyn, NY: Verso, pp.1–18.

Srnicek, N., 2017. *Platform Capitalism*. Cambridge; Malden, MA: Polity (Theory redux).

Stallabrass, J., 2012. Elite art in an age of populism. In: A. Dumbadze and S. Hudson, eds. *Contemporary Art*. Hoboken, NJ: Wiley-Blackwell, pp. 39–49.

Stavrakakis, Y., 2014. The return of "the people": populism and antipopulism in the shadow of the European crisis. *Constellations*, 21 (4), pp. 505–517.

Stavrakakis, Y. and Katsambekis, G., 2014. Left-wing populism in the European periphery: the case of SYRIZA. *Journal of Political Ideologies*, 19 (2), pp. 119–142.

Stewart, K., 2007. *Ordinary Affects*. Durham, NC: Duke University Press. doi: 10.1215/9780822390404.

Susser, I., 2021. "They are stealing the state": Commoning and the Gilets Jaunes in France. In: M. Ege and J. Moser, eds. *Urban Ethics: Conflicts over the Good and Proper Life in Cities*. Abingdon; Oxon; New York: Routledge, pp. 277–294.

Sutcliffe-Braithwaite, F., 2018. *Class, Politics, and the Decline of Deference in England, 1968–2000*. Oxford; New York: Oxford University Press.

Sutter, O., 2016. Alltagsverstand. Zu einem hegemonietheoretischen Verständnis alltäglicher Sichtweisen und Deutungen. *Österreichische Zeitschrift für Volkskunde*, (1–2), pp. 42–70.

Walsh, R.A., 2014: Frank Capra. In: A. Kindell and E. Demers, eds. *Encyclopedia of Populism in America*. Santa Barbara: ABC-CLIO, pp. 109–111.

Watkins, S., 2021. Paradigm shifts. *New Left Review*, 128 (March/April).

Watson, J., 2019. Rural-urban imagery in country music video: identity, space, and place. In: L. Burns and S. Hawkins, eds. *The Bloomsbury Handbook of Popular Music Video Analysis*. London: Bloomsbury Academic, pp. 277–296.

Williams, J.C., 2016. What so many people don't get about the U.S. working class. *Harvard Business Review* [online] November 10. Available at: https://hbr.org/2016/11/what-so-many-people-dont-get-about-the-u-s-working-class [Accessed 13 March 2017].

Worsley, P., 1969. The concept of populism. In: G. Ionescu and E. Gellner, eds. *Populism: Its Meanings and National Characteristics*. London: Macmillan, pp. 212–250.

Zuboff, S., 2019. *The Age of Surveillance Capitalism: The Fight for a Human Future at the New Frontier of Power*. London: Profile Books.

Filmography

Bad Banks, 2018 – present. [series] Directed by Christian Schwochow, GER.
Coup pour Coup, 1972. [film] Directed by Marin Karmitz, FRA
Dirty Harry, 1971. [film] Directed by Don Siegel, USA.
Élite, 2018 – present. [series] Directed by Dani de la Orden et al, ESP.
Hat Wolff von Amerongen Konkursdelikte begangen?, 2004. [film] Directed by Gerhard Friedl, GER.
Hillbilly Elegy, 2020. [film] Directed by Ron Howard, US.
Itaewon Class, 2020. [series] Directed by Kim Sung-yoon, KOR.
Joker, 2019. [film] Directed by Todd Phillips, USA.
Meet John Doe, 1941. [film] Directed by Frank Capra, USA.
Mr. Smith Goes to Washington, 1939. [film] Directed by Frank Capra, USA.
Nine to Five, 1980. [film] Directed by Colin Higgins, USA.
Norma Rae, 1979. [film] Directed by Martin Ritt, USA.
Paracelsus, 1943. [film] Directed by Georg Wilhelm Pabst, GER.
The Hunger Games, 2012–2015 [film series] Directed by Gary Ross, USA.
The Riot Club. 2014. [film] Directed by Lone Scherfig, USA.

"AGAINST THE ELITES!"

AN ANTI-ELITE MOMENT

PART I

JOHN CLARKE, PAOLO GERBAUDO, JENS WIETSCHORKE

2 Anti-elitism, populism and the question of the conjuncture

John Clarke

Anti-elitism forms a significant element of contemporary politics and is central to the varieties of populist discourse and politics affecting many places across the globe. However, I argue that analysis of these developments needs to treat them as conjunctural formations and not as an epochal shift. Rather than a new era of populism or anti-elitism, we confront a moment of many *-isms* in which cultural and political repertoires are assembled and put to work in conditions of disruption, dislocation and disaffection. The core of the chapter develops an account of the problems and possibilities of conjunctural analysis as a way of making sense of the present. In doing so, it draws on examples from the long moment of "Brexit" in the UK.

Anti-elitism has a long and complex history as a political discourse, and many elites have been named and shamed by popular political movements. Elites have been imagined and represented in diverse ways: from the "power elite" in the US identified by C. Wright Mills (Wright Mills, 1956) to the so-called global Jewish conspiracy and from the current cosmopolitan liberal elite to the rapacious Conservative elites who were identified as ruling many Latin American societies. The *elite*, in this sense, is a figure of political discourse – always imagined in an antagonistic relationship to its others: the rest of society, ordinary decent people, the poor and the *people*. This diversity points to an important double movement at the heart of anti-elitism: On the one hand, dominant social groups recurrently colonise resources, wealth, symbolic capital and forms of power and invest considerable political effort in securing the resulting inequalities. On the other hand, the *elite* is a polyvalent term in political discourse whose content and particular referent (which groups are identified as the elite) changes across place and time. The idea of the elite is particularly central to the workings of populist discourse (see, *inter alia*, Laclau, 2002; Müller, 2016). The couplet of anti-elitism and populism has been central to current political reformations in many settings, from the UK's Brexit moment through Trump's election to Orbán's rule in Hungary.

However, this focus on populism and anti-elitism leaves other *-isms* standing in the shadows, despite their contemporary salience. A short list might include nationalism, nativism, racism, colonialism (or neocolonialism), authoritarianism and some other tendencies lacking the *-ism* ending but no less significant: for example, xenophobia, homophobia and misogyny. Rather

DOI: 10.4324/9781003141150-3

than assuming that these dispositions are merely the natural fellow travellers of anti-elitism or populism, it may be important to consider how they are – selectively and unevenly – connected together or articulated. For me, that would imply thinking of such formations not as essences (what is the core of nationalism?) but asking what enables their assembling together with other dispositions in a chain of meaning-making – *nationalism+authoritarianism+misogyny+racism* – while knowing that each of the terms is itself capable of different valences or registers (there is no one nationalism, nor one racism). The process, I suggest, is better understood as articulation rather than aggregation (the plus signs are misleading), because in the process of articulation, the meaning of each element is inflected by the others (on articulation see, *inter alia*, Clarke, 2015; Slack, 1996).

Articulation and the challenge of thinking conjuncturally

Thinking conjuncturally involves overcoming a series of alternatives that promote ways of dealing with the present in more singular terms. One of the stumbling blocks is the undoubted pleasure of knowing "how things work": The certainty that each concrete moment is best understood as one typical example or effect of a larger process whose dimensions and dynamics are already well-known. Many contemporary phenomena have been treated as the latest instance of globalisation or neo-liberalism (or even neo-liberal globalisation). I do not mean to suggest that the processes of globalisation or neo-liberalisation are not significant, but to trace everything back to them – including the rise of populism and anti-elitism – short-circuits questions of how these phenomena come into being in particular places at specific moments. Such reductionist accounts treat the concrete phenomenon as an instance, rather than, as Marx argued, seeing the concrete as the "complex synthesis of multiple determinations" (Marx, 1857, p.101). By contrast, conjunctural analysis demands attention to the multiplicity of determinations in play.

There is a second problem to be overcome, one that emerges from a double dynamic. At root, the drive to name the moment as part of a wider historical shift is an understandable intellectual desire (I remember reading many books on the transition from feudalism to capitalism). It is also an effect of a certain academic dynamic, whereby naming the shift (announcing this as the age of populism, nationalism, rage, etc.) is a form of claim-making. Both dynamics contribute to the identification of a singularity – the age/era of X (which succeeds and is different from a different age/era). The argument against this way of framing critical analysis was beautifully put by Raymond Williams:

> In what I have called epochal analysis, a cultural process is seized as a cultural system, with determinant dominant features: feudal culture or bourgeois culture or a transition from one to the other. This emphasis on dominant and definitive lineaments is important and often, in practice, effective. But it then happens that its methodology is preserved for the very different function of historical analysis, in which a sense of

> movement within what is ordinarily abstracted as a system is crucially necessary, especially if it is connected with the future as well as the past. In *authentic historical* analysis it is necessary at every point to recognize the complex interrelationships between movements and tendencies both within and beyond a specific effective dominance. It is necessary to examine how these relate to the whole cultural process rather than only to the selected and abstracted dominant system.
>
> (1977, p.121; emphasis in original)

I do not mean to suggest that there are no problems about conjunctural analysis as an approach. On the contrary, there are several, not least the absence of any definitive statement of the approach or a clear method for conducting such an analysis (though see Grossberg, 2019). At its core, though, two critical principles seem to me to be at stake in approaching conjunctural analysis: the commitment to thinking heterogeneity, and of a conjuncture as unstable and contingent.

Thinking heterogeneity

In line with Williams' resistance to epochal views, conjunctural analysis points to the risks of identifying singularities rather than thinking multiply across a range of focal points. It then becomes important to resist simplifying accounts of the singular cause of our present condition. Instead, conjunctural analysis examines multiple causes that intersect and become condensed and entangled in the present – these might include globalisation, financial capitalism or neo-liberalism, but never alone. Similarly, we might need to move from thinking about social forces in the singular (usually classes and, even more usually, capital and labour) to paying attention to heterogeneous social forces, which are unevenly mobilised (or demobilised) as political forces in this particular conjuncture. Thirdly, there is the challenge of moving beyond thinking of the present as organised around *a crisis* towards considering the dynamic intersections of multiple, divergent crises that serve to unsettle social, political, economic and ecological formations. For me, this would imply tracing the failures of neo-liberalisation; the crises of globalisation (ranging from displacement to expropriation and the degradation of different populations); the crises of Western social democracy (and the exhaustion of consent); the unresolved crises of the post-colonial period; the crises of social reproduction; and, not least, the looming crises of the global environment. Each of these has a specific effectivity in constituting the present but they are collectively entangled in complex and unpredictable ways.

Similarly, analysis might move from a focus on a singular contradiction to a consideration of the multiple contradictions, antagonisms and points of division and conflict that are in play (not least because of the implications for the possible lines of political affiliation and articulation that they create). This leads inexorably to the challenge of moving from thinking of politics as a site of singular divisions (classes) or political projects (nationalism, populism, etc.) to considering the heterogeneous issues, divisions and potential alignments that they enable. The conjuncture, then, is the setting in which politics involves the assembling – or articulation – of social groups into political blocs. It is the site of *political work* – the construction and maintenance of social authority, leadership or even hegemony. This is an appropriate point to return to Williams' view of how historical analysis needs to move beyond attention to a "specific effective dominance" (1977, p.121). In political–cultural terms, Williams argued that such dominant formations were always accompanied by others: particularly the *residual*, which he described as the persistence of questions that cannot be answered in the terms of the dominant, and the *emergent*, which he identified as the rise of new questions and new demands that were, however, always at risk of being absorbed into the dominant. This triangulation of the political–cultural field (dominant–residual–emergent) is a useful discipline for thinking about the heterogeneity of the conjuncture and avoiding limiting the analytic focus to the dominant alone. Characteristically, Williams makes it more demanding still by suggesting analysis needs to be attentive to "recognize the complex interrelationships between movements and tendencies" (1977, p.121). That is, the *dominant* is itself a contested formation as different elements, tendencies and forces contend to lead it. At the same time, the dominant formation is always engaged in the work of subordination: trying to actively residualise the residual ("this is just old-fashioned thinking"), while trying to either marginalise or incorporate the emergent.

There are many conceptual vocabularies through which one might think about heterogeneity, but rather than a fully systematic conceptual apparatus, I have found it easier to hang on to a few phrases that provide me with an orientation when trying to think about the conjuncture in these ways. My starting point (borrowed, like so much else, from Stuart Hall) is a view of things (the objects of attention) as "unities in difference" rather than simple or expressive totalities (Hall, 1996, p.141). Hall developed this conception from Marx's reflections on the "circuit of capital" in the *Grundrisse* (1857; Hall, 2003). Although formulated by Marx as a way of talking about the circuit of capital as a unity of differences (of practices, of place and of time), the idea helps (me, at least) to think about a variety of objects from social formations through political blocs to discursive strategies. A second phrase then comes into play when thinking about the structuring of these unities-in-difference, borrowed from Louis Althusser, who talked about a "teeth-gritting harmony" between the different state apparatuses: "It is the intermediation of the ruling ideology that ensures a (sometimes teeth-gritting) 'harmony' between the repressive State apparatus and the Ideological State Apparatuses, and between the different State Ideological Apparatuses" (1971, p.24). I have always liked the phrase for its very physical sense of a harmony (unity) that may be painful and discordant, even as it is held together. That sense of strain seems important to me in the face of overly integrated or smoothly functioning conceptions of states, formations and ruling blocs.

The third phrase is borrowed from Antonio Gramsci who suggested that the life of the state can be conceived as "a series of unstable equilibria" (1973, p.182). This sense of a dynamic trajectory is fruitful, leading away from overly institutionalised conceptions of the state towards the shifting political forces in which it is enmeshed and to the oscillations between moments of settlement and their unsettling. In this important sense, political settlements are only ever temporary: always likely to be undone either by their inability to contain their internal contradictions or by emergent new forces, challenges and crises. I have taken up this idea of settlements in work on welfare states and suggested that different settlements might be distinguished (political, economic, social and organisational: see, *inter alia*, Clarke and Newman, 1997). It is the unstable/unsettled aspect of this dynamic that is critical for thinking conjuncturally. Settlements and their destabilisation create the conditions of crisis and possibility as new alliances are formed, new strategies are formulated, new blocs are assembled in support and attempts are made to forge (and stabilise) new settlements – although with no guarantee of success. This is not a systematic conceptual vocabulary but a set of orienting propositions for attending to the heterogeneous forces, conditions and pressures that make up a conjuncture – that are, perhaps, best described as being *condensed* in a conjuncture. The image of being condensed hints at the intersecting, entangled and enmeshed pressures that are at work within the conjuncture.

Locating the conjuncture

One of the unresolved questions for conjunctural analysis is how to find and demarcate the conjuncture. The banal starting point may be that the (current) conjuncture is to be discovered in the "here and now" – but both place and time are problematic locations. The authors of *Policing the Crisis* (Hall et al., 1978), for example, confidently asserted that the conjuncture (and the shift towards an exceptional state that they traced) involved several interlinked crises:

> First it is a crisis of and for British capitalism [...]
>
> Second, then, it is a crisis of the "relations of social forces" engendered by this deep rupture at the economic level – a crisis in the political class struggle and in the political apparatuses [...] at the point where the political struggle issues into the "theatre of politics," it has been experienced as a crisis of "Party" [...].
>
> (Hall et al., 1978, p.317)

It is notable that this conjuncture was framed in national terms, even though these authors knew well that Britain was not a closed space (the territorial box of the nation) but was a colonial/post-colonial space. Nevertheless, it was British capitalism, a British "theatre of politics" and, eventually, a British state that formed the focus of the book. Where might we now locate the conjuncture if studying the moment of, for example, Brexit? I would be less confident about the *British* character of British capitalism: given patterns of transnational ownership (for example, of British manufacturing), the transnational production and distribution of material objects and the global reach (and global self-conception) of the leading sector of capital – the finance capital based in the "World City" of London (Massey, 2007). Wrestling with this sort of spatial formation requires ways of thinking that escape the traps of "methodological nationalism", on the one hand (the territorial enclosures of nation and state) and "methodological globalism", on the other (Clarke et al., 2015, p.23). At the same time, questions of the state might point towards a more multilevel or multiscalar understanding of governing institutions and apparatuses, despite the problems of reification in such scalar thinking (see, *inter alia*, Ferguson and Gupta, 2002; Isin, 2007).

I can only hint here at two ways of addressing these challenges. The first follows Doreen Massey (2005) in treating place as relationally produced. Refusing the subsumption of space under time (and the singular narrative of modernity and progress), she argued that:

> 'Recognising spatiality' involves (could involve) recognising coevalness, the existence of trajectories which have at least some degree of autonomy from each other (which are not simply alignable into one linear story) [...] On this reading, the spatial, crucially, is the realm of the configuration of potentially dissonant (or concordant) narratives. Places, rather than being locations of coherence, become the foci of the meeting and the nonmeeting of the previously unrelated and thus integral to the generation of novelty.
>
> (Massey, 2005, p.71)

This argument guides me in thinking about the return of "nation" in the current conjuncture, both as a relational formation and a potent political imaginary. Brexit (and its contemporary echoes in other places, such as the USA, Hungary and France) invoked that impossible object of desire – the nation in all its glory. Brexit has enabled the renewal of imperial fantasies: Not least, the belief that a mercantilist Britain, capable of leading the world as a sovereign power, can both direct and benefit from a new phase of global free trade. This is a profoundly spatial and scalar *imaginary* that aims to reconstruct (some of) the spatial relationships that have produced the nation. A celebration of national identity is at stake in this British/English revitalisation of the nation: Britishness has historically been triangulated externally through contradictory relationships of distance and desire in relation to the (imagined and material) entities of America, Empire and Europe. These three poles form the *imagined* others of Britishness – entangled in complicated connections of desire, loss, anxiety and fear – and they continue to shape ideas about who *we* (the British) are, who *we* were and who *we* might become (see Clarke, 2009 for a fuller discussion). It should be noted that there are *internal* relations that are subject to increasing strain as

Britishness becomes dominated by Englishness, displacing and subordinating the other nations of the more or less *United* Kingdom (see, *inter alia*, O'Toole, 2018). I have argued elsewhere that this dynamic of displacement and subordination reveals the different nations and nationalisms that are in play in the moment of Brexit (Clarke, 2021). In the process, subordinate anti-elitisms (notably Welsh and Scottish) have challenged a dominant *English* elitism.

The second hint emerges from recent work around topological approaches to studying space and spatiality. In a thought-provoking article, Allen and Cochrane (2010) take up the question of what Neil Brenner (2004) called "new state spaces". They argue against a scalar understanding (structured by a conception of "height") and, instead, make the case for a topological understanding of "reach":

> We start from a topological account of state spatiality, one that draws attention to the spatial reconfiguration of the state's institutional hierarchies and the ways in which a more transverse set of political interactions holds that hierarchy in place, but not in ways conventionally understood through a topographical lens. In contrast to a vertical or horizontal imagery of the geography of state power, what states possess, we suggest, is reach, not height. Topological thinking suggests that the powers of the state are not so much "above us" as more or less present through mediated and realtime connections, some direct, others more distanciated.
>
> (Allen and Cochrane, 2010, pp.1072–1073)

This different approach to analysing relationships of proximity and distance seems productive for thinking about the shifting sites, spaces, scales and forms of states (and state-like agencies) in which state powers have been (often simultaneously) decentralised and centralised, dispersed and concentrated, contracted out and turned into relationships of meta-governance as states have been reformed, reconstructed, reinvented and reconstituted. There is more to say about how the spatial character of the conjuncture may be determined and analysed, but these two approaches provide starting points for thinking about the challenge of escaping methodological nationalism and moving towards a more relational grasp of place and power in the conjuncture (Clarke, 2018). But space is only one of the constitutive conditions of the conjuncture – the other is the question of temporality.

Working on Brexit's conditions, causes and consequences, I have been recurrently struck by the confusing sense of time that surrounds it. Arguments about Brexit, for example, have started from the immediate, relatively short-term political calculations that made it possible: the then Prime Minister David Cameron deciding that a referendum would resolve the fissures over *Europe* within the Conservative Party. His failure created a new setting for short-term political calculation and careerism. At the other extreme, commentators have pointed to the long drawn-out and unfinished business of Empire and its hold over the affective economy of Britishness and the antagonisms towards its others (see, *inter alia*, Younge, 2016; Dorling and Tomlinson, 2019). This issue underlines Gilroy's (2004) argument about the condition of "postcolonial melancholia" as the constitutive fracture in British politics and struggles over British identity. But between the political short term and the *longue durée* of the post-colonial period, there are many other temporalities in play: the dynamics of Britain's insertion into the maelstrom of neo-liberal globalisation and its consequences for the economic and social uneven development of the UK; Britain's flailing struggles to be in, but not of, the European Union (EU); the 2007–2009 financial crisis and its transmutation into the politics and policies of *austerity* and the slow decline of classical social democracy (or at least its Labourist variant) as it attempted to become the "best possible shell" for neo-liberalism in the UK. Somewhere within those transformations, there is a need to account for the devastation of the public realm and its infrastructural conditions in the agencies, resources and contradictions of a welfarist state. There are other medium-term dynamics, too, including the social and cultural transformations around gender, race and sexuality that eventually came to be denounced as "cosmopolitan liberalism" by those who felt displaced or undermined by such partial shifts in cultural and political recognition. The conjuncture is created out of this multiplicity of temporalities – with their different rhythms and tempos. Each contributes to the sense of dislocation, disruption and disaffection that came to characterise the moment of Brexit (discussed at greater length in Clarke and Newman, 2017; Clarke, 2019). However, it is important to consider that the conjuncture may be more than a multiplicity of different temporalities. They do not just coexist, running alongside one another. They intersect, interact and interfere with one another, generating tensions, pressures and contradictions. That is to say, they are conjoined or condensed in the making of the conjuncture (Michele Filippini [2017, p.109] makes a similar point about the importance of multiple temporalities in Gramsci's work).

We might then think of the conjuncture as the site of political–cultural struggles to "tell the time": to order these different temporalities, to treat some as significant, while discounting or denying others, and to find narratives of the desired trajectory through the conjuncture that commands political assent (of which the restoration of lost greatness may be the most common in the current conjuncture). That theme was, of course, central to the politics of Brexit and I now turn to the question of anti-elitism and its place in those peculiar politics.

Which elite is this? The paradoxical populism of Brexit

In many respects, the campaigns to leave the EU which drove the Brexit vote in 2016 were marked by many of the classic tropes of populist discourse, counterposing a virtuous people and their oppressors and betrayers – the cosmopolitan, metropolitan, liberal elite. This was summed up in the statement celebrating the Leave victory by Nigel Farage (then leader of UKIP – the United Kingdom Independence Party):

> We have broken free from a failing political union. We have managed, the little people, the ordinary people who have ignored all the threats that have come from big business and big politics and it has been a huge, amazing exercise in democracy.
>
> (Dunn, 2016)

There are many reasons to doubt this description, not least, the complicated deployment of money, power, influence and technology during the campaign that point to something other than an "amazing exercise of democracy" (see, *inter alia*, Cadwalladr and Graham-Harrison, 2018; McGaughey, 2018). However, that is not my main concern here; instead, it is worth considering how this peculiarly British populism was articulated. Although the "liberal elite" was a recurrent focal point for the Leave campaigns, with attacks on "big business", the "political class", "liberals" and "experts" being among the variants, the People's Others took three different forms. Alongside the liberal elite (London-based and out of touch with the country as a whole), the Leave campaigns identified two other targets: The first was "Brussels" (as a summative image of the EU which combined "foreign" politicians, especially Angela Merkel, and the "bureaucrats" who ruled over the UK). The second was named "Migrants", the beneficiaries of lax immigration policies and "liberal" social policies. These two "enemies" were linked together through the EU's commitment to the free movement of labour, although the "Migrant" in British populist discourse shape-shifted with bewildering fluidity between the "Polish plumber", the Romanian fruit picker, the Muslim "terrorist" and the African-Caribbean migrants, the latter who continued to be treated as "out of place" after 70 years in the UK. This multiplication of enemies is not unknown in studies of populism (Barker, 2006, pp.103–104); more recently, it has been conceived as the doubling of vertical (anti-elite) and horizontal (anti-ethnic minority) axes of antagonism, in which fusions of populism and nationalism or nativism can be generated (Breeze, 2018; Hameleers, 2018). The Brexit trinity, however, positions "Brussels" as a crucial hinge between the vertical and horizontal axes since it doubles as both a hierarchical elite and the foreign disruptors of a British "way of life".

Perhaps the strangest aspect of Brexit populism was the composition of the Leave campaigns' leaderships, notably the public school and Oxbridge educated "posh boys" (such as Nigel Farage, Boris Johnson and Michael Gove). Farage had certainly mastered a particular self-presentation as a "man of the people", performing political "bluntness" as leader of UKIP, especially about immigration. This style, combined with photo opportunities crafted around him holding a pint of beer and a cigarette, tended to obscure his public school education and former career as a stockbroker. By comparison, Johnson and Gove (Oxbridge educated, Conservative Party cabinet ministers, journalists writing for the *Telegraph* and the *Times*, respectively) representing themselves as insurgent "outsiders" challenging the elite was deeply disconcerting. This very peculiar populism has had equally peculiar political consequences, dislocating and disturbing many established party political alignments and affiliations, most notably in the Conservative Party itself. It remains unclear how much the Brexit vote was inspired by this populist repertoire and to what degree it was derived from deep political disaffection and disenchantment. Brexit was certainly able to crystallise this sense of disaffection, even as it derived from multiple sources (economic dislocation, social crisis, cultural anxiety and multiple senses of loss: Clarke, 2019). The referendum enabled a sense of collective empowerment – the capacity to just say "No" allied to the promise to "take back control" – that underscored how disconnected everyday politics had become from many citizens. This is one of the recurrent promises of populism and forms the underpinning for the continuing sense of frustration and rage that has accompanied the limping progress of making Brexit come true (at the time of writing – mid-October 2020 – the issue is still unresolved). There are, of course, other frustrations and feelings of loss and betrayal on the part of those who wished to remain within the EU (despite its contradictions). There are other political and governmental frustrations across the divide of Leave versus Remain that were not given voice in Brexit populism: the global environmental crisis; the collapse of public services (including the failures of private enterprises contracted to supply them); the "hostile environment" for migrants (created by Theresa May); the crisis of social housing (painfully dramatised in the Grenfell Tower fire of 2017); and the continuing saga of UK manufacturing decline. Such politically neglected dislocations and frustrations also point to the possibilities of further political realignment within the UK.

Animating a people; assembling a bloc

The preceding section points to the value of the concept of *articulation* for exploring politics in the conjuncture in two senses. Firstly, there is the practice of "bringing to voice": how particular clusters of popular sentiments, desires and frustrations are – selectively – given political expression as if they are collective common sense (Clarke, 2019). Secondly, there is the practice of assembling a bloc: the equally selective mobilisation of social groups into a (temporary) political unity. The Brexit vote has been the subject of considerable debate, some of which has tended to a simplifying conception of class. This has pointed attention to working-class disaffection and disenchantment (see, *inter alia*, McKenzie, 2018). It is also a view of class difference in voting (working-class Leave and middle-class Remain) that is reflected in some of the data available. Skinner and Gottfried (2016), for example, claim that working-class voters were more likely to have voted Leave than middle-class voters (AB 59% Remain; C1 52% Remain, C2 62% Leave, DE 64% Leave). However, Dorling argues that

> because of differential turnout and the size of the denominator population, most people who voted Leave lived in the South of England. Furthermore, of all those who voted for Leave 59% were middle class (A, B or C1), and 41% were working class (C2, D or E). The proportion of

> Leave voters who were of the lowest two social classes was just 24%. The Leave voters among the middle class were crucial to the final result. This was because the middle class constituted two thirds of all those who voted.
>
> (2016, p.1)

These arguments point to a significant puzzle about the relationship between class and Brexit voting, one that is further complicated by Skinner and Gottfried's (2016) figures about voting dispositions related to employment status: Those in work were more likely to vote Remain (53% of those in full- or part-time employment), while those unemployed, retired or not working were all more likely to vote Leave. This takes us closer to thinking about some of the regional distribution of Brexit voting with Leave voters more strongly clustered in non-metropolitan rural and urban settings in England and Wales (both Scotland and Northern Ireland had majorities for Remain). The employed/not working or retired also links with the role that age played (over 45s were more likely to vote Leave in contrast with under 45s, who were more likely to vote Remain, with the difference increasing along each arm of the age axis). Gender was not a particularly significant dimension, with slightly more women voting to Remain than men. The difference was largest in the 18- to 24-year-old group, where 60% of men voted to Remain, compared with 80% of women (Statista Research Department, 2016).

At the same time, the simplifying conception of class is not helpful for thinking about the middle classes, where the assumption is that they formed a "cultural elite" characterised by a cosmopolitan liberalism that prompted them to vote Remain. This ignores a substantial population of what might be called the "traditional" middle classes, an older *petit bourgeoisie* in generational, occupational and cultural senses, who might be understood as the occupants of the "suburbs and shires". Less urban, less cosmopolitan and invested (materially and affectively) in a "way of life" that appeared threatened, these middle-class strata seemed prone to anxieties about the corrosive effects of "change". While demography is certainly not destiny (McQuarrie, 2017), a more differentiated analysis of the social groups in play enables attention to be paid to the coalition that was assembled around Brexit.

The two forms of articulation (voice and bloc-making) point to the unstable combination of people and politics that was mobilised to vote Leave. It was put together under the leadership that combined idiosyncratic embodiments of capital (see, *inter alia*, Arron Banks), populist leaders denied a leadership role (Nigel Farage of UKIP) and a cast of Conservative Party opportunists (notably Boris Johnson and Michael Gove) who saw personal and political opportunities in the referendum. A very contingent coalition of the frustrated, angry and outraged was assembled around them. As I have tried to indicate, this coalition cut across classes in strange ways, nicely expressed in Cochrane's description of the strange alliance forged from the twin nostalgias emerging from the post-imperial Home Counties and the post-industrial heartlands (2020). Those who voted for Remain formed an equally complex and contingent coalition: leading fractions of industrial and especially financial capital; much of the broadsheet press, as opposed to most of the tabloids who supported Leave; core sections of most of the political parties; and the public sector-based and more socially liberal fractions of the middle classes, and some sections of the working class, particularly those in employment (all again shaped by age dynamics).

Putting anti-elitism to work

In short, conjunctural analysis means considering how blocs and coalitions are put together, not least the political practices through which groups of people come to see themselves as political (or not). It might help us to avoid mistaking these moments for epochal shifts, for example, as marking a structural change from "politics as usual" to an era of populism. At the same time, this mode of analysis keeps open the question of whether these blocs and coalitions can be stabilised as the basis for further political moves or whether new alignments may be created. In the UK context, conjunctural analysis makes visible the specific ways in which anti-elitism has been put to work in the formation of a political project that has (more or less) delivered Brexit and reshaped the terrain of British politics, not least the recomposition of the Conservative Party itself (Clarke, 2020; Gamble, 2019; Schwarz, 2019). As I noted earlier, anti-elitism was a powerful theme in the campaigns for leaving the EU, combining antagonisms towards British metropolitan–cosmopolitan–liberal elites and the EU as forces that had ignored or suppressed the British people. Despite the 2016–2019 hiatus under Theresa May, Boris Johnson reframed the Conservative project around a commitment to "getting Brexit done" and as a response to the "liberal elites" in the UK and Europe who stood in the way of that accomplishment. This anti-elitism came to centre on attacks on the judiciary and the processes of judicial review of government decisions. Johnson led the Conservatives to a large victory in the 2019 General Election, including taking a significant number of seats historically held by Labour.

Since the beginning of 2020, this Conservative project has struggled: Progress on Brexit trade talks proved slow and painful and the arrival of COVID-19 exposed multiple forms of government incapacity and incompetence. Despite a recurrent drum beat of nationalist claims to being "world beating", a lengthening series of failures have seen both Johnson personally and the Conservative Party itself decline in opinion polls. One response has been to intensify the British/English version of anti-elitism: attacking liberals, "lefty lawyers" (especially those working for migrant/refugee/asylum seeker rights), the BBC and the Labour Party as "traitors" and "anti-British". This version of anti-elitism features a UK mobilisation of the "culture wars", denouncing those who would criticise or, even worse, change "our" history in response to Black Lives Matter and anti-colonial movements (Malik, 2020). These "culture wars" mobilise a characteristic populist combination of nationalism and authoritarianism through articulating a particular chain of equivalences: the people=the Conservative Party=the government=the state=the nation. This attempted fusion of identities aims to locate all dissent or opposition outside of this chain – not popular, not patriotic, not responsible, not tolerable and certainly not "really British".

Whether these strategies can sustain the Conservative political project remains open to question. Indeed, this is a recurrent feature of conjunctures, characterised as they are by multiple antagonisms and shifting balances of political forces. But, for my purposes here, it is the assembling and articulating processes that are most significant. Anti-elitism has formed one potent strand of this reworking of British Conservatism – a major achievement for a Party that has historically been seen as a party of the elite, even until recently when the Cameron governments were attacked as a group of "toffs" and "posh boys". The current Conservative Party and government are no less "posh" but have, temporarily at least, managed to recast themselves as the people's friends. However, this anti-elitism has been carefully articulated with other strands – a potent British/English nationalism, a revitalised anti-Europeanism, a xenophobic and racist fear of the world, all alongside a commitment to an imagined past of imperial greatness. This mixture fleshes out the British anti-elitism in specific ways: it populates this anti-elitism with distinctive heroes and villains and then mobilises the whole as the foundation for a state of national outrage and grievance (O'Toole, 2018). The elements of this complex "structure of feeling" (Williams, 1977: 134ff) are familiar from other places and times, but it remains distinctive in its specific combinations of these elements. It is this mixture of familiarity, connection and distinctiveness that conjunctural analysis can help us to grasp – and to see ways in which these articulated unities may yet come apart and be remade.

Bibliography

Allen, J. and Cochrane, A., 2010. Assemblages of state power: topological shifts in the organization of government and politics. *Antipode*, 42 (5), pp. 1071–1089.

Althusser, L., 1971. *Lenin and Philosophy and Other Essays*. Translated from French by Ben Brewster. London: New Left Books.

Barker, R., 2006. *Making Enemies*. Basingstoke; Palgrave Macmillan.

Breeze, R., 2018. Positioning "the people" and its enemies: populism and nationalism in AfD and UKIP. *Javnost – The Public: Journal of the European Institute for Communication and Culture*, 26, pp. 89–104.

Brenner, N. 2004. *New State Spaces: Urban Governance and the Rescaling of Statehood*. Oxford: Oxford University Press.

Cadwalladr, C. and Graham-Harrison, E., 2018. Revealed: 50 million Facebook profiles harvested for Cambridge Analytica in major data breach. *The Guardian* [online] 17 March. https://www.theguardian.com/news/2018/mar/17/cambridge-analytica-facebook-influence-us-election [Accessed 25 January 2021].

Clarke, J., 2009. Making national identities: Britishness in question. In: S. Bromley, J. Clarke, S. Hinchcliffe and S. Taylor, eds. *Exploring Social Lives*. Milton Keynes: The Open University, pp. 203–246.

Clarke, J., 2015. Stuart Hall and the theory and practice of articulation. *Discourse: Studies in the Cultural Politics of Education*, 36 (2), pp. 275–286.

Clarke, J., 2018. Finding place in the conjuncture: a dialogue with Doreen. In: M. Werner, J. Peck, R. Lave and B. Christophers, eds. *Doreen Massey – Critical Dialogues* (Economic Transformations). London: Agenda Publishing, pp. 201–213.

Clarke, J., 2019. A sense of loss? Unsettled attachments in the current conjuncture. *New Formations*, 96–97, pp. 132–146.

Clarke, J., 2020. Building the "Boris" bloc: angry politics in turbulent times. *Soundings: A Journal of Politics and Culture*, 74, pp. 118–135.

Clarke, J., 2021. Which nation is this? Brexit and the not-so-United Kingdom. In: I. Corut and J. Jongerden, eds. *Beyond Nationalism and the Nation-state: Radical Approaches to Nation*. London/New York: Routledge, pp. 98–116.

Clarke, J., Bainton, D., Lendvai, N. and Stubbs, P., 2015. *Making Policy Move: Towards a Politics of Assemblage and Translation*. Bristol: Policy Press.

Clarke, J. and Newman, J., 1997. *The Managerial State: Power, Politics and Ideology in the Remaking of Social Welfare*. London: Sage Publications.

Clarke, J. and Newman, J., 2017. "People in this country have had enough of experts": Brexit and the paradoxes of populism. *Critical Policy Studies*, 11 (1), pp. 101–116.

Cochrane, A., 2020. From Brexit to the break-up of … England? Thinking in and beyond the nation. In: M. Gunderjan, H. Mackay and G. Steadman, eds. *Contested Britain: Brexit, Austerity, Agency*. Bristol: Policy Press, pp. 161–174.

Dorling, D., 2016. Brexit: the decision of a divided country. *British Medical Journal*, 2016, p. 354. http://www.dannydorling.org/wp-content/files/dannydorling_publication_id5564.pdf [Accessed 25 January 2021].

Dorling, D. and Tomlinson, S., 2019. *Rule Britannia: Brexit and the End of Empire*. London: Biteback Publishing.

Dunn, J., 2016. Nigel Farage calls for a new bank holiday to mark UK's "Independence Day" as his life's work comes to fruition as the country votes for Brexit. *Mailonline* [online] 24 June. https://www.dailymail.co.uk/news/article-3657627/What-Nigel-Farage-s-life-s-work-comes-fruition-stunning-Brexit-vote-lauds-new-dawn-Britain-outside-EU.html#ixzz4Cngl7Zj9 [Accessed 25 January 2021].

Ferguson, J. and Gupta, A., 2002. Spatializing states: towards an ethnography of neoliberal governmentality. *American Ethnologist*, 29 (4), pp. 981–1002.

Filippini, M., 2017. *Using Gramsci: A New Approach*. London: Pluto Press.

Gamble, A., 2019. The realignment of British politics in the wake of Brexit. *The Political Quarterly*, 90 (S2), pp. 177–186.

Gilroy, P., 2004. *Postcolonial Melancholia*. New York: Columbia University Press.

Gramsci, A., 1973. *Selections from the Prison Notebooks*, G.N. Smith and Q. Hoare, eds. London: Lawrence and Wishart.

Grossberg, L., 2019. Cultural studies in search of a method, or looking for conjunctural analysis. *New Formations*, 96–97, pp. 38–68.

Hall, S. 1996. On postmodernism and articulation: an interview with Stuart Hall (L. Grossberg). In D. Morley and K.-H. Chen, eds. *Stuart Hall: Critical Dialogues in Cultural Studies*. London: Routledge, pp. 131–150.

Hall, S., 2003. Marx's notes on method: a "reading" of the "1857 introduction". *Cultural Studies*, 17 (2), pp. 113–149.

Hall, S., Critcher, C., Jefferson, T., Clarke, J. and Roberts, B., 1978. *Policing the Crisis: Mugging, the State and Law and Order*. London: Macmillan.

Hameleers, M., 2018. A typology of populism: toward a revised theoretical framework on the sender side and receiver side of communication. *International Journal of Communication*, 12 (2018), pp. 2171–2190.

Isin, E., 2007. City state: critique of scalar thought. *Citizenship Studies*, 11 (2), pp. 211–228.

Laclau, E., 2002. *On Populist Reason*. London/New York: Verso Books.

Malik, N., 2020. *We Need New Stories: Challenging the Toxic Myths behind Our Age of Discontent*. London/New York: Verso.

Marx, K., 1857. Introduction to a contribution to the critique of political economy. *Grundrisse: Foundations of the Critique of Political Economy*. Translated and edited by M. Nicolaus, 1973. Harmondsworth: Penguin, pp. 81–114

Massey, D. 2005. *For Space*. London: Sage Publications.

Massey, D., 2007. *World City*. Cambridge: Polity Press.

McGaughey, E., 2018. Fraud unravels everything: Brexit is voidable and Article 50 can be revoked. *London School of Economics* [online] April 19. https://blogs.lse.ac.uk/politicsandpolicy/fraud-unravels-everything-brexit/ [Accessed 25 January 2021].

McKenzie, L., 2018. We don't exist to them, do we? Why working class people voted for Brexit. *London School of Economics* [online] January 15. http://blogs.lse.ac.uk/brexit/2018/01/15/we-dont-exist-to-them-do-we-why-working-class-people-voted-for-brexit/ [Accessed 25 January 2021].

McQuarrie, M., 2017. The revolt of the Rust Belt: place and politics in the age of anger. *British Journal of Sociology*, 68 (S1), pp. 120–152.

Müller, J-W., 2016. *What Is Populism?* Philadelphia: University of Pennsylvania University Press.

O'Toole, F., 2018. *Heroic Failure: Brexit and the Politics of Pain*. London: Head of Zeus.

Schwarz, B., 2019. Boris Johnson's conservatism: an insurrection against political reason? *Soundings*, 73, pp. 12–23.

Skinner, G. and Gottfried, G., 2016. How Britain voted in the 2016 EU referendum. *IpsosMORI* [online] 5 September. https://www.ipsos.com/ipsos-mori/en-uk/how-britain-voted-2016-eu-referendum [Accessed 21 January 2021].

Slack, J., 1996. The theory and method of articulation in cultural studies. In: D. Morley and K-H. Chen, eds. *Stuart Hall: Critical Dialogues in Cultural Studies*. London: Routledge, pp. 112–127.

Statista Research Department, 2016. Distribution of EU Referendum votes in the United Kingdom (UK) in 2016, by age group and gender. *Statista* [online] September. https://www.statista.com/statistics/567922/distribution-of-eu-referendum-votes-by-age-and-gender-uk/ [Accessed 25 January 2021].

Williams, R., 1977. *Marxism and Literature*. Oxford: Oxford University Press.

Wright Mills, C., 1956. *The Power Elite*. Oxford: Oxford University Press.

Younge, G., 2016. Brexit: a disaster decades in the making. *The Guardian* [online] 30 June. https://www.theguardian.com/politics/2016/jun/30/brexit-disaster-decades-in-the-making [Accessed 25 January 2021].

3 The betrayal of the elites

Populism and anti-elitism

Paolo Gerbaudo

Introduction

The elite, the controller of power, culture and money, towers over contemporary political discourse as a sort of Great Villain against which political polemic is targeted. The term "elite" has become a byword of populist movements, which, regardless of their ideological orientation, share a strong anti-elitism, an enmity towards the elite. Anti-elitism is a common leitmotiv of many groups described as populist, from Donald Trump in the US attacking the "swamp" of Washington politics and the mainstream news media represented as "fake news" and the likes of Nigel Farage ranting against the European Union and the Brussels bureaucracy to protest movements, such as the Gilets Jaunes, expressing discontent at entrepreneurs, economists and journalists, to continue with many examples of left-wing populist movements and their enmity against the rich and corporations. Indeed, enmity against the elites has often been taken as a defining element of populist movements, a condition for their interpellation of the people.

The elite is a foundational element of all forms of populism because the elite plays the role of what Ernesto Laclau and Chantal Mouffe, building on the Schmittian opposition between friend and enemy, has called the "constitutive outside" within populist discourse (Laclau, 1990; Mouffe, 2000). This is an external entity which, through its presence, offers a sort of inverse mirror against which the Self of the people is constituted. The constitutive outside is the role played by the "Other" as an external entity, which, through its presence, offers a sort of inverse mirror against which the Self of the people is constituted. Populist movements almost invariably appeal to a majority that is seen as having been wronged by a minority of people, who are deemed to be in a position of power – be it political, economic or cultural – that sets them at odds with the interest of the people (Canovan, 1999). However, the identity of these "few", who act as a necessary counterpart of "the many", is far from being as immediate and intuitive as populist discourse would let us believe. In order to understand anti-elitism in contemporary populism, it is necessary to adopt a relational and intersubjective view of politics, accounting for the way in which the "Self of the People" is defined in opposition to the "Other of the Elite". In other words, if we are to understand the spirit of

DOI: 10.4324/9781003141150-4

contemporary populism, it is essential to clarify the nature of its antagonist and to explore what is actually meant by the term elites: Who are the elites and what are the different categories to which they belong? Who makes up the elite antagonised in populist discourse? Is there one elite or many? What are the elite groups that different sorts of populists tend to antagonise? And from where does this enmity towards the elite ultimately stem?

I will argue in the course of this chapter that different elites exist and that they are antagonised by different brands of populism for different political purposes. On the Right, anti-elitism is mostly geared towards "cosmopolitan elites", a shadowy collective made of academics, journalists, leftist teachers and other people working in what Althusser called the "ideological state apparatuses" (2014). These people are deemed to have adopted a cosmopolitan viewpoint and custom which sets them apart from the community, towards which they harbour feelings of superiority and disdain. On the Left, instead, anti-elitism is mostly developed as a sort of adaptation and extension of the classic framework of class struggle. In this context, the elites who are mostly antagonised are the elites of the wealthy, the capitalist and rentier class, and the way in which they have consolidated a condition of power that insulates them against redistributive demands coming from ordinary citizens. Finally, as we shall see, there is a third form of anti-elitism that concentrates mostly on the role played by the political class, often represented as a self-serving "caste", as seen in the discourse of the Movimento 5 Stelle (Five Star Movement) in Italy and Podemos in Spain.

Regardless of these specific declinations of anti-elitist discourse, common characteristics exist, particularly the perception of the illegitimacy of the power wielded by the elites, which is sometimes described in populist discourse as the "betrayal of the elites". The elites are seen as actors whose power is illegitimate because it runs against the interests of the community of which they are part, mostly conceived as the national community, binding together people who share a common citizenship and the different rights and duties that are connected to that membership. The elite is perceived as being outside the pale of society, or hovering above it, being aloof and disconnected, hence, living in a state of contradiction to membership of a civic community. While the elites are, by definition, "the elected ones", the reason for populist antagonism towards them seems to be down to the fact that it is not clear exactly who elected them in the first place and to what extent this process of selection was democratic and not just meritocratic. This applies not only to cultural and economic elites, who are, by nature, non-elective, but also to political elites, who are increasingly seen not as the legitimate representatives of the citizenry but as the manifestations of a self-referential political class amid a post-democratic situation (Crouch, 2004), in which choices for citizens are reduced to the minimum. Anti-elitism, in this sense, stems from the perception that the presence of the elites militates against the basic principle of equality, the idea that all citizens are equal, irrespective of their condition or upbringing, and that those in power should be chosen and controlled by the collective will of the people.

This chapter develops a theoretically informed overview of anti-elitism in populist movements, drawing on the discourse analysis of populist movements in Europe and the US. The chapter begins by exploring the notion of the elite and its definition in political science and sociology. It continues by exploring different kinds of elites and anti-elitism in different strands of contemporary populism. The chapter will conclude with a discussion of the theoretical and political implications of populist anti-elitism. I will argue that it is necessary to take the motivations behind populist anti-elitism seriously. The political solution, however, is not doing away with the elites. Instead, a progressive democratic populism, in addition to denouncing illegitimate neo-liberal elites, should also set itself the task of constructing alternative elites that are responsive to the interests of the popular classes.

The people vs. the elite

Discussions of anti-elitism and populism call for some definitions, starting from what is meant exactly by the elite. The elite, from the perfect particle Latin term *eligere* (to elect; hence meaning "the elected" or "chosen ones"), can be defined at the outset as a group of people possessing exceptional power and influence which sets it apart from the general population. It is a selected group of people, an "elected few", whose membership is highly exclusive. According to the Cambridge Dictionary, the elite comprises "those people or organizations that are considered the best or most powerful compared to others of a similar type". The question of the elites is not altogether new. It has traversed social and political thought in modernity.

The existence of elites for elite theorists, such as Mosca, Pareto and Michels, was a fundamental fact that was becoming apparent in mass societies in which the equality of mass suffrage was contradicted by the fact that power – cultural, social, economic and political – was concentrated in a few hands (Mosca, 1939; Pareto, 1991; Michels, 1915). The elite tendency of society became naturalised in Pareto's power distribution law, asserting that the concentration of resources in a few units of a given system was a natural tendency of all societies. Mosca argued in *The Ruling Class* (1939) that all societies throughout history had been marked by the dominance of a class of power-holders. These authors were eagerly read by Italian fascists, who, also informed by the *Übermensch* suprematism of Nietzsche, saw these theories as a means to justify that the dominance of one group of people was only a natural tendency. An element of the elite was also attributed to Jews, especially in Nazi discourse, which painted the Jewish community as almighty and engaged in an international conspiracy, as famously suggested by the forged document "The Protocols of the Elders of Zion". The alleged influence of the Jewish elite (in its economic and cultural ramifications) in Nazi discourse had to be displaced by asserting the power of an Aryan elite, organised in highly hierarchical and militarised structures such as the Schutzstaffel (commonly abbreviated to the SS), and acting as a guide for the Germanic race (Neumann, 1944, pp.475–476). The elitism of fascist movements went hand in hand with a despite of the masses, seen as a passive and irrational element that had to be repressed and dominated. This stance was informed by the reactionary theory of the crowd in the late 19th and 20th century and the work of Gustave Le Bon (1897). In what one Italian fascist leader described as his favourite book, Le Bon depicted the crowd as callous and mindless, incapable of coherent action without the presence of a leader.

The elite is a growing phenomenon in the present day and age, as a reflection of the growing role of finance, the need for advanced knowledge for the operation of a complex and technological society and, finally, the technocratic transformation of politics and public administration (Rothkopf, 2008; Esmark, 2020). The growth in the division of labour and functional differentiation of complex post-industrial society calls for the presence of "experts" of all kinds, whose composition is, by its nature, elitist and anti-democratic. This new elite is defined more in meritocratic terms, i.e. based on the exceptional skills and capacity that are attributed to it. However, meritocracy has also become a polemical target in recent years and has been seen as a tendency which lessens democratic participation and equality, while justifying capitalist competition (Littler, 2017; Lind, 2020). Interestingly, at its very inception, the term meritocracy was seen in opposition to populism. It was coined by British sociologist and politician Michael Young in *The Rise of Meritocracy* (1958), where he described a dystopian society dominated by a class of technicians. In Young's vision of the future, this class would be antagonised by populists, appealing to the growing masses of the poor, whose conditions compare badly to that of pampered elites. Regardless of the negative implications in Young's use of the term, meritocracy has been adopted positively by neo-liberal ideologues to affirm the vision of a society based on competition, opportunity and upward social mobility. Yet, as Young predicted, this cult of merit has been accompanied by a worsening of living conditions for the many, who do not qualify for entry into elite circles (Milanovic, 2016). It is in the light of this vindication of Young's sociological prophecy that we can understand the reasons behind populist anti-elitism.

Populism has often been described in political science as a political orientation that centres on the opposition to elites. As Daniele Albertazzi and Duncan McDonnell stated, it "pits a virtuous and homogeneous people against a set of elites and dangerous 'others' who were together depicted as depriving (or attempting to deprive) the sovereign people of their rights, values, prosperity, identity, and voice" (2007: p.3).

> A thin-centred ideology that considers society to be ultimately separated into two homogenous and antagonistic camps, "the pure people" versus "the corrupt elite," and which argues that politics should be an expression of the volonté générale (general will) of the people .
>
> (Mudde 2004, p. 543)

The presence of the elite in these and other classic definitions of populism is seen as a necessary component of populist discourse and practice (Moffitt and Tormey, 2014; Moffitt, 2016). Amidst the present "populist zeitgeist" (Mudde, 2004), this enmity of populist movements against an elite felt to be a threat to a community has only become more central. It involves the mobilisation of a basic democratic principle: the superiority of the majority over the minority. In this sense, populism comes very close in its logic to the old democratic adage of *hoi polloi* vs. *hoi oligoi* (Vergara, 2020). Populism, in this sense, is anti-oligarchic; it is a political logic based on opposition to the power of the few: a fundamental democratic leitmotiv. The elite are resented because they are an "elected few", often without having been elected, or in a context in which not even the electoral procedure (as in the case of the political class) is seen as a criterion endowing the elite with legitimacy.

But who actually are these elites that the populists antagonise? As seen in Table 3.1, elites can be classified based on their field of activity. In this regard, it can be said that three types of elites exist: cultural, economic and political. Cultural elites comprise such figures as academics, journalists, researchers, designers, architects, filmmakers, writers and artists. These figures possess cultural influence, as they have the power of shaping the values, beliefs and ideas that are dominant in society. Hence, they are often suspected of having an illegitimate power to shape the way people think. Economic elites comprise people whose wealth is well above the population average and who have positions of power in industrial and financial companies and economic institutions. A list would include people with large personal wealth, company owners, shareholders with large holdings, landlords with many properties, wealthy merchants, financiers and brokers. Finally, the political elite comprises people who are part of the political class proper, i.e. elected representatives, and the political personnel employed by political parties and government apparatuses, including consultants and spin doctors.

Table 3.1 Different elites, different anti-elitisms

	Cultural elite	Economic elite	Political elite
Main antagonist	*Populist Right*	*Populist Left*	*Populist Centre*
Examples	Academics, journalists, showbiz, NGOs, creative class; scientists; doctors; news media and digital media	The wealthy; entrepreneurs; bankers; brokers; landlords; managers and highly paid technicians/lawyers	Politicians; bureaucrats, civil servants; technocrats; supranational institutions; government consultants and experts
Reasons for enmity	Perception of betrayal of tradition and popular sentiments; imposition of cosmopolitan and liberal world-view	Exploitation of workers; tax avoidance; environmental degradation; interference with political decisions	Corruption; lack of transparency; laziness and wastefulness; distortion of the popular will; vote rigging

This classification of different elites is useful to make sense of the differences in contemporary anti-elitism because it corresponds to different strands of populism. Right-wing populism mainly takes aim at cultural elites, centrist populism at political elites and left-wing populism at economic elites. While some of these enemies overlap, this simplified classification goes a long way to demonstrate the differences which exist in different kinds of populist anti-elitism. Each of these populisms carries very different notions of what the elites are and why their influence is problematic. These will be described in what follows by focusing on three narratives that condense how different populists understand the betrayal of the various elites they antagonise: the dictatorship of the intelligentsia, the greed of the super-rich and the corruption of the political caste.

The dictatorship of the intelligentsia

One of the most despised elite in contemporary politics is no doubt the cultural elite, a category that has often come under attack by populists, especially on the Right. One of the leitmotivs of the populist Right is that the intelligentsia exercises a great degree of control over culture and opinion in our society in ways that are deemed to be nefarious to the said society. Popular targets of populist Right propaganda have included academics, judges and the law, NGOs (especially those rescuing migrants in the Mediterranean), the Pope, politicians, film actors/actresses, journalists, civil servants (especially in the central bureaucracy who are considered "privileged"); the Green elite (68ers) in Germany; the *soixante-huitards accompli* (well-to-do 68ers) in France; and the Covid-19 elites made up of virologists and epidemiologists. As this list shows, many of these figures are intellectuals in one way or another, people whose power is fundamentally connected with the production and distribution of knowledge.

For many on the populist Right, we live under what Santiago Abascal's Spanish extreme-right party Vox has described as a "*dictadura progre*" (progressive dictatorship) enforced by the dominance of the cultural elite. This and similar terms are often used when talking about the "liberal establishment" or the "radical Left", as proposed in the US by Trump and his acolytes, or the "*professoroni*" (big professors) antagonised by Matteo Salvini, leader of the Lega party in Italy. He has often vilified "radical chic" academics, journalists and teachers who are out of touch with reality and hate their own culture and religion, arguing that they are forcing alien values onto Italians. In France, Marine Le Pen has equally taken aim at the news media, journalists and academics accused of excessive tolerance towards the Muslim community and having no pride of their own country. She has often ranted against the supposed ostracism suffered by her views and reclaimed strongly a Christian and Western identity which she considers as having been attacked by multiculturalism and the embracing of other values and beliefs. In the UK, during the Brexit campaign and its aftermath, similar accusations were seen being directed towards the "metropolitan elites" accused of being in cahoots with the Brussels bureaucracy. A strong anti-intellectualism has also been on display in the discourse of Eastern European right-wing populists, such as Viktor Orban in Hungary and Jaroslaw Kaczyński in Poland, who have embraced virulent attacks on intellectual elites and universities accused of pushing anti-national interests and being too hospitable towards migrants.[1]

In this narrative, the reasons for anti-elite enmity revolve around the perception that the cultural elite illegitimately occupies its function of shaping society's world-views and values and is not responsive to the desires of the community of which they are part, understood as the national community defined by common citizenship. A link is often drawn in right-wing populist propaganda between intellectuals and the Left, based on the suspicion that the Left has monopolised cultural posts in society. Intellectuals are accused of peddling a dangerous mix of cultural Marxism and queer politics that are corrupting the demos and contributing to low fertility rates. This attitude has been compounded during the coronavirus crisis by aspersions aimed at doctors, virologists and epidemiologists accused by groups on the far Right of creating an atmosphere of excessive fear and paternalism and being in cahoots with pharmaceutical corporations. This was most pronounced in the context of conspiracy theories by "no-vax" and "no-mask" groups, which accused experts of rushing in a "health dictatorship" with no regard for people's freedom and livelihood.

This antagonism vis-à-vis the cultural elites is strongly reminiscent of that anti-intellectualism which Richard Hofstadter had already famously retrieved as a key motive in US politics (1963). Hofstadter argued that anti-intellectualism was not unified but that it encompassed different strands. Specifically, he listed anti-rationalism, populist anti-elitism and unreflective instrumentalism as distinct tendencies within anti-intellectualism. However, he did not anticipate the scale of anti-intellectualism targeting academics and the news media. Yet, this is precisely one of the key elements in contemporary anti-intellectualism. As theorist of populism Margaret Canovan highlights, "Populist animus is directed not just at the political and economic establishments but also at opinion-formers in the academy and the media" (1999, p.3). In a society that is increasingly mediatised, i.e. dependent on news media and opinion and analysis furnished by various media institutions for its functioning, journalists come to be perceived as having a power that many ordinary people do not possess. This dominance of mainstream news media causes populist movements to mobilise in their pursuit of progressive or reactionary alternative news media, including social media, seen as providing a better channel for the "vox populi" (Gerbaudo, 2018). On the Right, anti-intellectualism targets those intellectuals seen as promoting ideas that go against traditional world-views. The beliefs and values which are seen as setting them apart from the people include a wide range of behaviours based on the social and cultural inclination of different forces on the Right.

An example is the Right's criticism of "gender ideology" (i.e. the acceptance of sexual and gender freedom), which is presented as imposing ideas on people and depriving them of certainties. Similarly, scientists are accused of having turned science into an idol, while disregarding people's views. Furthermore, the cultural elite is lambasted for its lifestyle, for which it stands accused of being fundamentally out of touch with reality, egotistical and self-serving. The intellectuals, it is deemed, have isolated themselves from the surrounding reality and are not cognisant of the living conditions of the person on the streets. What is attributed to intellectuals, in other words, is a mix of sanctimony, lack of patriotism, hypocrisy and sheer venality since they are sometimes even pictured as agents of a conspiracy of global finance in ways that are sometimes reminiscent of anti-Semitic discourse in the 20th century. Conspiracy theories, often mobilised by right-wing supporters, go as far as imagining that the cultural elites are involved in satanic rites or planning the substitution of local population with migrants, as in the so-called Kalergi Plan and the theory of "great replacement" proposed by French far Right author Renaud Camus.

The anti-elitism of the Right projects a very particular kind of people. It is a people that is imagined as suspicious of affectation and sophistication, a body of liberty-loving citizens who do not accept others telling them what to do. It is a people that is seen as animated by outrage at the fact of having had its culture and self-awareness hijacked by a small number of highly educated progressives who want to force their view of the world on society. Intellectuals are presented as an oligarchic force who pretend to have the right to speak in the name of truth without speaking in the name of the people, despite the fact that, as argued by Pierre Bourdieu, the intellectuals, while powerful and rich in cultural capital, are in fact a "dominated dominant class", subject to those comprising the super-rich and big entrepreneurs (1984). In fact, this is not by chance. The entire point of the Right's targeting of intellectuals seems to revolve around the attempt to divert attention away from other kinds of elites, particularly the elite of the wealthy.

The greed of the super-rich

The anti-elitism harboured by the Left has a radically different orientation to the one seen on the Right, focusing on the elite of the rich. The examples of this anti-rich spirit have been made manifold in recent years. Take, for example, the slogan: "Billionaires should not exist", used by Bernie Sanders during the 2019–2020 US primary campaign, unveiling the proposal for a wealth tax on the richest Americans. He proclaimed after his win in New Hampshire in February 2020:"We're taking on billionaires and we're taking on candidates funded by billionaires". These anti-rich tirades, which are reminiscent of the populism of the American People's Party in the late 19th century, have been echoed by many other left-wing populist leaders, such as Jeremy Corbyn, Jean-Luc Mélenchon, Pablo Iglesias and Alexandria Ocasio-Cortez. The likes of Jeff Bezos, Mark Zuckerberg, Elon Musk, the Koch brothers in the US, Richard Branson in the UK, Bernard Arnault in France, the Benetton family in Italy and Amancio Ortega in Spain have become familiar targets.

This "soak the rich" sentiment has become a defining element of the New Left that emerged in the aftermath of the 2008 crisis, and it has only intensified in recent years. These attacks reflect a shift in public opinion over the last few years: The extraordinary enrichment of the wealthy elites and the huge inequalities that have resulted have attracted widespread outrage in the population. Even members of the billionaire class, such as Warren Buffett, have admitted the enormity of the imbalance in wealth by famously saying, "There's class warfare, all right, but it's my class, the rich class, that's making war, and we're winning." Indeed, looking at the data on the distribution of wealth and income in society (Milanovic, 2016) and the growing share that is going to reward capital rather than labour, it is quite apparent how the elite of the wealthy wields enormous and perhaps unprecedented power in our society. Hence, there should be little surprise that they have been identified as a key elite responsible for the betrayal of ordinary people.

Part of the adoption of populist discourse and strategy on the Left can be seen as an attempt to recuperate this anti-elitist motive, as a means to recentre public attention on elites that are much more deserving of disdain than the cultural elites. The post-1960s New Left has often been accused by the populist Right and conservative intellectuals as being the source of the progressives' cultural elitism and suspicion of workers and the common people. Yet, at its source, it contained a strong anti-elitist streak. This is visible perhaps most prominently in C. Wright Mills' famous book *The Power Elite*, in which the sociologist who inspired the New Left reconstructed the composition of the elite in US society, from the central apparatus of the state and large industries down to the American small town with its local elites (1956). He argued that US society was controlled by few tens of thousands of people, strongly united by their participation in common social events and institutions, who wielded great power in determining the course of society.

The questions raised by Mills became of great interest to a number of sociologists, especially those concerned with technocratic power in ever more complex and secularised societies. Alain Touraine, for example, in France, saw May 1968 as prefiguring a new class conflict opposing the popular classes against the technocratic power of the state and organised capitalism (1971). In the present day and age, suspicion of the oligarchic power of economic elites is directed strongly towards corporate power: the way in which multinational corporations and digital companies in particular have accrued enormous power. These anti-elitist motives have continued to resonate at different levels in the radical Left and social movements, as seen in the mobilisations against economic globalisation and corporate power or in the 99 % vs. 1 % meme developed by Occupy Wall Street protestors. Amid the pandemic, the growth in the wealth of figures such as Jeff Bezos, the founder of Amazon, and Elon Musk the CEO of Tesla have been similarly denounced as an affront to ordinary people. If anything, it is surprising that, to date, this growing inequality has not attracted even fiercer denunciation and anger. Ultimately, as argued by economist Dani Rodrik, the only way for the Left to counter the Right's cultural populism is by developing an economic populism in which the rich are the obvious target of attack (Rodrik, 2018).

The corruption of the political caste

The third elite that is antagonised by populists is the political. In recent years, in fact, a lot of anger has been directed at the behaviour of politicians and experts working for governments, variously accused of corruption, lack of respect for the democratic will and, finally, a tendency to look with disdain at the demands raised by the citizenry. My argument is that this elite enemy corresponds most clearly to what is perhaps the least discussed brand of populism, i.e. centrist populism, while also acting as a shared enemy for all forms of populism, regardless of their political orientation. In fact, as argued by Ernesto Laclau, populism stands in opposition to political institutions accused of being unable to satisfy popular demands (2005). Therefore, it is evident that the agents identified with political institutions, and with "institutionalism", i.e. the defence of institutions and their necessity regardless of their actual satisfaction of popular demands, can easily end up in the crosshairs of populist movements.

Perhaps the most evident case of populist contestation of the political class and main evidence of the centrist character of this type of populism is provided by the Italian Five Star Movement and its lambasting of the so-called political "*casta*" (caste). The Five Star Movement has, since its inception, criticised politicians, accusing them of being corrupt and of seeing politics as merely a career through which to earn salaries and an opportunity for pilfering public resources, and highlighted the need for public probity at a time when trust in the political class is at a historical low. In addition to the Five Star Movement, other centrist formations have mobilised populist criticism of the political class. These include the Spanish party Ciudadanos that combines a liberal economic platform with a strong criticism of the corruption of existing parties, including the Popular Party, from whose electorate it drew many of its supporters, and partly also Emmanuel Macron's La Republique En Marche, which is often understood as being strictly neo-liberal, and hence, at loggerheads with populism but has mobilised using typical populist themes, such as a criticism of existing parties and the way in which they are seen as representing a bottleneck in the political system.

Opposition to the political class can be seen as the form of anti-elitism that best corresponds to centrist populism as manifested by the Five Star Movement. However, it is also visible on the Left and Right brands of populism. On the Right, Donald Trump has made the anti-politics element a key component of his rhetoric. In the 2016 election, he challenged the establishment of the Republican Party, starting with the Bush family and the former governor of Florida, Jeb Bush, and came out victorious. During his first mandate as president, he promised to "drain the swamp" of Washington in order to fix problems in the federal government. This discourse and the QAnon conspiracy theory strongly informed the Capitol Hill rioters who invaded Congress on January 6, 2021, precisely on the back of the perception that the political class had betrayed them. In Italy, the rise of Lega leader Matteo Salvini, a career politician who entered politics when he was just

20 years old, has largely been fuelled by his ability to appropriate some of the Five Star Movement's anti-politics polemic. This can be seen in the way he lambasted the Conte II government, formed by the Five Star Movement and the Italian Democratic Party, as a "government of armchairs", to express the fact that it was formed solely to avoid an election and maintain power. This motive is reminiscent of the way in which, in the UK, Boris Johnson turned the 2019 national elections into a "people vs parliament" contest when he called people to support him to overcome what he denounced as the obstructionism of parliament. A further example is the tirades against experts that constituted a central rhetorical component of the Leave camp in preparation for the Brexit referendum of June 2016.

The Left is not entirely alien to this criticism of the political class. As we have seen in the foregoing section, new left parties and candidates that have emerged in the aftermath of the 2008 financial crisis have recuperated an anti-elite frame that had already been formulated at the inception of the New Left. Within this narrative, links are often drawn between the power of the economic elite and their alliance with the political elite. This is the argument developed by UK commentator Owen Jones in *The Establishment*, where he takes aim at the complicity between capitalists and politicians (Jones, 2015). Similarly, Podemos, after initially adopting the Five Star Movement's jargon of the "*casta*", later turned to the term "*trama*" (plot) to denounce the imbrication of political and economic interests, while in France, Jean-Luc Mélenchon has often turned to a similar rhetoric in his tirades against French President Emmanuel Macron, often branded as the president of the rich. Even politicians with a long history in their respective parliaments, such as Jeremy Corbyn in the UK and Bernie Sanders in the US, did not shy away from attacking the political class, and many of their credentials derived from the fact that while being career politicians, they were perceived as different from the rest and having been long marginalised.

The perception of a betrayal of the interests of ordinary people lies at the heart of this polemic against the political class. This perception is not merely the product of irrational propaganda smearing the political class. It reflects an objective state of affairs, the rise of a "post-democratic society", in which many decisions have been removed from public scrutiny and collective decision-making (Crouch, 2004). The existence of technocrats and growing power in government seems to contradict the basic democratic principle, according to which rulers should be chosen by citizens. To return to the etymology of "elite" mentioned previously, technocrats are perhaps the most despised type of elite because while acting as an "elected few", they have not, in fact, been elected by anyone. In this sense, the alleged meritocracy based on which technocrats are *selected* (rather than elected) is seen as having created a gap between the class of technicians working for government and ordinary citizens, with no possibility for the latter to control the former. This sense of betrayal is amplified in the case of technocratic governments that are not only manned but actually led by a technocrat, as in the Italian case of the 2011–2013 Mario Monti government and the new government formed by

Mario Draghi in February 2021. But this denunciation of aloofness is not restricted to technocrats. It also applies more generally to the political class, which is suspected of having, by and large, internalised many of the tendencies of this technocratic transformation of society and having broken the representative linkage with the citizenry, as stated in the theory of "cartel parties" (Katz and Mair, 1995).

While the criticism of the political class stems from an objective reality, i.e. the growing dominance of technocrats, and the "cartelisation" of political parties, the risk is that this stance may turn into a nihilistic denunciation of the political system, which only ends up reinforcing neo-liberal common sense and the view of politics as inherently corrupt. Blaming politicians and institutions for all the wrongs can result in making business look relatively good in comparison. Measures informed by an anti-politics orientation actually risk making the issue of the corruptibility of the political class even worse, as in the case of the Five Star Movement's battle against the power of money in politics and the elimination of public financing for the funding of political parties. The net result of this measure in Italy has been making politicians more reliant on rich donors, while the reduction in the number of members of parliament, another flagship Five Star Movement policy, has been fiercely criticised for its weakening of parliament. Therefore, enmity against political elites stands to reveal a more general limit of all brands of anti-elitism; while it is informed by a well-motivated criticism of the concentration of power in a few hands, it does not provide any clear remedy.

What use does anti-elitism have?

Anti-elite politics comes in many shapes and forms, as can be seen in the sketch of different types of elites and forms of anti-elitism. The key question, however, concerns the political implications of anti-elitism in contemporary populism. As we have seen in the course of the chapter, growing anti-elitism reflects the condition of a society in which cultural, economic and political power has become ever more concentrated and the way in which "meritocracy", a fundamental principle of selection in neo-liberal societies, has ended up colliding with democracy and its principles of equality and popular sovereignty. The predictions of Michael Young on the enmity between technicians and populists in advanced capitalist societies have been vindicated. In these days, this diagnosis is echoed by a number of authors, such as American democratic nationalist Michael Lind, who has argued that a managerial elite has become a powerful new actor in contemporary society, one caught in a class struggle with the mass of the lowly educated (2020). However, there are serious questions about the viability of anti-elitism as an effective discourse for the Left. The general perception to date is that the nationalist Right has been far more effective than the socialist Left in mobilising elites as enemies and utilising them as their constitutive outside. This is particularly evident when confronting the effectiveness of the Right's cultural populism vis-à-vis the Left's economic populism (Rodrik, 2018). Part of this problem seems to stem from the way in which neo-liberal common sense, while providing legitimacy for the business elites, does not provide the same line of defence for the cultural elite.

While it is evident that the elites are often resented for good reasons, this does not mean that anti-elitism should be accepted uncritically, nor that the Left should feel compelled to adopt it indiscriminately as a means to reconnect with the people. Anti-elitism, in fact, can turn out to be a means of political scapegoating that, while emphasising the sins of the elite, at the same time, exculpates citizens from the responsibilities they have towards society. It can lead to the opposite of the ethic of virtue and responsibility recommended by republicanism: Citizens can end up feeling exculpated by laying all the blame on the political class. Furthermore, it needs to be accepted that some elites, particularly political elites, will always be necessary in complex and highly technological societies, marked by a minute division of labour and complex cognitive and supervisory tasks. Rather than wallowing in the hyper-populist fantasy of doing away with all elites altogether, the pragmatic political question should instead be how the elite can be made more democratic at different levels. First and foremost, this would mean improving the access to elite circles not only in meritocratic but also in democratic terms, guaranteeing that it does not become a self-reproducing body but that it actually draws from different classes and particularly from lower socio-economic tiers. Secondly, it means making sure that the elite is not detached from the interests of the popular classes, also as a consequence of *esprit de corps*, but that measures are taken to guarantee that it remains responsive to the citizenry and its actions remain the object of democratic control and scrutiny. In other words, the real challenge is that only way to mediate, if not resolve, the opposition between the people and the elite is by creating "popular elites", which, similar to Gramsci's organic intellectuals (2007), have to remain, by definition, rooted in a broader society.

Note

1 This analysis of right-wing populist discourse draws on the discourse analysis conducted in my book *The Great Recoil* (Gerbaudo, 2021).

Bibliography

Albertazzi, D. and McDonnell, D., eds., 2007. *Twenty-first Century Populism: The Spectre of Western European Democracy*. Berlin: Springer.

Althusser, L., 2014. *On the Reproduction of Capitalism: Ideology and Ideological State Apparatuses*. London; New York: Verso Trade.

Bourdieu, P., 1984. *Distinction: A Social Critique of the Judgement of Taste*. Cambridge, MA: Harvard University Press.

Canovan, M., 1999, Trust the people! Populism and the two faces of democracy. *Political Studies*, 47 (1), pp. 2–16.

Crouch, C., 2004. *Post-democracy*. Cambridge: Polity Press.

Esmark, A., 2020. *The New Technocracy*. Cambridge: Policy Press.

Gerbaudo, P., 2018. Social media and populism: an elective affinity? *Media, Culture & Society*, 40 (5), pp. 745–753.
Gerbaudo, P., 2021. *The Great Recoil: Politics after Populism and Pandemic*. London; New York: Verso.
Gramsci, A., 2007. *Selections from the Prison Notebooks*. Durham, NC: Duke University Press.
Hofstadter, R., 1963. *Anti-intellectualism in American Life*. New York: Vintage Books.
Jones, O., 2015. *The Establishment: And How They Get Away with It*. Brooklyn; London: Melville House Publishing.
Katz, R.S. and Mair, P., 1995. Changing models of party organization and party democracy: the emergence of the cartel party. *Party Politics*, 1 (1), pp. 5–28.
Laclau, E., 1990. *New Reflections on the Revolution of Our Time*. London; New York: Verso.
Laclau, E., 2005. *On Populist Reason*. London; New York: Verso.
Le Bon, G., 1897. *The Crowd: A Study of the Popular Mind*. Reprint 2002. North Chelmsford, MA: Courier Corporation.
Lind, M., 2020. *The New Class War: Saving Democracy from the Managerial Elite*. London: Penguin.
Littler, J., 2017. *Against Meritocracy: Culture, Power and Myths of Mobility*. Abingdon: Taylor & Francis.
Michels, R., 1915. *Political Parties: A Sociological Study of the Oligarchical Tendencies of Modern Democracy*. New York: Hearst's International Library Company.
Milanovic, B., 2016. *Global Inequality: A New Approach for the Age of Globalization*. Cambridge, MA: Harvard University Press.
Mills, C.W., 1956. *The Power Elite*. Afterword by A. Wolfe in reprint 2000. Oxford: Oxford University Press.
Moffitt, B., 2016. *The Global Rise of Populism: Performance, Political Style, and Representation*. Stanford: Stanford University Press.
Moffitt, B. and Tormey, S., 2014. Rethinking populism: politics, mediatisation and political style. *Political Studies*, 62 (2), pp. 381–397.
Mosca, G., 1939. *The Ruling Class (Elementi di Scienza Politica)*. Reprint 2015. New York; London: McGraw-Hill Book Company, Inc.
Mouffe, C., 2000. *The Democratic Paradox*. London; New York: Verso.
Mudde, C., 2004. The populist zeitgeist. *Government and Opposition*, 39 (4), pp. 541–563.
Neumann, F.L., 1944. *Behemoth*. New York: Oxford University Press.
Pareto, V., 1991. *The Rise and Fall of Elites: An Application of Theoretical Sociology*. Abingdon: Routledge
Rodrik, D., 2018. Populism and the economics of globalization. *Journal of International Business Policy*, 1 (1), pp. 12–33.
Rothkopf, D., 2008. *Superclass: The Global Power Elite and the World They Are Making*. New York: Farrar, Straus and Giroux.
Touraine, A., 1971. *The Post-industrial Society: Tomorrow's Social History: Classes, Conflicts and Culture in the Programmed Society*. New York: Random House.
Vergara, C., 2020. Populism as plebeian politics: inequality, domination, and popular empowerment. *Journal of Political Philosophy*, 28 (2), pp. 222–246.
Young, M.D., 1958. *The Rise of the Meritocracy*. Reprint 1994. Piscataway, NJ: Transaction Publishers.

4 The *transclasse* and the *common people*

Autosociobiographies and the anti-elitist imaginary

Jens Wietschorke

Introduction: elite as a relational concept

The term elite has had a chequered career. In its semantic flexibility and reliance on context, it is nothing less than a chameleon of German and European social history. "Elite" established itself in the transforming estatist societies of the 18th century as a political battle cry, directed *from the bottom up*, against the supremacy of the nobility and the Church (Hartmann, 2004, p. 9). Those who spoke of the *élite* in the context of the French Revolution, for instance, used the term to mean a bourgeois functional elite that was chosen and legitimised not by dynastic and familial ancestry, but by the principle of merit. Over the course of the 19th century, the meaning of the term elite then shifted decisively: "elite" now referred less and less to the idea of a bourgeois democratic meritocracy, but, rather, functioned as a formula for the intellectual leadership of the popular masses. The direction of view, from the perspective of the "elite", was now *top-down*. The trope of "the masses and leadership" – and, with it, the imagining that the population always needs to be led by elites – thus shaped large parts of the political and social science discourse of the early 20th century.[1] For, it was around the turn of the century that three of the classic theoretical blueprints of elite sociology were developed: Gaetano Mosca's theory of the "ruling class", Vilfredo Pareto's "circulation of elites" and Robert Michels' reflections on the principles of oligarchy (Hartmann, 2004, 13–42). Under National Socialism, "elite" became a concept of *völkisch*-racist selection linked to the Führer principle. After 1945, the concept then underwent further transformations in Western societies: Where, under the impact of war and National Socialist tyranny, ethically based ideas of an elite had dominated, later, a functionalist understanding of elite prevailed. The term now mainly referred to positional or functional elites, who were, likewise, always understood as groups of experts, as sector and sectional elites, and who, in this sense, no longer denoted a claim to leadership in society as a whole, though the "social magic of individual selection" was at work in them almost unchanged (Reitmayer, 2009, 377). Since the 1970s, in light of the public negotiation of the concept of the elite, it has again, ultimately, been possible to detect tendencies towards a "more dichotomous worldview", in which the "elites" are

DOI: 10.4324/9781003141150-5

primarily subject to political and moral criticism: In the public discourse, they are a synonym for the rich and powerful, who occupy the key positions of control in society, for those "out of touch with reality", who live in a "parallel world with their own rules" (Hartmann, 2018, 7). Anti-elitist positionings and affects are currently booming in the context of right- and left-wing populist movements. The "Against the elites!" theme of crisis addressed in this volume is thus found at many junctions of the political field: in parliament and on the street, in party conference speeches and protest movements. At the same time, it can be seen that in protest rhetoric, economic, political and intellectual elites are often lumped together wholesale as a "power bloc". In this sense, the "elites" represent a *top* of society that is barely subject to any differentiation.

While the brief history of terms sketched out above only maps the rough lines of concepts of the elite over the ages, it is necessary to take a more detailed look at the concrete contexts and dynamics of the relational formula of "elites/the people", in which political constellations and conjunctures are illustrated. On the basis of two such specific contexts, this chapter examines what one might call the *anti-elitist imaginary*. It takes advantage of the relative openness to interpretation and flexibility of elitist and anti-elitist positionings and uses this as a starting point to show how anti-elitism functions in different specific contexts as a pattern of thought that establishes relations. In anti-elitist speech acts, actors not only assume a position themselves but also assign positions to other actors in the field – in each case, within the framework of what can be said discursively and the cultural figures available at that specific time. These dynamics can be characterised as a play with the assignments of *top* and *bottom*. Adopting the position of the *bottom*, and thereby speaking for the *common people*, can, at the same time, bring situational gains in authenticity that change the perception of the orders of social space in an "imaginary" way. There is a special focus here on the spaces in between that are created by the biographies of the so-called *transclasses* (Jaquet, 2018) and which can be read as spaces for manoeuvre for the anti-elitist imaginary. This essay outlines its argument in two loosely connected, as it were, rhapsodic steps: In the first part, a scene is presented and analysed that was shown on a German television talk show in 1982, in which the then Federal Minister of Finance, Hans Matthöfer, and Fritz Teufel, an original 1968er and former "*Spaßguerillero*" ("fun guerrilla") of *Kommune* 1, encountered one another in a remarkable semantic game of *top* and *bottom*. The second part offers some observations from the field of current autosociobiographical literature. Didier Eribon's *Retour à Reims*, the *Hillbilly Elegy* by JD Vance and comparable stories of social climbers meet with an enduringly strong reception on the international book market because they seem to link perspectives of *top* and *bottom* in a specific way. As this piece argues, they serve both the anti-elitist imaginary *and* the ascension narratives of the meritocracy at once. This makes them ambivalent texts that negotiate the question of configurations of elites/the people in a way that may be illuminating for the analysis of the current conjuncture.

Good behaviour: an experimental arrangement

When, in February 1982, a round table of psychologists, dance teachers, the *Schlager* singer Abi Ofarim, Federal Minister of Finance Hans Matthöfer and Fritz Teufel sat down together to discuss the topic of "good behaviour" on television under the direction of hosts Marianne Koch and Wolfgang Menge, no one could anticipate that a brilliant sociological lesson would shortly unfold. The composition of the round table altogether followed the principle formulated by the local public service broadcaster Radio Bremen to mark the occasion of the 30th anniversary of its talk show "3 nach 9": "Bring people into the studio who couldn't be more different and simply wait and see what happens."[2] The scene dealt with here begins with Fritz Teufel's statement on the topic of the programme:

> When I was asked if I wanted to say something about "genteel behaviour", I did give thought to it. But I don't come… I don't get any point of contact. I feel here like how it sometimes used to go in court; they speak a different language […], the language of dance teachers. And the problems are not really so much my problems either, how to open bottles, for example.[3]

Fritz Teufel's distancing of himself from the problems of "the dance teachers" was, of course, to have been expected and factored in by the makers of the show because, quite obviously, the experts for genteel behaviour had been invited as symbolic representatives of an "elite", which is subject to no further definition, whereas Teufel had been invited as the left-wing revolutionary who was supposed to toss barbs at this elite. What then followed, however, was no longer part of the plan. Teufel initially continued his remarks as follows:

> In my worldview, genteel behaviour is, for one thing… tenderness. This maybe only applies to people who resist though, because that is simply a necessity, because otherwise you… those who conform cannot actually be tender; that's my experience. The other side of it, though, is that genteel behaviour is by necessity unconventional and criminal in resistance. He, for example [Teufel points to Matthöfer], would like to shake hands with everyone; I would just like to wet a federal minister.

At that, Teufel pulled a water pistol filled with magic ink out of his jacket and squirted it on Matthöfer's shirt and suit. After the minister had sat there for about 20 seconds with a stony look on his face, he reached for his wine glass and flung its contents with resolute vigour at Fritz Teufel, who was sitting next to him. The audience acknowledged this action with violent applause – a close-up from the programme shows two viewers in the studio who, obviously outraged by the magic ink attack, clap with downright dogged enthusiasm. The reaction of the etiquette experts in attendance, however, was significantly different. Immediately after the exchange of wet onslaughts, the dance teacher Hinrich Wulff piped up:

> Ladies and gentlemen, even at the… at the risk of being misunderstood here, I have to say, I didn't think very much of that. I didn't think very much of it, you see, because Mr Teufel here was using a water pistol that doesn't leave stains, but red wine stains. So, that's why I didn't think much of that.

As the discussion went on, Hans Matthöfer was accused of having allowed himself to be provoked by Teufel to an emotional reaction. The general tenor was that a seasoned minister ought not to allow himself to get carried away into such an open counterattack. Fritz Teufel himself drew on this point, incidentally, when he appeared pleased to have "broken through the reserve of the man who is jointly responsible for top-security wings and such affairs". At first, Matthöfer politely apologised, pointing out that he did not know that it was just magic ink, and that his reaction was therefore based on the assumption that his shirt and suit had been "chemically soiled" – a response that the sociologist Hans Haferkamp has assessed as "self-negating" (Haferkamp, 1983, 60). A little later, however, Matthöfer justified himself with a remark that crucially shifted the symbolic markings of *top* and *bottom* within this conversation, and with which he definitively won large parts of the audience over to his side: "I am a working-class lad. I don't put up with things like that."[4]

The shifting of the symbolic markings

Spontaneous, approving applause in the audience made it clear that a key moment had been reached in the entire television discussion about "genteel behaviour". For, here, the "ungentlemanly behaviour" of the Federal Minister Hans Matthöfer, who, in an act of no self-control, had hurled his glass of red wine at fellow discussion participant Fritz Teufel, was suddenly being negotiated on a completely different level than before. The brief reference to Matthöfer's working-class origins suddenly reversed the polarities: The lack of composure was now re-interpreted as class-specific obstinacy. Matthöfer went from the contrite minister of finance, whose "reserve" had been "broken through", to a self-assured Social Democrat who can be proud of his biography of ascension and who does not let a bohemian get the better of him. All of a sudden, Matthöfer had, as it were, the core values of the proletarian class habitus on his side: quick-wittedness, directness, honesty and class pride. The inversion of the symbolic markings is astonishing: It was now no longer ministers (*top*) and revolutionaries (*bottom*) opposing each other, but the working class (*bottom*) and the educated middle class (*top*) or – to use the words of Luc Boltanski and Eve Chiapello – "social critique" and "artist critique" (Boltanski and Chiapello 2007). While Teufel's demeanour articulated the values of "freedom, autonomy and creativity" (Boltanski and

Chiapello, 2007, 37), the minister now symbolically represented exactly what was previously missing in the group: namely, the voice of the *common people* and of *common sense*.[5] In that one moment, Matthöfer fulfilled what Sebastian Dümling has termed the "multiple desire for the *common people*". This desire is dependent on a "game of referents": "When it comes to the 'common people', this is primarily about politics of reference and the enforcement of reference" (Dümling, 2020, 17).

This example shows that it was not just the hidden structures of social inequality that co-determined the communication here; rather, something like a temporary, imaginary recoding of the situation also took place. By referring to the values of the working class, Matthöfer succeeded in assigning the other participants present in the discussion group very specific speaking positions of the *top*. Like in a force field or magnetic field–this metaphor is often used, for example, by Bourdieu in the context of his version of field theory to characterise the relationality and interdependence of positions within a field – Matthöfer's new positioning also caused the positions of the others to change: The morals of the dance teachers now suddenly appeared as a formalistic code that was detached from lived reality and that necessarily fails in conflict situations, and Fritz Teufel suddenly appeared as an arrogant, brattish middle-class son and screwed-up academic who is only interested in the spectacle. As a result of the recontextualisation of the situation against the background of Matthöfer's biography of ascension from working-class lad to federal minister, Fritz Teufel unexpectedly switched for a moment to the side of the "establishment", towards which the anti-elitist effects of the audience were now directed. An otherwise rather conservative Social Democrat in the style of Helmut Schmidt, Matthöfer now used his "proletarian" gesture of assertiveness to harness the anti-elitist imaginary for his own gain. A punchline of this gesture was also that adopting this attitude of "taking no nonsense" and "defending himself" on the principle of "tit for tat" suddenly also guaranteed him a distinct position in relation to the assembled dance teacher "elite", with their antiquated, bourgeois-aristocratic ideals of "discipline", "composure" and "moderation".

Matthöfer was able to plausibly refer to the position of the "working-class lad" because he did actually have a distinctive biography of ascension to show. Measured against the structural conditions of his background, his career was, indeed, meteoric. Hans Matthöfer grew up in a working-class family in Bochum, which, at times, had to live in very precarious circumstances. His father was an unskilled labourer of Polish origin who made his way as a casual and temporary worker in the 1920s; his mother came to the Ruhr region from the Eifel as a working woman. Although he was a good student and a diligent reader, Matthöfer did not attend grammar school but completed his schooling at a school for basic primary and secondary education (Abelshauser, 2009, 26–34). Matthöfer's biographer, Werner Abelshauser, notes that the future politician probably suffered from not having enjoyed any upper secondary education. He therefore always criticised the low social penetrability of the West German school system, and, thus, "even the later academic was not always able to keep himself completely free of deep-seated resentment against the bourgeois academic upper class" (Abelshauser, 2009, 33). Matthöfer was only able to start studying without a secondary school leaving certificate and higher education entrance qualification due to an exemption clause that had been introduced by the Social Democratic government of Hessen in order to enable war veterans to study economics. Over the course of his life, Matthöfer passed through many stages that outline a steep rise: he was

> student of a school for compulsory basic education, apprentice, soldier, black marketeer, language teacher, student leader, publicist, economics and automation expert, diplomat, head of the education department at IG-Metall, member of the Bundestag, campaigner against the Franco dictatorship, parliamentary state secretary to the Federal Minister for Economic Cooperation, research minister, finance minister, postal minister, treasurer of the SPD, head of the trade union holding company BGAG, member of multiple supervisory boards and international economic advisor.
>
> (Abelshauser, 2009: 9)

At the time of the television debate on "3 nach 9", Hans Matthöfer was at the zenith of his political success: The introduction of the federal budget in 1982 became his "shining hour in parliament", and Matthöfer was thereafter repeatedly called "Crown Prince" and tipped as possible successor to Helmut Schmidt as Federal Chancellor of the Bundesrepublik Deutschland (Abelshauser, 2009, 478–492).

While Matthöfer's family background did not conform with "any of the conceivable patterns of mobility at that time in the circuit of the elites" (Abelshauser, 2009, 27), the path to university for Fritz Teufel, son of a senior employee of the chemical group Boehringer-Ingelheim, was almost predestined. "The eight Teufels wanted for nothing," writes Marco Carini in his biography (Carini, 2003, 11), and so there was never any doubt that Fritz Teufel could commence his studies in German, journalism and theatre studies at the Freie Universität after obtaining his secondary school leaving certificate – a path that concurrently led him to Kommune I with its "radically subject-oriented understanding of politics" (quoted from Reichardt, 2014, 100). One has to know these biographical backgrounds in order to be able to understand the anti-elitist game that was played during the television discussion. Nevertheless, it is characteristic of the situation that the actual social status of the two participants did not really play a role. What was instead crucial was the *reference* to the "common people" and the dynamic of the anti-elitist imaginary, which made it possible for the sitting Federal Minister of Finance to adopt a *bottom-up* speaking position towards the radical left-wing dropout. The brief reference to his biography as a class transitioner from poor circumstances, who had, nevertheless, retained the down-to-earth attitude and resilience of his proletarian milieu of origin, was enough at that moment to clearly decide the mood of the room in his favour. By bringing the topic of social background and social ascent into the discussion, he secured himself gains in recognition for his achievement in life. Incidentally, the imaginary nature of the assignments of *top* and *bottom* is also made clear from the fact that the pianist Gottfried Böttger, who was present in the television studio, claimed to have remembered things that cannot be seen in the video recording of the programme: While Fritz Teufel had worn an "ancient, egg-stained jumper", the Federal Minister of Finance appeared in "white tie and tails". After the red wine attack by Matthöfer, Teufel was "awfully indignant". The recording shows neither the tie and tails nor the indignation – in his recollection, Böttger turned the scene from February 1982 into an exemplary encounter of *haute-volée* and *underdog*.[6]

The Bremen talk show of February 1982 can be read as a historical conjunctural snapshot of a time in which certain configurations of "elite/the people" could be called up discursively, so to speak. At that time, left-wing anti-elitism was circulating in the Federal Republic of Germany in two prominent forms above all: on the one hand, as the radical "anti-authoritarian" position of the late "non-parliamentary opposition", represented here by Fritz Teufel, and, on the other, as the social democratic trade union position of the "ordinary people".[7] Although Matthöfer, as Federal Minister of Finance, had long since switched sides, so to speak, his working-class background enabled him to reclaim this position at short notice. For a moment, this created a new space for symbolic positions and figurations: The elite figures of the dance teachers and etiquette experts present, on the one hand, and the minister as a representative of the state, on the other, were now joined by a third figure: The academic, intellectual and "egghead", who, with his post-pubescent political ideas, understands nothing of the "real" life of the "common people". It is important to note that these cultural figures and figurations of *top* and *bottom*, the "elite" and the "people", maintain an exceedingly complicated relationship to social reality. They are symbolic representations that denote real positions and thus contribute to the order and interpretation of the social sphere. Conflicts between these positions are therefore always at once real and imaginary – but, above all, they are always relational: They make clear "that […] 'culture' is not about a system of identities, but about an ensemble of differences".[8]

Background and "complexion": the hour of autosociobiography

"I am a working-class lad. I don't put up with things like that." This brief reference by Federal Minister of Finance Hans Matthöfer to his social background positioned him as a class transitioner, as "transclasse" for the purposes of the French philosopher Chantal Jaquet.[9] In her work on the "non-reproduction of social power", Jaquet spelled out the figure of the class transitioner in theory. According to this, the "transit between the classes" (2018, 213) creates a state of "complexion" that takes on the "dynamic form of a permanent de- and reconstruction" and establishes a "position that fluctuates between distancing and turmoil" (2018, 214–215). What is interesting in this context is less the complex socio-philosophical derivation of the "complexion", as presented by Jaquet in her work, and rather the expectations that are placed on the figure of the class transitioner in the contemporary public discourse – because books in which the author's social background is addressed in an autobiographical or autofictional manner and which tell of the social ascension of their protagonists are currently being published in rapid succession. Didier Eribon's book *Retour à Reims*, especially, has caused a sensation in France since its publication in 2009 and, as of 2016, became a bestseller in its German translation as well (2009, 2016). This book seemed to promise to understand the shift to the right in France from the perspective of formerly left-voting working people who had now voted for the *Front National*. German readers expected Eribon to explain the rise of *Alternative für Deutschland*, and Eribon also became a shooting star internationally – the stage version of *Retour à Reims* was even performed in Manchester and New York.

It is not difficult at this juncture to mention a number of other prominent autosociobiographies from the past few years – from *En finir avec Eddy Bellegueule* by Édouard Louis to Darren McGarvey's memoir *Poverty Safari*, with the descriptive subtitle "Understanding the Anger of Britain's Underclass", to Sarah Smarsh's *Heartland* and Tara Westover's *Educated* (Louis, 2014; McGarvey, 2017; Smarsh, 2018; Westover, 2018). In Germany, the genre is represented, for example, by the autosociobiographies of Daniela Dröscher and Christian Baron (Dröscher, 2018; Baron, 2020). The question that arises here is whether the boom in autosociobiography might perhaps articulate a significant shift in the political discourse of the last ten years: To what extent does the new popularity of all the first-person stories about social background and social ascension point to a new way of thinking about the relationship between class and identity, between "elites" and the "people"? How are elitist and anti-elitist motifs entangled in these books and in their history of reception in different national contexts? Because one thing is obvious: Those writing here are social climbers, to greater or lesser degrees of success, telling their lives' journeys from humble backgrounds in the circle of academically educated elites. The authors are the ones who have made it; at the same time, they return for a moment to the milieus of their childhood, grapple with the history of their families and criticise condescending arrogance and the hermetic insularity of elitist circles based on their history of social injury. This creates a specific perspective that oscillates between the positions of the milieu of origin and that of arrival, and thus between *top* and *bottom*. The reproduction of social inequality is illuminated from the standpoint of "non-reproduction" (Jaquet, 2018, 29–102), and the rule, by the exception.

In the second part of my essay, I would like to pursue these questions with the aid of a book that shines a light on both current public discourses on the *transclasse* and the ambivalent role of anti-elitist narrative motifs and the anti-elitist imaginary: *Hillbilly Elegy* by the Silicon Valley capital manager JD Vance, published in 2016 (Vance, 2016). I will show how the neoliberal narrative of the self-made man Vance is so interwoven with anti-elitist motifs that it was able to be celebrated – at least for a short time – as a central contribution towards overcoming social divisions. For, this book, in particular, has been credited as displaying an enormous degree of skill in social diagnostics: The *New York Times* named *Hillbilly Elegy* as one of "six books to help understand Trump's win" (NY Times 2016). *The Economist* said it was "the most important recent book about America". David Brooks wrote that Vance's "description of the culture he grew up in is essential reading for this moment in history". "Couldn't have been better timed," said the *National Review*,[10] and the *Frankfurter Allgemeine Zeitung* also joined in: The book leads us "into the world of Trump voters". In short: "This is how to understand America" (Hochgeschwender, 2017). The background to all these reviews is the US-American discourse on social division, as has been conducted in an intensified form since the financial crisis of 2007–2009, but, above all, since the surprising election success of presidential candidate Donald J Trump: In this discourse of division, cosmopolitans and communitarians, liberal "anywheres" and conservative "somewheres", democrats and republicans, globalisation winners and losers face each other irreconcilably (cf. Wietschorke, 2020). It was therefore natural to read *Hillbilly Elegy* as the book of the moment that tells the story of this schism from its two poles, so to speak, and connects both perspectives – that of Silicon Valley capitalism and that of the "left-behind" province. Without in any way referring to *Retour à Reims* or other examples of the autosociobiographical genre, Vance suddenly appeared in the public reception as a kind of American Eribon, who had made a central contribution to understanding the social dislocations and the rise of right-wing populist parties.

Hillbilly elitism: *top* and *bottom* between rupture and narrative reconciliation

The story that JD Vance tells in his bestseller is a story of survival: The focus is on the feeling "that [he] had survived decades of chaos and heartbreak and finally come out on the other side" (2016, 188). This perspective from the safety of dry land characterises the entire book. It is a retrospective view of the author's childhood and youth in Middletown/Ohio and Jackson/Kentucky in the 1990s with parents and grandparents in extremely difficult social circumstances. Drug consumption, financial worries and violence dominated everyday life; the cohesion of the family was – at least in the maternal line – at times the only thing that could offer any kind of stability. This story is told at a time when JD Vance has long since had a law degree from Yale University in the bag and is working as a capital manager in Silicon Valley, where billionaire and Trump supporter Peter Thiel is his mentor. By now, however, the motif of the return also defines the biography of the author, who after the media success of bestseller *Hillbilly Elegy* moved back to Ohio.

The reference to the American Dream is explicit in this book:

> I want people to understand how upward mobility really feels. And I want people to understand something I learned only recently: that for those of us lucky enough to live the American Dream, the demons of the life we left behind continue to chase us.
>
> (Vance, 2016, 2)

Vance also explicitly avows himself a member of the white working class:

> I identify with the millions of working-class white Americans of Scots-Irish descent who have no college degree. To these folks, poverty is the family tradition – their ancestors were day laborers in the Southern slave economy, sharecroppers after that, coal miners after that, and machinists and millworkers during more recent times. Americans call them hillbillys, rednecks or white trash. I call them neighbors, friends, and family. [...] To understand me, you must understand that I am a Scots-Irish hillbilly at heart.
>
> (Vance, 2016, 3)

A *hillbilly* "at heart": This social positioning is the point of departure in the book for a story of ascension that always looks back. Yet, it is also – and above all – a cultural positioning. This is because, for Vance, the difficulties faced by many families in his two regions of origin are "problems of family, faith, and culture" (Vance, 2016, 238). There is no talk of the capitalist economy, politics, social structure and the replication of social inequality through the education system. "What separates the successful from the unsuccessful are the expectations that they had for their own lives. Yet the message of the right is increasingly: It's not your fault that you're a loser; it's the government's fault." (Vance, 2016, 194). The figure of the "loser" functions here as a central tool of understanding, which the author JD Vance can only position because he himself had every chance of becoming one. The political message of *Hillbilly Elegy* is therefore extremely problematic: Its view of the American underclass is radically culturalistic; its view of the possibilities of social ascension is radically individualistic.

In its political dimension, *Hillbilly Elegy* fundamentally differs from the most famous French examples of the autosociobiographical genre, the books by Didier Eribon and Édouard Louis. With Eribon, the question of political socialisation and class consciousness is not only ever present but forms the actual reason for existence of *Retour à Reims*, which revolves around the problem of the formerly socialist-voting parents of Eribon having turned to Le Pen and the then *Front National* in the 2000s. Louis, on the other hand, especially in his second book, *Qui a tué mon père*, openly accuses the political leadership of the French Republic, whom he holds directly responsible for the physical injury of his father. There are sentences like missiles: "Jacques Chirac and Xavier Bertrand destroyed your intestines", "Nicolas Sarkozy and Martin Hirsch were breaking your back". "Hollande, Valls, and El Khomri asphyxiated you" and "Emmanuel Macron is taking the bread out of your mouth" (Louis, 2019). There is nothing of the kind from Vance, who tells the story of his family, who has always lived precariously, and especially that of his drug-addicted mother, not as an effect of a failed social policy or as a product of social conditions in general but as a family story. His book elicits sympathy for the protagonists – above all, for "Mamaw", who, as a resolute grandmother, fights to the last for the well-being of the family – if need be, using her shotgun – and whom JD ultimately has to thank for being able to reach "dry land" at all.

Just as *Hillbilly Elegy* is shaped by a critical but consistently empathetic view of the rural underclass in the Rust Belt and the Appalachians, so the view of the academically educated East and West Coast elites is pragmatic and critical. Vance makes a point of always maintaining symbolic distance from them and *not* being one of them. "Sometimes," says Vance in his book, "I view members of the elite with an almost primal scorn" (2016, 253). Even the move to university created biographical ruptures, as are found in many reports of class transitioners from the working class (cf., e.g., Dews and Leste, 1995, Ryan and Sackrey, 1996, Muzzatti and Samarco, 2006). Vance writes: "At Yale Law School, I felt like my spaceship had crashed in Oz" (Vance, 2016, 204). In contrast to Eribon, however, and many other autosociobiographical authors, the problem of the "split habitus" is not addressed any further, and the question of the complicated relationship between elitist and anti-elitist affects is not discussed. JD's recipe for success is "optimism" – and thus a quality that "contrasted starkly with the pessimism of so many of [his] neighbors" (Vance, 2016, 188). Since the moment when JD talked himself into believing he could make it, it was apparently clear that he would, indeed, make it. In the words of the sociologist and Appalachian expert Dwight B. Billings, the story thus becomes "an advertisement for capitalist neoliberalism and personal choice" (Billings, 2019, 38). No wonder then that it was precisely conservatives like the journalists Reihan Salam and David Brooks, investor Peter Thiel and "tiger mother" Amy Chua who were among the first admirers of *Hillbilly Elegy* (see Hutton, 2019, 22). Thiel even explicitly sees the book as offering a refreshing counter-position to the academic exploration of social inequality: "Elites tend to see our social crisis in terms of 'stagnation' or 'inequality'. JD Vance writes powerfully about the real people who are kept out of sight by academic abstractions."[11]

The great paradox of *Hillbilly Elegy* is the ambivalent role of the anti-elitist imaginary in this story. On every page of this book, as it were, it is clear that real life, the life that really matters, revolves around family and social background. This is where true feelings have their home, where the values of loyalty, solidarity and cohesion are what count. Yale University, Silicon Valley – from the book's perspective, these are not the places that really matter to the heart, but the places where you do your job, where you get access to a better life. The view of the world of the elites, however, remains hard-headed: "I have to give it to them: Their children are happier and healthier, their divorce rates lower, their church attendance higher, their lives longer. These people are beating us in our own damned game." (Vance, 2016, 290–291). It becomes abruptly clear here how *Hillbilly Elegy* works as a narrative. The book references the world of the *common people* as a world of "us", with a warmth and empathy that the "other side" in Yale or San Francisco – with the exception of Usha, the girlfriend and wife of JD Vance – is never afforded. At the same time, however, the book is pervaded by discourses of social contempt, as the historian TRC Hutton highlighted in his apt analysis in *Jacobin*: "Vance's personal story permits him to claim the term 'hillbilly', then scold his fellow hillbillies for their cultural and moral failings" (Hutton, 2019, 25). *Hillbilly Elegy* thus falls in line with the more recent culturalist discourse about white trash, as has been conducted in the circuit of the national conservative magazine *National Review* for years.[12] The historian Anthony Harkins, who has propounded a history of the hillbilly figure in American pop culture, identifies the function of this figure in its enabling a "non-rural, middle-class, white, American audience" to "imagine a romanticized past, while simultaneously enabling the same audience to caricature the negative aspects of premodern, uncivilized society" (Harkins, 2003, quoted from Hutton, 2019, 27). This hits on the targeted direction of the book: It is a modern fairy tale in which the "forgotten" and "left behind" apparently get their own voice – but in which this voice then turns out to only be that of the neoliberally conditioned social climber. The *ordinary people* are left behind in history, but for a brief moment, satisfy the longing for the "real", the "popular" – in short: the "multiple desire for the *common people*". At the same time, the anti-elitist identification with the "hillbillies" functions here as a prerequisite for the elitist narrative: the strange and paradoxical "Hillbilly Elitism" (Hutton, 2019) of the author JD Vance.

Underclass as a family story: the "Elegy" on Netflix

The figure of JD Vance as a paradigmatic American *transclasse*, as a wanderer between the social worlds, has meanwhile also been milked by the medium of film. In November 2020, a Netflix film production of *Hillbilly Elegy* was released with a star cast. The actresses Amy Adams and Glenn Close played the roles of Mom and Mamaw, thereby embodying two characteristic figurations of the "underclass": the drug-addicted mother who can handle neither her own life nor bringing up her children, and the grandmother whose indomitable "hillbilly" identity ultimately constitutes the young JD's salvation. The film shows notorious images from the shabby, left-behind province, packed full of clichés, though social romanticism is only found in very sparing doses. Loved and beaten up as a child, JD is exhorted to perform at school under the care of his grandmother and is thus faced with a decision: "But you, you got to decide: you want to be somebody or not".

Fourteen years later, JD – now played by Gabriel Basso – is seen sulking at the sink in a restaurant kitchen, working to earn money for his studies at Yale Law School. One cut later, he's already there, preparing for an important dinner during which legal positions will be assigned. JD admits to his girlfriend that he is nervous because: "[He's] going to shine up against the triple Ivy league types". A blue plaid shirt points to the origins of the protagonist from the left-behind province, who is otherwise shown to be a sensitive, romantic, all-round likeable young man. The following scene is a key scene for the picture of the "elitist" outlined here. A darkened room, men and women in suits. They are giving each other business cards and making appointments. Choosing the right wine and the order in which to use the cutlery pose such problems for JD that he has to call his girlfriend at one point to get her advice. Just when dinner is underway, JD receives a call from his sister Lindsay, who tells him that their mother has just been taken to hospital for a heroin overdose. In his mind's eye, he sees images from his childhood pass by, which mark the greatest possible contrast to the dignified society at Yale. Back at the table, JD is asked to tell "his story", whereupon he proudly reports that his grandparents in Kentucky belonged to a kind of "hillbilly royalty". When, eventually, one of the prominent lawyers at the dinner party speaks jokingly of "rednecks", JD coolly replies: "We don't really use that term". The lawyer starts to explain the differences between Yale and the Appalachians and stresses: "I mean, you are at one of the top educational institutions in the world". JD's reply: "My mother was salutatorian of her high school. Smartest person I've ever met. Probably smarter than anyone in this room". The awkward silence is finally broken again by a joke from one of the top lawyers present: "It sounds like maybe we should be offering your mom a position". Ultimately, of course, JD leaves to help his family – the scenes that follow mix childhood memories with episodes of current events in the hospital. "Can't you come home?" – the tear-choked question of his sister prevails; roots are stronger than wings. JD takes care of his mother and runs the risk of missing interview week at Yale.

The film also makes the complexion of the class transitioner clear in its editing: It tells the story of JD's childhood, his career at Yale and his return to Middletown not in chronological succession but as interwoven episodes. It thereby reproduces what Chantal Jaquet calls the "complexion" of the class transitioner in its dramaturgical adaptation. JD appears in the Netflix drama as "someone torn apart, who cannot really succeed in uniting the incompatible – also because the irresolvable contradiction of his self is an inexhaustible source of shame" (Spoerhase, 2018, 233). Here, the film blends origins and arrival, mirrors the two in each other and thus unfolds the central narrative of the book *Hillbilly Elegy*, which shows identity as a resource and ascension as a self-earned achievement. "Where we come from is who we are, but we choose every day who we become". This is how JD sums up the central life lesson of his grandmother "Mamaw" in the film – the rule that she gave him to take with him on his way, and which brought him to Yale. This fundamental rejection of Bourdieu inherently sets the memoir *Hillbilly Elegy* apart from the French examples of the genre, from *Retour à Reims* to *Qui a tué mon père*. And, of course, the film ends with the – presumably – successful job interview with one of the prominent lawyers from the circle connected to Yale Law School.

Conclusion: the figure of the *transclasse* and the referencing of the "people"

There are several reasons for why the figure of the *transclasse* is currently receiving such massive media attention. Against the background of an intensive discourse about social divisions, this figure seems to allow a view from both sides: if the diagnosis is correct that the USA has long since disintegrated into two political echo chambers that hardly know anything about each other (Lütjen, 2017); if the alienation between the social and the political camps has grown so large that they seem separated by "empathy walls" (Hochschild, 2016, 5) and are heading for a "hyperconflict" (West, 2019), then the *transclasse* harbours a fascinating promise: He knows both camps, he knows the *deep story* that connects the two of them – and he can possibly re-start the dialogue between them both. Autosociobiography becomes an "analysis of society" (Spoerhase, 2017) and possibly contributes to an answer to the question that Thomas Frank already posed in 2004: "What's the Matter with Kansas?" (Frank, 2004). Even before that, the position of the *transclasse* has repeatedly been interpreted as an epistemologically privileged position from which the mechanisms of the social can be seen through particularly well. Pierre Bourdieu pointed out several times in his "Sketch for a Self-Analysis" that it was precisely his meteoric rise from humble circumstances to the upper echelons of the French education system that allowed him to develop his sociological perspective: for a start, through the "deep refusal of the scholastic point of view […] to which the relationship to the social world associated with certain social origins no doubt predisposes" (Bourdieu, 2008, 41) and, beyond this, through empirical accuracy and attention to apparently incidental objects of investigation: "Perhaps the fact of coming from what some like to call 'modest' origins gives in this case virtues that are not taught in manuals of methodology." (Bourdieu, 2008, 103). Chantal Jaquet also underlines the "precious opportunity to take a step back and gain distance" (2018, 146) as a chance for insight. In this respect, the autosociobiographies à la *Hillbilly Elegy* also deliver the promise of speaking from a third position: a position between the elite, whose mechanisms the authors see through due to the travails of their biographies of ascension, and the *common people*, whom they understand, but among whom they no longer belong.

However, the *transclasse* runs the risk of losing sight of the structural relationships between *top* and *bottom* because his individual story drives him away from the *bottom* in a specific way. One can only properly understand this genre in the political context of the present if one examines the complex constellations between the "elites" and the "people" that are handled within it. In this essay, I have tried to illuminate two of these constellations on the basis of examples and to work out the relationality of the references to *top* and *bottom*. Both examples involve referencing the *ordinary people* in a way that changes the context of a situation or a narrative. Although they have long since advanced into the ranks of the political and economic elite, both Hans Matthöfer and JD Vance operate using the anti-elitist imaginary of their own social backgrounds. The affect towards the *top* in society is a key part of their social self-positioning but fits into social democratic or neoliberal narratives of progress of the time. Matthöfer and Vance both appear as class transcenders who obtain a specific gain in authenticity from their popular stories of origin. They make use – albeit in very different ways – of the "performative gesture of speaking on behalf of the common people, of representing the common people, or even of simply knowing where the common people can be found" (Dümling, 2020, 19).

The *transclasse* can be understood from this viewpoint as a cultural figure that mobilises specific symbolic resources and has gained in importance, especially in the contemporary discourse. Following on from the classic subculture studies of the Birmingham Centre for Contemporary Cultural Studies, in which subcultures are read as "imaginary solutions" to class conflicts (Clarke et al., 1993), many of the currently popular biographies of social climbers can be understood – according to my thesis – as imaginary solutions to the cultural cleavage of the present, which, in a narrative way, solve the diagnosed problem of a cultural division of society into cosmopolitans and communitarians, or "elites" and "those left behind", by seeming to mirror the perspectives in one another. In the case of JD Vance, as least, it becomes clear that, here, the representation of the "people" ultimately supports the neoliberal narrative of *pull-yourself-up-by-your-bootstraps*. The reference to the *common people* once again proves to be a relational formula that appears in very different historical constellations and configurations. All the more need for a cultural analysis that shows itself to be sensitive to all these contexts of use, in which it is only through mutual relation that the elitist, the anti-elitist and the ordinary constitute their meanings.

Translated from German by Josephine Draper

Notes

1 On the discourse of the 1920s, in particular, see Berking, 1984, Nolte, 2000, 118–127. For an overview of the intellectual history of the figure of the "masses", see Gamper, 2007.
2 The following section is in part based on: Wietschorke, 2007, quotation 4.
3 A ten-minute excerpt from the broadcast can be seen on www.youtube.com. That section essentially contains the scenes discussed here.
4 This sentence is not included in the aforementioned YouTube excerpt. I am quoting from my notes on the rerun of the programme on NDR television from February 23, 2007, and these also form the basis of the earlier piece mentioned (Wietschorke, 2007).
5 On this, see also the brilliant intellectual history of *common sense* in Rosenfeld, 2011.
6 What is also interesting about Böttger's recollections is that after Fritz Teufel had laid his water pistol on the table, a special police unit with sub-machine guns allegedly moved into position backstage. "People didn't see them, but the whole studio was on the floor, and I was lying behind the piano." Carini 2007, 236.
7 On the political culture of that time in West Germany, cf. the illuminating sections in Schildt and Siegfried, 2009, 277–302 and 365–385. Incidentally, the union positioning in this game is notably ambivalent: In characteristic manner, the figure of

the union official stands at odds with the dichotomy of business elites and employees. In the first third of the 20th century, in particular, the Social Democratic "party" or "trade union big shot" was always heavily criticised by the party base. See Stein, 1985.

8 Michael Frank, quoted from Lindner, 2003, 180.
9 Matthöfer also emphasises his class transition in the unpublished manuscript of his autobiography: The working title of the text is "Vom Kohlenpott in den Bundestag" ("From the Ruhr coalfield to parliament") (Abelshauser, 2009, 21).
10 The press quotes are printed in the title pages of Vance, 2016.
11 Printed in the title pages of Vance, 2016.
12 On the history of discourse on *white trash*, cf. Isenberg, 2016.

Bibliography

Abelshauser, Werner (2009). *Nach dem Wirtschaftswunder. Der Gewerkschafter, Politiker und Unternehmer Hans Matthöfer*. Bonn: J.H.W. Dietz.

Baron, Christian (2020). *Ein Mann seiner Klasse*. Berlin: Claassen.

Berking, Helmuth (1984). *Masse und Geist. Studien zur Soziologie in der Weimarer Republik*. Berlin: WAV.

Billings, Dwight B. (2019). Once Upon a Time in "Trumpalachia": *Hillbilly Elegy*, Personal Choice, and the Blame Game. In: Anthony Harkins and Meredith McCaroll (Eds.), *Appalachian Reckoning. A Region Responds to* Hillbilly Elegy. Morgantown: West Virginia University Press, 38–59.

Boltanski, Luc and Ève Chiapello (2007). *The New Spirit of Capitalism*. London: Verso.

Bourdieu, Pierre (2008). *Sketch for a Self-Analysis*. Chicago: University of Chicago Press.

Carini, Marco (2003). *Fritz Teufel. Wenn's der Wahrheitsfindung dient*. Hamburg: Konkret Literatur Verlag.

Clarke, John, Stuart Hall, Tony Jefferson and Brian Roberts (1993). Subcultures, Cultures, and Class. In: Stuart Hall and Tony Jefferson (Eds.), *Resistance through Rituals: Youth Subcultures in Post-War Britain*. London: Routledge, 9–74.

Dews, C.L. Barney and Law Carolyn Leste (Eds.) (1995). *This Fine Place So Far from Home: Voices of Academics from the Working Class*. Philadelphia: Temple University Press.

Dröscher, Daniela (2018). *Zeige Deine Klasse. Die Geschichte meiner sozialen Herkunft*. Hamburg: Hoffmann und Campe.

Dümling, Sebastian (2020). "I want to live like common people" – Populismus und das multiple Begehren nach den "einfachen Leuten". Eine einführende Skizze. In: Sebastian Dümling and Johannes Springer (Eds.), *Die "einfachen Leute" des Populismus – Erzählungen, Bilder, Motive*. Schweizerisches Archiv für Volkskunde 1, 9–19.

Eribon, Didier (2009). *Retour à Reims*. Paris: Fayard.

Eribon, Didier (2016). *Rückkehr nach Reims*. Berlin: Suhrkamp.

Frank, Thomas (2004). *What's the Matter with Kansas? How Conservatives Won the Heart of America*. New York: Henry Holt.

Gamper, Michael (2007). *Masse lesen, Masse schreiben. Eine Diskurs- und Imaginationsgeschichte der Menschenmenge, 1765–1930*. Munich: Wilhelm Fink.

Haferkamp, Hans (1983). *Soziologie der Herrschaft. Analyse von Struktur, Entwicklung und Zustand von Herrschaftszusammenhängen*. Opladen: Westdeutscher Verlag.

Harkins, Anthony (2003). *Hillbilly. A Culture History of an American Icon*. New York: Oxford University Press.

Harkins, Anthony and Meredith McCaroll (2019). *Appalachian Reckoning. A Region Responds to* Hillbilly Elegy. Morgantown: West Virginia University Press.

Hartmann, Michael (2004). *Elitensoziologie. Eine Einführung*. Frankfurt am Main: Campus.

Hartmann, Michael (2018). *Die Abgehobenen. Wie die Eliten die Demokratie gefährden*. Frankfurt am Main: Campus.

Hochgeschwender, Michael (2017). Diesen Weißen geht es so wie den Schwarzen. *Frankfurter Allgemeine Zeitung*, 08 May 2017, URL: https://www.faz.net/aktuell/feuilleton/buecher/rezensionen/sachbuch/hillbilly-elegie-von-j-d-vance-fuehrt-in-welt-der-trump-waehler-14960625.html.

Hochschild, Arlie Russell (2016). *Strangers in Their Own Land. Anger and Mourning on the American Right*. New York: The New Press.

Hutton, T.R.C. (2019). Hillbilly Elitism. In: Anthony Harkins and Meredith McCaroll (Eds.), *Appalachian Reckoning. A Region Responds to* Hillbilly Elegy. Morgantown: West Virginia University Press, 21–33.

Isenberg, Nancy (2016). *White Trash. The 400-Year Untold History of Class in America*. New York: Viking.

Jaquet, Chantal (2018). *Zwischen den Klassen. Über die Nicht-Reproduktion sozialer Macht*. Konstanz: Konstanz University Press.

Lindner, Rolf (2003). Vom Wesen der Kulturanalyse. *Zeitschrift für Volkskunde* 99, 177–188.

Louis, Èdouard (2014). *En finir avec Eddy Bellegueule*. Paris: Seuil.

Louis, Èdouard (2019). *Who Killed My Father*. London: Harvill Secker (EBook).

Lütjen, Torben (2017). Die grosse Entzweiung. Wie Amerika in politische Echokammern zerfiel. *Aus Politik und Zeitgeschichte* 67 no. 18, 9–15.

McGarvey, Darren (2017). *Poverty Safari. Understanding the Anger of Britain's Underclass*. Edinburgh: Luath Press.

Muzzatti, Stephen L. and C. Vincent Samarco (Eds.) (2006). *Reflections from the Wrong Side of the Tracks. Class, Identity, and the Working Class Experience in Academe*, Lanham: Rowman & Littlefield.

New York Times. (2016). 6 books to Help Understand Trumps's Win, November 9. URL: https://www.nytimes.com/2016/11/10/books/6-books-to-help-understand-trumps-win.html

Nolte, Paul (2000). *Die Ordnung der deutschen Gesellschaft. Selbstentwurf und Selbstbeschreibung im 20. Jahrhundert*. Munich: C.H. Beck.

Reichardt, Sven (2014). *Authentizität und Gemeinschaft. Linksalternatives Leben in den siebziger und frühen achtziger Jahren*. Berlin: Suhrkamp.

Reitmayer, Morten (2009). *Elite. Sozialgeschichte einer politisch-gesellschaftlichen Idee in der frühen Bundesrepublik*. Munich: R. Oldenbourg.

Rosenfeld, Sophia (2011). *Common Sense. A Political History*. Cambridge: Harvard University Press.

Ryan, Jake and Charles Sackrey (Eds.) (1996). *Strangers in Paradise: Academics from the Working Class*. Lanham: Rowman & Littlefield.

Schildt, Axel and Detlef Siegfried (2009). *Deutsche Kulturgeschichte. Die Bundesrepublik – 1945 bis zur Gegenwart*. Munich: Hanser.

Smarsh, Sarah (2018). *Heartland: A Memoir of Working Hard and Being Broke in the Richest Country on Earth*. New York: Simon and Schuster.

Spoerhase, Carlos (2017). Politik der Form. Autosoziobiographie als Gesellschaftsanalyse. In: *Merkur. Gegründet 1947 als Deutsche Zeitschrift für europäisches Denken* 71, 818, 27–37.
Spoerhase, Carlos (2018). Aufstiegsangst: Zur Autosoziobiographie des Klassenübergängers (afterword). In: *Chantal Jaquet: Zwischen den Klassen. Über die Nicht-Reproduktion sozialer Macht*. Konstanz: Konstanz University Press, 231–253. https://www.k-up.de/9783835391048-zwischen-den-klassen.html.
Stein, Gerd (1985). Vorwort. In: Stein, Gerd (Ed.), *Lumpenproletarier – Bonze – Held der Arbeit. Verrat und Solidarität. Kulturfiguren und Sozialcharaktere des 19. und 20. Jahrhunderts*, vol. 5. Frankfurt am Main: Fischer, 9–19.
Vance, J.D. (2016). *Hillbilly Elegy. A Memoir of a Family and Culture in Crisis*. London: HarperCollins.
West, Darrell M (2019). *Divided Politics, Divided Nation. Hyperconflict in the Trump Era*. Washington, DC: The Brookings Institution.
Westover, Tara (2018). *Educated: A Memoir*, New York: Random House.
Wietschorke, Jens (2007). Gutes Benehmen – oder: Der Teufel steckt im Detail. Kulturanalytische Randnotizen zu einer Talksendung vor 25 Jahren. *Kuckuck. Notizen zur Alltagskultur* 2, 4–6.
Wietschorke, Jens (2020). The Politics and Poetics of Cultural Cleavage. Notes on a Narrative. *Journal for European Ethnology and Cultural Analysis* 5, 102–118.

"AGAINST THE ELITES!"

POLITICS, ECONOMY, INEQUALITY

PART II

REBECCA BRAMALL, STEFANIE HÜRTGEN, OLGA REZNIKOVA

5 What are we going to do about the rich?

Anti-elitism, neo-liberal common sense and the politics of taxation

Rebecca Bramall

Introduction

It's early Spring 2019, and I'm listening to a new song by the Pet Shop Boys on YouTube. The song's lyrics itemise and complain about the extravagant habits and behaviours of the super-rich, from buying football clubs to acquiring media outlets, before moving inexorably to their tax arrangements: "They're avoiding paying taxes/While the welfare state collapses [...] What are we gonna do about the rich?".

In the long decade since the financial crisis, tax issues have gained considerable political salience in the UK and around the globe. Stories exposing the avoidance of tax by celebrities, politicians and business leaders regularly circulate in the press and social media, and opinion polls indicate that the topic is of significant public interest.[1] In defining "the rich" as tax avoiders, the Pet Shop Boys' song is exemplary of a broader popular culture in which the issue of tax avoidance has helped to bring wealthy elites into focus. Taxation has served as a terrain on which "the elite" have been imagined and defined and an antagonism between *them* and *us* has been played out. It is not surprising that this construction of the rich as tax avoiders has been adopted and bolstered by movements on the left. Political parties who have pursued left populist strategies in the 2010s – specifically Podemos in Spain and the UK Labour Party under the leadership of Jeremy Corbyn – have espoused this view. What has been more remarkable is the proliferation of complaints against elite tax avoidance right across the political spectrum. Senior leaders of the UK Conservative Party have declared that those who evade taxes are "leeches on society", that aggressive tax avoidance is "morally repugnant" and that individuals and businesses "must pay their fair share" (BBC, 2013; Osborne, 2011). There are even some sections of the rich who have aligned themselves against their tax-avoiding peers. In 2020, a group of super-rich individuals from around the globe published an open letter noting that "tax avoidance and tax evasion have reached epidemic proportions" and demanding higher taxes on millionaires and billionaires (Millionaires Against Pitchforks, 2020).

For decades, "common-sense" thinking about taxation has been dominated by neo-liberal ideas (Hall and O'Shea, 2013) which align the interests of the rich with the public interest and posit that taxation inhibits the

DOI: 10.4324/9781003141150-7

entrepreneurial risk-taking of society's wealth creators. Yet these recent popular cultural and political developments are suggestive of a certain shift in common-sense thinking about taxation and the rich – or, at the very least, of the availability of a potential resource for the critical contestation of neo-liberal tax regimes. My aim in this chapter is to consider the political implications of the new visibility of tax-avoiding elites and, in particular, the extent to which it creates opportunities to challenge the neo-liberal ideas that have become embedded in common sense about taxation. Although my discussion is centred on the UK, neo-liberal ideas about taxation circulate around the globe, making it useful to reference other national contexts, including the US, Spain, Ireland and France.

There are four parts in this chapter. In the first part, I review the neo-liberal ideas about taxation that have become sedimented in common sense over the last four decades. I go on to identify some of the key frames through which "the rich" have been discursively constructed in the crisis years of the 2010s. In the third part, I discuss how the characterisation of elites as tax avoiders has been exploited by political parties on the left, who have sought to use tax issues to enable "the people" to recognise that neo-liberal economic policies privilege elites and are designed to serve their interests. In the last part of the chapter, I evaluate the new visibility of elites as tax avoiders and their activation as adversaries in left politics and consider the opportunities that the political salience of tax avoidance in the 2010s presents to left political actors. This final part of the chapter moves through three stages in which I assess the specificities of the elites, the people and the political demand that have emerged on the terrain of debate about tax avoidance. I argue that while the construction of elites as tax avoiders has helped to foreground the ways in which neo-liberal capitalism is designed to favour the economic interests of the few, it has not supported the consolidation of diverse grievances against the current economic system into a general political demand. The chapter contributes to our understanding of the "performative dimension" (Laclau, 2007, p.14) of ideas about taxation in contemporary political debate and popular culture.

Neo-liberalism and common sense about taxation

It is essential in any discussion of neo-liberalism to begin by noting the heterogeneous nature of this complex political, economic and ideological formation, the "elasticity of neoliberal norms and principles" (Slobodian and Plehwe, 2020, p.11) and the interpenetration of those norms with competing philosophies (Cooper, 2017). It is also helpful to distinguish between neo-liberal economic policy and practice, neo-liberal "free market" ideology and the neo-liberal ideas that have become sedimented in common sense – although these three planes are closely imbricated and often mutually reinforcing.

Low taxation is recognised as a key orientation and destination in the neo-liberal agenda, and so tax policies are often used to identify the neo-liberal quality of past and ongoing economic practice. When critics characterise neo-liberal economic policies, tax reduction for wealthy individuals and corporations invariably appears near or at the top of a list which also includes deregulation, privatisation and marketisation (Brown, 2016; Jessop, 2015). Ronald Reagan's tax cut in 1981 is seen, more than any other policy intervention, as "the most important instance of American neo-liberalism" and as a central pillar in the rise of the market society (Prasad, 2012, p.352). If the goal of neo-liberalism is to promote a competitive order in which individuals are encouraged to behave as free market actors, a commitment to low taxation is understood to support that goal by placing limits on the state's capacity to intervene in the functioning of the market, for example, through policies which seek to redistribute wealth and to be consistent with the ambition to free entrepreneurial actors from these interferences.

In this chapter, I am primarily interested in the third plane of neo-liberalism: the "neo-liberal" elements of "common sense". By "common sense", I refer to the Gramscian concept of popular understanding and knowledge defined by Stuart Hall and Alan O'Shea as "a form of 'everyday thinking' which offers us frameworks of meaning with which to make sense of the world" (2013, p.8). Common sense contains "[b]its and pieces of ideas from many sources", thus, the frameworks of meaning that enable people to make sense of taxation are not identical to neo-liberal ideology. These frameworks are also composed of elements from past political projects and philosophical traditions, such as welfare capitalism, some of which may support insights guided by "good sense" (Hall and O'Shea, 2013, p.10). Nevertheless, common sense about taxation is often judged to be very strongly shaped by neo-liberal ideas and lacking in alternative resources to counter those ideas. While other aspects of the neo-liberal project have modulated over time, the orientation towards low taxation both in neo-liberal ideology and economic practice has been comparatively consistent, helping to secure its sedimentation in common sense. Writing about "hegemonic common sense about taxation", Doreen Massey submits that "[i]n the unthought assumptions of everyday speech, tax is a (necessary) evil" (2016b, p.161).

Turning specifically to the subject of rich elites and taxation, there are a series of interlinked neo-liberal ideas which are strongly embedded in common sense and are in turn linked to other powerful explanatory narratives. Firstly, the idea that the rich have worked hard for their money – that they have earned it through legitimate entrepreneurial activity – is supported by a wide consensus that we live in a "meritocratic" society. As Jo Littler (2018) has demonstrated, the notion of meritocracy has become absolutely central to the legitimation of plutocratic neo-liberal capitalism. Wealth that is seen as legitimately accrued in a meritocratic system is insulated from debates about inequality because the latter tends to be justified and naturalised (Harvey et al., 2015). Neo-liberalism further legitimates the unequal distribution of wealth through the idea that the innovation and enterprise which produce economic growth derive from the risk-taking and shrewd investments made by entrepreneurs (Mazzucato, 2013). Wealth is said to "trickle down" to everyone else in that society through the production of jobs and economic activity.

Neo-liberalism holds that it is essential to keep taxes low, so as not to inhibit the productivity of these entrepreneurial individuals. Indeed, according to one powerful common-sense notion, there is an ideal level of taxation that will optimise incentives to work, innovation and productivity. Taxing individuals above this ideal rate is said to disincentivise risk-taking or, even worse, lead to the removal of assets from the jurisdiction in question. We have become accustomed in the UK to warnings that the super-rich will "leave the country" if tax rates are increased. In the run-up to the 2019 general election, for example, the billionaire founder of a mobile phone company vowed to leave if the Labour Party formed the next government, claiming he would "just go and live in the south of France or Monaco" (Neate, 2019). These threats illustrate that the capacity for an individual to organise their affairs so as not to pay too much tax has come to be regarded as a legitimate feature of the lives of celebrities and the super-rich (Urry, 2014). Neo-liberalism frames tax avoidance by the wealthy as the fault of governments who have designed uncompetitive tax policy, while the rich are cast as "deserving" of their wealth and shrewd in their use of legal loopholes. In a debate with Hillary Clinton during the run-up to the 2016 US presidential election, Donald Trump famously claimed that avoiding federal income tax made him "smart" (Diaz, 2016).

These ideas about ideal levels of taxation and trickle down derive largely from supply-side economics and discredited theories such as the Laffer curve (Prasad, 2012). Yet they have become deeply embedded in common sense and they provide powerful frameworks for understanding debates about taxation and the economy in general. The popular non-fiction book *What Everyone Needs to Know About Tax* delivers an "entertaining and informative look at the UK tax system" and promises to take the reader beyond the "media hype" (Hannam, 2017). Adopting a rational, pragmatic tone, the book reiterates key neo-liberal ideas about taxation as established truths: "A common suggestion to meet the government's need to raise more money is to tax the rich. Sadly, things are a bit more complicated than that", the author explains (Hannam, 2017, p.10). Reviewing the effort to maintain higher rates of taxation on the super-rich in the 1970s, he claims that "[m]any of the most talented individuals just left the country. [...] When the tax burden is heavy, it drives them out of the country". The author ends this section of the book by concluding that high marginal rates on the rich are "economically perverse" (Hannam, 2017, p.12).

There is a final neo-liberal idea which does not relate directly to the rich and wealthy but does have an important bearing on the topic of this chapter. This idea is uncritically repeated in media discourse and by actors across the political spectrum on a routine basis. Over the last forty years, the "taxpayer" has become a significant subject position in neoliberal culture. Emerging as a discursive figure within Thatcherism (Hall, 1988, p.49), the taxpayer was later promoted within the New Labour variant of neoliberalism (Hall, 2011, p.715) and is also central to political discourse in the US (Williamson, 2018) and many other countries. The figure of the taxpayer became a focal point in the context of the extension of competitive markets into domains of social life previously serviced by welfare states. Opportunities for individuals to opt for private provision of health, education and other services increased in the 1980s and 1990s. These opportunities position citizens as "customers" ready to exercise freedom of choice between the options made available to them by private markets. As the middle classes in particular become habituated to providing for themselves in this way, the payment of taxes – which provide "generalized support to the community as a whole" (Streeck, 2017, p.12) – become open to greater contestation and scrutiny. Not all citizens are included when "taxpayers" are addressed: on the contrary, the concept has an exclusionary and divisive function, promoting social antagonism towards citizens who are fashioned as welfare beneficiaries and "non-taxpaying others" (Hackell, 2013, p.134). Writing about the New Zealand context, Melissa Hackell notes that "the taxpayer subjectivity condenses a range of human attributes [...] that come under the rubric of competitive individualism" (2013, p.134).

Constructing elites as tax avoiders

For decades, then, common sense thinking about taxation has been dominated by neoliberal ideas. The global financial crisis of 2007–2008 delivered opportunities to contest this neo-liberal consensus, and a wide range of social actors – including politicians, NGOs, campaigners and activists – played a role in producing tax avoidance as an issue of concern. The impacts of fiscal austerity measures are regarded as particularly important among the many factors that gave rise to the increased political salience of tax avoidance. These impacts are understood to have prompted citizens in the UK and elsewhere to take "a different kind of interest" in taxation and spending (Cobham in Burgis, 2016) and to have delivered opportunities for civil society actors to problematise tax avoidance in terms of national revenue (Birks, 2017; Vaughan, 2019). This argument was, for example, highly salient in the UK context in the wake of interventions by the grassroots activist organisation UK Uncut, which brought to light alleged tax avoidance by multinationals in order to critique austerity economics (Bramall, 2016). The era of post-crisis austerity also provided a context for debate about the threat to fiscal sovereignty posed by the development of the offshore world in the globalised neo-liberal era (Christensen and Hearson, 2019; Urry, 2014).

Beginning in the earlier part of the decade and continuing for at least five years, the tax affairs of well-known individuals began to receive significant levels of critical media attention (Bramall, 2018). The factors determining this focus of attention are not identical to those motivating civil society actors. Campaigners have tended to focus their efforts on tax avoidance by multinationals, while the UK tabloid and right-wing press has focused on cases in which celebrity misdemeanours can be brought to centre stage. This strand of reporting aligns with the orientation of tabloids towards "click bait" stories which celebrate the risk-taking, "frontier" existence of celebrities

(Rojek, 2012, p.37). In 2016, the UK broadsheets provided extensive analysis of the Panama Papers leak, and news organisations have also conducted many smaller-scale investigations into elite individuals and celebrities of interest to their readers. This critical attention has developed into a broad and diverse discourse on elite tax avoidance, extending well beyond the press and news media.

Media discourse does not directly reflect or constitute the workings of common sense, as there are many other domains of our social lives through which "everyday thinking" is fostered and secured. Media institutions and forms are, however, key sites of "hegemonic work" (Clarke, 2010, p.350), and media discourse is one of the more accessible "repositories" of common sense (Hall and O'Shea, 2013, p.9). It is a site where enduring and emergent frameworks of meaning become concretised and are available to be identified and evaluated. The media discourse which posits rich elites as "tax avoiders" has been iterated through an identifiable set of devices, practices and frames, and three particularly persistent characteristics can be noted.

Firstly, a rhetoric of exposure – of shining a light on "hidden" activities, behaviours and practices – consistently animates these stories. The activities and motivations of the International Consortium of Investigative Journalists, who investigated and reported on the Panama Papers, can be placed in a long tradition of journalistic revelation and exposure (Inglis, 2010). While this genre of journalistic exposé tends to adopt conventions of objectivity and impartiality, it has fuelled a "growing culture of naming and shaming" (Barford and Holt, 2013), in which revelation invites condemnation. Commentators across the political spectrum have welcomed "tax shaming" as a significant device in the fight for tax justice. They argue that in lieu of the government action needed to simplify the tax code and close loopholes, the threat of public exposure and condemnation serves to deter would-be tax avoiders. In one financial journalist's words, "there are few more powerful weapons than a really high-profile public shaming" (Ford, 2012).

The Panama Papers exposé also firmly established the idea that intermediary organisations, such as corporate law firms, consultants, business registries and corporate services providers, hold vast stores of documents about the tax affairs of rich elites, and that leaks from these organisations provide only a glimpse of arrangements that have become globally pervasive. Other unseen documents – specifically politicians' tax returns – have become objects of attention within this rhetoric of exposure. While some politicians have made a practice of publishing their tax returns on an annual basis, Donald Trump's refusal to make his affairs public became the subject of a major, long-running court battle and media controversy (Doerer, 2020), which culminated in an investigation published in the *New York Times* (Buettner et al., 2020). Commentators on both sides of the Atlantic speculated about whether or not these revelations would "sink" Trump (Smith, 2020; Yokley, 2020).

The rhetoric of exposure has been central to the construction of elites as tax avoiders, yet this configuration of the "elite" has not been produced solely through journalistic discourse and news reporting. Paulo Gerbaudo

(2018) argues that social media provide channels in which the interests of "ordinary people" can be invoked as against the "establishment", a tendency he describes in terms of an "elective affinity" between social media and "populism". Social media are certainly important sites for the dissemination and discussion of stories about tax avoidance by wealthy elites, and platforms such as Twitter and Facebook provide spaces in which memes, jokes and satirical hot takes can be shared.[2] Yet analysis of this material quickly reveals its connectedness to other media forms – not just news journalism, but film, television and popular culture.[3] Rather than focusing solely on journalism and/or social media, it is more accurate to emphasise the hybridity and interpenetration of the media systems, practices and genres that have contributed to the construction of elites as tax avoiders. Secondly, then, it is a context of media convergence and connectivity, in which media users "move across multiple platforms and engage with a diverse array of content types" (Cupples and Glynn, 2020, p.178), which has enabled and sustained the construction of elites as tax avoiders.

A third tendency has been consistently present in the "elite tax avoiders" discourse. Media reports about tax avoidance invariably reference a "public" that is said to be "angry" about this issue. An opinion piece in the UK political magazine the *New Statesman* opens with the assertion that "[p]ublic anger over tax avoidance is palpable" (Rowney, 2015). Where journalists provide evidence of this public anger, they usually refer to one of the many opinion polls that have addressed this topic. The charity Christian Aid, for example, asked participants to agree or disagree with the statement that tax avoidance "makes me feel angry" (Savanta: ComRes, 2013). Two related theoretical perspectives can be employed to interpret this referencing of a "public that is angry about tax avoidance". According to Nick Mahony and John Clarke, "publics as entities [...] are always mediated and always emergent, rather than being pre-existing" (2013, p.933). On this account, media reporting about tax avoidance has contributed to the construction and mediation of a "public" that is concerned about this topic, just as it has contributed to the construction of an "elite" engaged in such practices. As for the recourse to survey results, Mahony and Clarke argue that although public opinion polling relies upon and reproduces an idea of the public as a "pre-existing collectivity" (2013, p.935), it is best understood as another medium through which "subjects and objects of publicness" are "assembled" (Barnett, 2008, p.404; also see Hall and O'Shea, 2013).

This perspective on the mediation of publics can be productively integrated with Karin Wahl-Jorgensen's work on mediated anger in the press coverage of protests. Building on research that emphasises the discursive construction of anger through journalistic practice, Wahl-Jorgensen demonstrates that this emotion is attributed to protesters in order to provide an explanatory framework for collective grievances and a "barometer of the intensity of public feeling" (2018, p.2083). There is, she argues, "a spectrum of discursive constructions of the legitimacy of mediated anger", and it can be construed as rational and legitimate, or irrational and illegitimate (2018, p.2077). Wahl-Jorgensen's perspective indicates that the construction in the UK press of a public that is angry about tax avoidance is enabling of further claims about the legitimacy of the complaint, a point to which I will return later in this chapter.

Political activations

To summarise, the discursive construction of elites as tax avoiders has been shaped through a media discourse which has consistently adopted a rhetoric of exposure. This discourse has concurrently summoned a "public" that is "angry" about tax avoidance. Although the framing of stories about elite tax avoidance derives from a tradition of journalistic revelation and exposure, this discourse has circulated and been sustained in a context of media convergence. This media discourse has helped to bring wealthy elites into focus in the decade since the global financial crisis and is suggestive of a certain shift in common-sense thinking about taxation.

Left political parties and movements share a challenge, as Doreen Massey and others have put it, in constructing a political frontier: how should this frontier be characterised, and "who [or] what is the 'enemy'?" (2016a, p.11). This challenge is often discussed using the conceptual framework developed by Ernesto Laclau and Chantal Mouffe to interrogate populism (Laclau, 2007; Laclau and Mouffe, 2001; Mouffe, 2018). Laclau and Mouffe's post-Marxist framework enables us to discern that tax avoidance in recent years has played an important role in left populist efforts to communicate demands that establish a frontier between "the people" and their antagonists. These actors have, consequently, played a significant role in activating the status of elites as tax avoiders. In the UK, the Labour Party under Jeremy Corbyn (2015–2020) adopted such an approach. In the general election manifesto of 2017, ideas about fair taxation were used to construct the political subject of "the many" against "the few". As Mouffe points out, this was originally a New Labour slogan which was "re-signified [...] in an agonistic way" (2018, p.28). The manifesto argued that the richest in society – the top 5% of earners, earning more than £80,000 – needed to pay more in tax to fund public services. It also reiterated Labour's view that tax avoidance is a "social scourge", which it promised to tackle through the closing of legal loopholes (Labour Party, 2017). During this period, the Labour front bench consistently attempted to define the super-rich as "tax dodgers" depriving the nation of essential funds. On the occasion of the Paradise Papers exposé, John McDonnell (then Shadow Chancellor) released a short video on social media, in which he commented that:

> My neighbours will be getting up [...] and going to work and they will pay their taxes and those taxes will pay for our public services. [...] What is happening is that the super-rich are avoiding paying taxes [...] and as a result of that not funding our public services. We've gone through seven years of austerity [...] largely [...] because the super-rich are just not paying their taxes.[4]

In Spain, the political party Podemos has worked to construct a frontier between *la gente* ("the people") and *la casta*, a group Sirio Canos has defined as "the highly corrupt political and economic revolving-door elite" (Prentoulis et al., 2015, p.22). Podemos has, at different times, foregrounded various corrupt practices in order to characterise this elite, and the party's campaign launch in Valencia in May 2014 drew attention to tax avoidance and evasion (Sanders et al., 2017). Party leader Pablo Iglesias was shown completing his tax return in a campaign video during the 2014 European elections (Sanders et al., 2017), a performance which can be compared to Corbyn's practice of publishing his tax return on his constituency website.[5] The issue of legal avoidance in particular resonates strongly with Podemos's ambition to unveil the multiple ways in which the economic system is the outcome of political decision-making which favours the wealthy over the interests of "normal" people (Canos in Prentoulis et al., 2015). Rodrigo Stoehrel (2017, p.562) argues that the party's significant achievement has been "the production of a narrative that places [...] the whole current financial politics of corporate safety and favourable tax conditions for the wealthy, against the interests of the 'average Spaniard'".

In Spain, as in the UK and other European contexts, the frame of "austerity" has been a consistent reference point in debates about tax avoidance, providing opportunities for political actors such as John McDonnell to problematise tax avoidance in terms of national revenue. The argument is that closing tax avoidance loopholes will recoup lost revenue and mitigate the need for austerity. In the US, where the notion of austerity has not had the same ideological purchase, the debate about rich elites and taxation has taken a slightly different turn. The activation of this rhetoric has centred around the newly resurgent Democratic Left and, in particular, Senator Alexandria Ocasio-Cortez. In January 2019, she proposed that income above 10 million dollars should be taxed at 70%. Later that year, a three-word tweet – "Tax the rich." – was "liked" by over 280,000 users of the platform.[6] Unlike her associates on the left in the UK and Spain, Ocasio-Cortez tends not to refer to the rich as tax *avoiders* and instead focuses on the argument that they should simply pay more tax.

What kind of elite?

The political salience of the issue of tax avoidance in the decade after the global financial crisis has created opportunities for left political actors, who have responded in a variety of ways shaped by their national contexts. In the final sections of this chapter, I want to evaluate the new visibility of "elites as tax avoiders" and offer some critical observations about their activation as adversaries in left politics. What is useful and what is limiting about this formulation? And does it provide an opening to contest neo-liberal ideas about taxation? Before I embark on this discussion, it is worth clarifying that I am not interested here in contributing to a debate about populism. Following Laclau (2007, p.17), I understand "populism" as "a *political logic*", rather than as a "*type* of movement", and so a discussion of whether or not the Labour Party or Podemos are or were "populist parties" at the time of their activation of these ideas about tax avoidance is not my concern. Neither do I wish to examine the political value of populist political logic at an abstract level – that is, whether or not, in general, populist and anti-elitist strategies are effective in hegemonising left demands. Instead, the discussion aims to analyse populist logic in the concrete: in the long decade after the global financial crisis, what opportunities does the political salience of tax avoidance afford left political actors in the long decade after the global financial crisis? The formulation of elites as "rich tax avoiders" is only one of a number of competing ways currently in circulation of describing wealthy elites, and it can be distinguished from recently popular conceptions of elites such as "the 1%". Thus one way to address the question about the political salience of tax avoidance might be to consider the specificities of the "elite" that has surfaced on the terrain of debate about tax avoidance. This will be followed by a discussion of the kind of "people" that is constituted on this terrain and, finally, of the kind of political demand that has emerged.

While reporting on tax avoidance in the UK tabloid and right-wing press has tended to foreground celebrity misdemeanours, the media discourse which constructs rich elites as tax avoiders does not focus exclusively on the already visible wealthy. On the contrary, the rhetoric of exposure – which has revealed the existence of vast datasets of unseen, "secret" documents – has tended to confirm the ubiquity of tax avoidance amongst the rich and affluent, including the "faceless rich" (Littler, 2018, p.136) who have previously escaped prominence. The *Independent*, for example, carried a widely circulated data visualisation representing the occupations of people associated with the Panama Papers, which includes an extensive range of professions (Sheffield, 2016). The sheer volume of reporting based on the Paradise and Panama Papers has also tended to communicate the pervasiveness of tax avoidance by high-net-worth individuals all around the globe.

There has been a marked readiness in the British press to articulate tax avoidance to other markers of elite status. For example, there has been much greater scrutiny of philanthropy in recent years, with links made to the issue of tax avoidance through discussion of the benefits of tax relief to philanthropists (Vallely, 2020). In 2019, Rutger Bregman, a Dutch historian, made a widely reported intervention during the World Economic Forum at Davos, demanding that his influential audience "just stop talking about philanthropy and start talking about taxes. [...] That's it. Taxes, taxes, taxes. All the rest is bullshit in my opinion" (Matthews, 2019a). The connection was also lucidly expressed in the backlash against the French super-rich and their donations in the wake of the Notre Dame fire (Baker and Denis, 2019). In a similar vein, questions have been raised about whether individuals who avoid tax should be considered ineligible for commendation via the British honours system. David Beckham's nomination for a knighthood was reportedly blocked after he was "red flagged" by the UK tax authority (Booth and Grierson, 2017). In these kinds of stories, the tax avoiding practices of elite individuals have become articulated to other elite privileges and markers of status.

Media discourse has also enriched our sense of the spatial settings that global elites are imagined to inhabit and benefit from. Tax avoidance is a difficult phenomenon to visualise, and so stock images of Caribbean tax islands – featuring sandy beaches, palm trees and yachts – have become the go-to illustration for stories about the tax affairs of the global elite. These are not unfamiliar images, and their association with the super-rich is an established one. What has been clarified is their status as offshore locations that enable the super-rich to operate outside the jurisdiction of the tax authorities and nation states to which non-elite individuals are subject (Beaverstock et al., 2004). The increasing prominence of elite financial arrangements has contributed to a popular representation of the super-rich as inhabiting privileged spatial settings from which the majority of people are excluded. As Paula Serafini and Jennifer Smith Maguire put it, "there is a general sense of the deepening chasm between the very wealthy and 'the rest,' [...]. The super-rich appear increasingly isolated in a foreign land in which different tax regimes and life expectancy outcomes apply" (Serafini and Smith Maguire, 2019, p.2).

What is notable about the kind of elite that emerges out of the debate about tax avoidance is the extent to which it aligns with and makes visible many of the issues that have been foregrounded in the critical analysis of neo-liberalism. This is not to say that the elite animated in this debate is identical to that defined in critical scholarship, which, in any case, receives different emphases depending on the nature of the enquiry. William Davies's work on elite power in neo-liberalism, for example, foregrounds the financial elites who benefit extensively from finance-led capital, but are "characterized by an *absence* of public identity" (2017, p.229). These financial intermediaries have not remained completely invisible in the media discourse around elite tax avoidance – the disgraced lawyers Jürgen Mossack and Ramón Fonseca feature in Steven Soderbergh's loose fictionalisation of the Panama Papers scandal, for instance[7] – but they have not appeared front and centre. I mean to point instead to the way in which the "elites as tax avoiders" formulation foregrounds the mechanisms and structures that have been the focus of substantial research (Serafini and Smith Maguire, 2019, p.5). Critical assessments of global neo-liberalism invariably point to the role of tax policy (Ott, 2017) and the capacity of the super-rich to avoid paying tax as key factors in increasing levels of global inequality (Zucman, 2015). The terrain of debate about taxation clearly offers significant potential to illuminate the ways in which neo-liberal capitalism is designed to favour the economic interests of the super-rich, and to support the formulation of political demands that would address this systemic problem. The construction of elites as tax avoiders does not foreground a specific percentage of high-net-worth individuals – as per the identification of the elite as the "1%" by the Occupy movement (Matthews, 2019b). Instead, it confirms the universal and pervasive nature of the economic, legal and political advantages that the rich enjoy. Through the articulation of these advantages to other elite privileges and markers of status – such as practices of philanthropy or the UK honours system – the "elite" that emerges from this discursive formation is clearly positioned within an interconnected set of structural entitlements.

Furthermore, the foregrounding of capitalism's mechanisms and structures through the construction of elites as tax avoiders has facilitated critical challenge of dominant neo-liberal ideas. This is particularly evident in relation to neo-liberal ideas that legitimate wealth and are interconnected with ideas about taxation, such as the trickle-down effect. The elites as tax avoiders formulation provides resources to challenge the notion that wealth at the top benefits and makes its way to us all through investment and job creation. Instead, it has become possible to assert an imaginary in which wealth is amassed and sheltered in tax havens, "remote from the point where value is extracted" (Davis and Williams, 2017, p.11). In an opinion piece for the *New York Times*, the economist Gabriel Zucman (2018) used the example of Portuguese footballer Ronaldo's tax dodging to refute trickle-down economics and explain why "[s]ky-high incomes for star athletes are socially useless" (2018).

Certain cohorts of the global super-rich have played an important role in challenging neo-liberal ideas in this way in recent years. Representatives of groups such as "Millionaires for Humanity" and "Patriotic Millionaires" regularly feature in social media campaigns as well as publishing open letters calling for governments to permanently increase taxes on them. In a video shared in 2017, the US entrepreneur Nick Hanauer attacks Trump's recent tax cuts, describing them as "criminally stupid and totally corrupt". Hanauer addresses head on the Republicans' neo-liberal argument that tax cuts for entrepreneurs will lead to more jobs and higher wages for the average working American, concluding that "You got trickle down scammed, America!".[8] The rhetoric adopted by these rich tax advocates indicates that debate about taxation has created opportunities to cut through neo-liberal elements of common sense and activate a shared good sense about the production and circulation of wealth in societies.

What kind of people?

The discussion so far suggests that the political salience of tax avoidance affords significant opportunities for the left. Yet, there are other dimensions to the elite tax avoiders formulation which merit scrutiny and give significant pause for thought. As I have already suggested, the production of antagonism towards tax-avoiding elites does more than define the richest few in society. Media discourse has concurrently summoned a "public" that is "angry" about tax avoidance, and this construction sustains claims about the legitimacy of this complaint. Such claims – whether sympathetic or dismissive – invariably link public anger about tax avoidance to the rise of populism. The author and campaigner Richard Murphy, who is an influential voice in the debate about tax policy in the UK, argues that "the wave of political populism that is now sweeping through many countries is at least partly based on an awareness that tax havens threaten the well-being of most 'ordinary' people" (2017, p.2). Rhetorically, Murphy's mention of political populism helps to confer an urgency on his recommendations – if today's politicians do not take action, he implies, they are unlikely to be in power tomorrow.

The problem with the framing of tax avoidance as a populist issue is that it can – in less sympathetic hands – easily became a means of dismissing the complaint in question. Marco D'Eramo (2013) has drawn attention to a "negative revaluation" of populism which aims to reassert the political legitimacy of centrist politics. An increasing range of political actors come to be characterised as "populist" precisely "at the historical moment when the developed world is advancing into an oligarchical despotism" – when "anti-popular measures are multiplying" (D'Eramo, 2013, p.27). The ascription of anger to the public that is concerned about tax avoidance makes it possible to frame this matter as a populist issue. In turn, the charge of populism characterises complaints as uninformed and lacking in political legitimacy. A French lawyer quoted in a *Bloomberg* article about the European Union's response to tax dodging by tech giants warns that "tax populism and Google-bashing are on the rise among certain politicians" (Sebag, 2017). The notion of tax populism does the work of dismissing legitimate complaint by presenting it as popular but naïve and simplistic.

Political actors have also made use of the idea that the public is angry about tax avoidance. In the video I cited earlier, John McDonnell declares that "most people will be shocked and some will be outraged" at the information contained in the Paradise Papers. His statement moves rapidly on to assert that people like his neighbours go to work and pay taxes, which pay for public services. In this way, the activation of rich tax-avoiding elites as antagonists has also informed the way in which the "the people" are defined and positioned. If *la casta* are tax *avoiders*, the implication is that the *la gente* are *taxpayers*. This is more than just implied in the Labour Party (2017) and Podemos (2016) manifestos, where "the people" are addressed respectively as "taxpayers" or "*contribuyentes*". As I explained earlier in this chapter, there is an antagonism established in common sense between "taxpayers" and their others, who are defined as non-taxpayers and as *beneficiaries* of social protection. This pairing ("taxpayers" and "welfare beneficiaries") seems to be directly challenged by the alternative pairing proposed when left political actors establish a frontier between taxpayers and tax-avoiding elites. In her recent recapitulation of hegemony theory for the current conjuncture, Mouffe makes a number of references to Hall's discussion of the taxpayer in neo-liberalism and underscores the role of this signifier in the successful articulation of the "political idea of liberty [to] the economic idea of the free market" (2018, p.64). Her intervention is suggestive of the idea that there is scope to rework the "taxpayer" for radical democratic ends, and it is fair to assume that the mobilisation of the figure of the taxpayer by Labour and Podemos is informed by a deliberate strategy of re-signification.

Could such a strategy be successful? The question is whether the terrain of current debate about taxation offers fertile ground on which left political actors can successfully define taxpayers' others as elites rather than welfare beneficiaries. To answer this, we need to consider the deep sedimentation of neo-liberal ideas about taxation in hegemonic common sense or, to repeat Massey's phrase, the unthought assumption that "tax is a (necessary) evil". Reinforcing a notion of the taxpayer who works and pays taxes while others do not could simply serve to perpetuate the neo-liberal figure of the taxpayer in common sense – albeit as a subject position with two antagonists rather than one. Indeed, this triadic antagonism (Judis, 2016) offers a persuasive description of the dominant way in which the discourse around taxpaying currently operates, in which the meaning and significance of the "taxpayer" is produced in opposition to both a "free-riding super-rich" elite (Stanley, 2016, p.399) *and* an undeserving, non-taxpaying poor. In 2017, the British Social Attitudes survey found that while people in Britain disapprove of tax dodging, they disapprove even more of welfare fraud, leading the researchers to identify "a double standard in attitudes to tax avoidance and benefit manipulation: [...] benefit recipients are judged more harshly than tax offenders for what might be considered similar 'offences'" (NatCen Social Research, 2017). There is a risk, then, that the efforts of left populists to resignify the figure of the taxpayer end up strengthening the "underlying neoliberal sense of the individual's relationship with the state", rather than reactivating a residual concept of taxation as "a collective responsibility to society" (Birks, 2017, p.14).

What kind of demand?

Having considered the kind of elite and the kind of people that emerge on the terrain of debate about tax avoidance, I want to consider the *demand* that has (or has not) prevailed. Following Laclau (2007, p.74), popular complaint against tax avoiders represents the coalescence of a plurality of grievances against neo-liberal capitalism and particularly against an economic system which is rigged in favour of the wealthy. The articulation of these complaints into a "chain of equivalence" has enabled the construction of an "antagonistic frontier" separating "elite tax avoiders" and "the people" (the "taxpayers"). What is far less clear is the extent to which we can really speak of a "general demand" emerging from the cleaving together of diverse grievances against neo-liberalism and their coalescence in a complaint about elite tax avoiders. To put this in more simple terms: what would "the people" have us do about this problem?

I am not, of course, suggesting that the global left does not have a robust and compelling set of proposals about how to tackle the problem of tax avoidance. My point is rather that these solutions have not surfaced in the form of a general demand secured to the constitution of the people in question. There are a number of explanations for this problem. Firstly, it is clear that while convergent and connective media practices have enabled and sustained the construction of elites as tax avoiders, bringing them sharply into focus, media discourse has not supported the articulation of a general demand to tackle this issue. The predominance of journalistic practices of revelation and exposure has contributed to the construction of a public that is angry about tax avoidance, which has enabled tax avoidance to be framed as a "populist" issue. The charge of tax populism diminishes and obscures the demands of left political actors, merging and equating them with complaints arising from the political right.[9] The obfuscation of left political demands has meant that those challenges to neo-liberalism that have surfaced successfully in popular debate about tax avoidance have been "predominantly nationalist in form" (Birks, 2017, p.14) and have been articulated to the dissatisfaction with globalisation that has been powerfully harnessed by right-wing political actors.

Secondly, the articulation of a general demand about how to address tax avoidance has been severely constrained by the context of fiscal austerity in the UK and Europe, in which the complaint against tax avoidance was popularised. As we have seen in the decade after the global financial crisis, left political actors have tended to problematise tax avoidance in terms of national revenue. The statement in John McDonnell's video, cited above, is a good example of how the complaint against elite tax avoidance can be constricted and can fail to be articulated as a general demand. The limited demand here is for legal loopholes to be closed, rather than for a system which favours the economic interests of the super-rich to be dismantled. By contrast, the US Democratic Left's call simply to "tax the rich" represents a purer, more general demand, unconstrained by reference to a particular fiscal context, and unconfined to the specific problem of tax *avoidance*.

What are we going to do about the rich?

In the long decade since the global financial crisis, opportunities have emerged to contest the neo-liberal consensus in general and hegemonic common sense about taxation more specifically in the long decade since the global financial crisis. A new configuration of the rich has emerged on the terrain of debate about tax avoidance, and this has been animated through media discourse and political interventions.

Media reporting on this topic has consistently adopted a rhetoric of exposure and summoned a "public" that is "angry" about tax avoidance. These constructions of the elite and of an enraged public have circulated in a context of media convergence. Left political actors have activated these configurations in order to construct a political frontier and designate an "enemy". As a result, the issue of tax avoidance has become a way of differentiating "them" from "us".

The picture of the rich elite that emerges from this discursive formation builds on existing formulations, such as the 1%, but has certain distinctive characteristics. The practice of tax avoidance which defines this elite is understood to be ubiquitous, and so it encompasses the anonymous rich as well as those with a strong public identity. Thanks to the articulation of tax avoidance to other elite privileges, this elite is clearly positioned within an interconnected set of structural entitlements. The elite tax avoiders formulation helps to illuminate the ways in which neo-liberal capitalism is designed to favour the economic interests of the super-rich and foreground the mechanisms and structures that have been the focus of substantial criticism of neo-liberalism. Relatedly, this formulation facilitates the critical challenge of dominant neo-liberal ideas and particularly common-sense ideas that derive from supply-side economics. These outcomes demonstrate the political opportunities that can follow from the popularisation of debate about tax avoidance.

The production of antagonism towards tax-avoiding elites has concurrently summoned a public that is angry about tax avoidance. Public anger about tax avoidance has been designated a populist concern, which has enabled some political actors to dismiss a popular demand that action must be taken as simplistic and uninformed. The mobilisation of tax avoidance as a political frontier has also positioned "the people" as "taxpayers". Left political actors have apparently sought to resignify this neo-liberal discursive subject for progressive ends, with the aim of defining taxpayers' others as elites rather than welfare beneficiaries. However, resources to support an alternative conception of the citizen-taxpayer are thin on the ground, and there is a risk that this project could end up strengthening neo-liberal ideas about taxpayers and taxation embedded in common sense.

In the final part of this chapter, I considered the extent to which a general demand has emerged from the coalescence of diverse grievances against neo-liberalism in a complaint about elite tax avoiders. I argued that the charge of populism has tended to obscure left political demands, while the demands that have successfully surfaced have tended to be articulated to a right-wing nativist political agenda. The articulation of a general demand about how to address tax avoidance in Europe has also been severely constrained by the context of fiscal austerity, which has driven left political actors to problematise tax avoidance in terms of national revenue. While the elite tax avoiders formulation brings the rich sharply into focus and draws attention to their location in an economic system which perpetuates inequality, it does not tell us what to *do* about them.

However, it would be a mistake to conclude that taxation is therefore the wrong terrain on which to forge emergent antagonisms and formulate left political demands. On the contrary, debate about taxation delivers significant potential to illuminate the ways in which neo-liberal capitalism is designed to favour the economic interests of the super-rich. The challenge is that neo-liberal ideas about taxation are deeply embedded in common sense. This means that there are limited resources to support alternative ideas about taxation and alternative identities for the taxpayer – although each national context presents different opportunities and limitations in this respect. In particular, we lack alternative *fiscal imaginaries* that would support common-sense understandings of the importance of tax justice outside of a fiscal crisis. The political salience of tax avoidance has furthered the development of these resources over the last decade. It has created opportunities to cut through neo-liberal common sense and activate a shared good sense about the production and circulation of wealth in societies. These outcomes must be acknowledged and extended, even as we conclude that defining the rich as tax avoiders does not provide a simple fix to the problem of neo-liberal hegemony.

Notes

1 Tax Justice UK (2020), for example, found that 84% of those they polled want politicians to close loopholes to stop big companies and wealthy people avoiding paying tax, leading the campaigning organisation to assert that "the public hate tax avoidance".
2 For example: "I can't believe the BBC pay [singer] Gary Barlow £250,000, I mean after tax that's £250,000" (Twitter user, July 2017).
3 For example, members of the rock band U2 have a reputation in Ireland as tax dodgers (Van Nguyen, 2017) and photographs of graffiti on this theme have circulated on social media. See: https://twitter.com/Freewheeler12/status/887234275522871296.
4 See: https://twitter.com/johnmcdonnellMP/status/927499495679713280.
5 See: http://jeremycorbyn.org.uk/articles/jeremy-corbyn-my-tax-return-2/.
6 See: https://twitter.com/AOC/status/1184269930704916481.
7 The Laundromat (2019), a feature film distributed by Netflix.
8 See: https://twitter.com/attn/status/943910882953773056.
9 Sections of the political left and right (particularly but not exclusively in their nativist variants) do, indeed, have analogous concerns about globalisation and, more specifically, the threat to fiscal sovereignty posed by the development of the offshore world in the globalised neo-liberal era.

Bibliography

Baker, L. and Denis, P., 2019. As Notre-Dame money rolls in, some eyebrows raised over rush of funds. *Reuters* [online] 17 April. https://www.reuters.com/article/us-france-notredame-donations-idUSKCN1RT28Q [Accessed 26 January 2021].

Barford, V. and Holt, G., 2013. Google, Amazon, Starbucks: the rise of "tax shaming". *BBC News* [online] 21 May. http://www.bbc.co.uk/news/magazine-20560359 [Accessed 26 January 2021].

Barnett, C., 2008. Convening publics: the parasitical spaces of public action. In: K. Cox, M. Low, M. and J. Robinson, eds. *The Sage Handbook of Political Geography*. London: Sage, pp. 403–417.

BBC, 2013. Davos 2013: David Cameron calls for action on tax avoidance. *BBC News* [online] 24 January. https://www.bbc.co.uk/news/business-21176992 [Accessed 26 January 2021].

Beaverstock, J., Hubbard, P. and Rennie Short, J., 2004. Getting away with it? Exposing the geographies of the super-rich. *Geoforum*, 35(4), pp. 401–407.

Birks, J., 2017. Tax avoidance as an anti-austerity issue: the progress of a protest issue through the public sphere. *European Journal of Communication*, 32(4), pp. 296–311.

Booth, R. and Grierson, J., 2017. Publication of hacked David Beckham emails renders injunction worthless. *Guardian* [online] 6 February. https://www.theguardian.com/football/2017/feb/06/hacked-david-beckham-emails-renders-injunction-worthless [Accessed 26 January 2021].

Bramall, R., 2016. Tax justice in austerity: logics, residues, and attachments. *New Formations*, 87, pp. 29–46.

Bramall, R., 2018. A "powerful weapon"? Tax, avoidance, and the politics of celebrity shaming. *Celebrity Studies*, 9(1), pp. 34–52.

Brown, W., 2016. Sacrificial citizenship: neoliberalism, human capital, and austerity politics. *Constellations*, 23(1), pp. 3–14.

Buettner, R., Craig, S. and McIntire, M., 2020. The President's taxes: long-concealed records show Trump's chronic losses and years of tax avoidance. *New York Times* [online] 27 September. https://www.nytimes.com/interactive/2020/09/27/us/donald-trump-taxes.html [Accessed 26 January 2021].

Burgis, T., 2016. How fury over tax havens moved from the margins to the mainstream. *Financial Times* [online] 6 April https://www.ft.com/content/ebbf9556-fa72-11e5-8f41-df5bda8beb40 [Accessed 26 January 2021].

Christensen, R. and Hearson, M., 2019. The new politics of global tax governance: taking stock a decade after the financial crisis. *Review of International Political Economy*, 26(5), pp. 1068–1088.

Clarke, J., 2010. Of crises and conjunctures: the problem of the present. *Journal of Communication Inquiry*, 34(4), pp. 337–354.

ComRes, 2013. Christian Aid tax avoidance poll. ComRes interviewed the British public on behalf of Christian Aid. *ComRes* [online] 1 March. https://comresglobal.com/polls/christian-aid-tax-avoidance-poll-2/ [Accessed 26 January 2021].

Cooper, M., 2017. *Family Values: Between Neoliberalism and the New Social Conservatism*. Brooklyn: Zone Books.

Cupples, J. and Glynn, K., 2020. Neoliberalism, surveillance and media convergence. In: S. Springer, K. Birch and J. MacLeavy, eds. *Handbook of Neoliberalism*. Abingdon: Routledge, pp. 175–189.

D'Eramo, M., 2013. Populism and the new oligarchy. *New Left Review*, 82, pp. 5–28.

Davies, W., 2017. Elite power under advanced neoliberalism. *Theory, Culture & Society*, 34(5–6), pp. 227–250.

Davis, A. and Williams, K., 2017. Introduction: elites and power after financialization. *Theory, Culture & Society*, 34(5–6), pp. 3–26.

Diaz, D., 2016. Trump: "I'm smart" for not paying taxes. *CNN* [online] 27 September. https://edition.cnn.com/2016/09/26/politics/donald-trump-federal-income-taxes-smart-debate/index.html [Accessed 26 January 2021].

Doerer, K., 2020. Why do people want to see Donald Trump's tax returns? *ProPublica* [online] 1 July. https://www.propublica.org/article/why-do-people-want-to-see-donald-trumps-tax-returns [Accessed 26 January 2021].

Ford, J., 2012. Public shaming is the way to tackle tax cheats. *Financial Times* [online] 20 June. https://www.ft.com/content/b92601cc-bbb8-11e1-9436-00144feabdc0 [Accessed 26 January 2021].

Gerbaudo, P., 2018. Social media and populism: an elective affinity? *Media, Culture & Society*, 40(5), pp. 745–753.

Hackell, M., 2013. Taxpayer citizenship and neoliberal hegemony in New Zealand. *Journal of Political Ideologies*, 18(2), pp. 129–149.

Hall, S., 1988. The toad in the garden: Thatcherism among the theorists. In: L. Grossberg and C. Nelson, eds. *Marxism and the Interpretation of Culture*. Urbana, IL: University of Illinois Press, pp. 69–70.

Hall, S., 2011. The neo-liberal revolution. *Cultural Studies*, 25(6), pp. 705–728.

Hall, S. and O'Shea, A., 2013. Common-sense neoliberalism. *Soundings*, 55, pp. 8–24.

Hannam, J., 2017. *What Everyone Needs to Know about Tax*. Chichester: Wiley.

Harvey, L., Allen, K. and Mendick, H., 2015. Extraordinary acts and ordinary pleasures: rhetorics of inequality in young people's talk about celebrity. *Discourse and Society*, 26(4), pp. 428–444.

Inglis, F., 2010. *A Short History of Celebrity*. Princeton, NJ: Princeton University Press.

Jessop, B., 2015. Margaret Thatcher and Thatcherism: dead but not buried. *British Politics*, 10(1), pp. 16–30.

Judis, J.B., 2016. Rethinking populism. *Dissent* [online] Fall. https://www.dissentmagazine.org/article/rethinking-populism-laclau-mouffe-podemos [Accessed 26 January 2021].

Labour Party. 2017. *For the Many Not the Few: Labour Party Manifesto 2017*. https://labour.org.uk/wp-content/uploads/2017/10/labour-manifesto-2017.pdf [Accessed 26 January 2021].

Laclau, E., 2007. *On Populist Reason*. London: Verso.

Laclau, E. and Mouffe, C., 2001. *Hegemony and Socialist Strategy*. London: Verso.

Littler, J., 2018. *Against Meritocracy*. Abingdon: Routledge.

Mahony, N. and Clarke, J., 2013. Public crises, public futures. *Cultural Studies*, 27(6), pp. 933–954.

Massey, D., 2016a. Exhilarating times. *Soundings*, 61, pp. 4–16.

Massey, D., 2016b. Tax: a political fault line. *Soundings*, 62, pp. 161–163.

Matthews, D., 2019a. Meet the folk hero of Davos: the writer who told the rich to stop dodging taxes. *Vox* [online] 30 January. https://www.vox.com/future-perfect/2019/1/30/18203911/davos-rutger-bregman-historian-taxes-philanthropy [Accessed 26 January 2021].

Matthews, J., 2019b Populism, inequality and representation: negotiating "the 99%" with occupy London. *Sociological Review*, 67(5), pp. 1018–1033.

Mazzucato, M., 2013. *The Entrepreneurial State*. London: Anthem.

Millionaires Against Pitchforks, 2020. Davis sign on letter. *Millionaires Against Pitchforks* [online]. https://www.millionairesagainstpitchforks.com/ [Accessed 26 January 2021].

Mouffe, C., 2018. *For a Left Populism*. London: Verso.

Murphy, R., 2017. *Dirty Secrets: How Tax Havens Destroy the Economy*. London: Verso.

NatCen Social Research, 2017. Tax avoidance and benefit manipulation, *British Social Attitudes edition 34* [online]. http://www.bsa.natcen.ac.uk/latest-report/british-social-attitudes-34/tax-benefit-manipulation.aspx [Accessed 26 January 2021].

Neate, R., 2019. Super-rich prepare to leave UK "within minutes" if Labour wins election. *The Guardian* [online] 2 November. https://www.theguardian.com/news/2019/nov/02/super-rich-leave-uk-labour-election-win-jeremy-corbyn-wealth-taxes [Accessed 26 January 2021].

Osborne, G., 2011. Tax cheats have no hiding place under this coalition. *The Guardian* [online] 27 August. https://www.theguardian.com/commentisfree/2011/aug/27/tax-cheats-coalition-george-osborne [Accessed 26 January 2021].

Ott, J., 2017. How tax policy created the 1%. *Dissent* [online] 18 April. https://www.dissentmagazine.org/online_articles/tax-policy-history-by-for-1-percent [Accessed 26 January 2021].

Podemos, 2016. *Podemos 26J*. http://estaticos.elperiodico.com/resources/pdf/9/4/1465389843149.pdf [Accessed 26 January 2021].

Prasad, M., 2012. The popular origins of neoliberalism in the Reagan tax cut of 1981. *Journal of Policy History*, 24(3), pp. 351–383.

Prentoulis, M., Canos Donnay, S., Dubbins, S. and Massey, D.,. 2015. European alternatives: a roundtable discussion with Marina Prentoulis, Sirio Canos and Simon Dubbins, introduced by Doreen Massey. *Soundings*, 60, pp. 13–28.

Rojek, C., 2012. *Fame Attack: The Inflation of Celebrity and Its Consequences*. London: Bloomsbury.

Rowney, M., 2015. What's wrong with tax avoidance? *New Statesman* [online] 20 April. https://www.newstatesman.com/politics/2015/04/whats-wrong-tax-avoidance [Accessed 26 January 2021].

Sanders, K., Hurtado, M. and Zoragastua, J., 2017. Populism and exclusionary narratives: the "other" in Podemos' 2014 European Union election campaign. *European Journal of Communication*, 32(6), pp. 552–567.

Sebag, G., 2017. Google faces $1.3 billion French ruling amid "tax populism". *Bloomberg Quint* [online] 11 July. https://www.bloombergquint.com/politics/google-faces-1-3-billion-french-ruling-amid-rising-tax-populism [Accessed 26 January 2021].

Serafini, P. and Smith Maguire, J., 2019. Questioning the super-rich. *Cultural Politics*, 15(1), pp. 1–14.

Sheffield, H., 2016. Panama Papers bubble chart shows politicians are most mentioned in document leak database. *Independent* [online] 12 May. https://www.independent.co.uk/news/business/news/occupations-people-panama-papers-a7025736.html [Accessed 26 January 2021].

Slobodian, Q. and Plehwe, D., 2020. Introduction. In: D. Plehew, Q. Slobodian and P. Morowski, eds. *Nine Lives of Neoliberalism*. London: Verso, pp. 1–18.

Smith, D., 2020. Will the New York Times taxes report sink Donald Trump? *The Guardian* [online] 28 September. https://www.theguardian.com/us-news/2020/sep/28/will-the-new-york-times-taxes-report-sink-donald-trump [Accessed 26 January 2021].

Stanley, L., 2016. Legitimacy gaps, taxpayer conflict, and the politics of austerity in the UK. *The British Journal of Politics and International Relations*, 18(2), pp. 389–406.

Stoehrel, R., 2017. The regime's worst nightmare: the mobilization of citizen democracy. A study of Podemos' (aesthetic) populism and the production of affect in political discourse. *Cultural Studies*, 31(4), pp. 543–579.

Streeck, W., 2017. *How will Capitalism End?* London: Verso.

Tax Justice UK, 2020. Talking tax: how to win support for taxing wealth. *Tax Justice UK* [online] September. https://www.taxjustice.uk/tax-and-public-opinion.html [Accessed 26 January 2021].

The Laundromat, 2019. [film] Directed by Steven Soderbergh. New York: Netflix.

Urry, J., 2014. *Offshoring*. Cambridge: Polity Press.

Vallely, P., 2020. How philanthropy benefits the super-rich. *The Guardian* [online] 8 September. https://www.theguardian.com/society/2020/sep/08/how-philanthropy-benefits-the-super-rich [Accessed 26 January 2021].

Van Nguyen, D., 2017. Where the streets have no statues: why do the Irish hate U2? *The Guardian* [online] 12 July. https://www.theguardian.com/music/2017/jul/12/where-the-streets-have-no-statues-why-do-the-irish-hate-u2 [Accessed 26 January 2021].

Vaughan, M., 2019. Scale shift in international tax justice: comparing the UK and Australia from 2008 to 2016. *Social Movement Studies*, 18(6), pp. 735–753.

Wahl-Jorgensen, K., 2018. Towards a typology of mediated anger: routine coverage of protest and political emotion. *International Journal of Communication*, 12, pp. 2071–2087.

Williamson, V., 2018. Who are "the taxpayers"? *The Forum: A Journal of Applied Research in Contemporary Politics*, 16(3), pp. 399–418.

Yokley, E., 2020. Most voters expected Trump's business ties to impact how he governs. But more now say that's a bad thing. *Morning Consult* [online] 30 September. https://morningconsult.com/2020/09/30/trump-income-taxes-nyt-polling/ [Accessed 26 January 2021].

Zucman, G., 2015. *The Hidden Wealth of Nations: The Scourge of Tax Havens*. Chicago: University of Chicago Press.

Zucman, G., 2018. If Ronaldo can't beat Uruguay, the least he can do is pay taxes. *The New York Times* [online] 3 July. https://www.nytimes.com/2018/07/03/opinion/ronaldo-world-cup-tax-evasion.html [Accessed 26 January 2021].

6 Criticism of elites and subjective social agency

A look at the workers

Stefanie Hürtgen

Subalternity and elite critique

Criticism of decision-makers and leaders in companies and the state apparatus is crucial for the workers' movement and culture[1] in a comprehensive sense. In essence, it is about the structural ruthlessness of capitalist political economies towards work-related human and social concerns. Workers have experienced brutal ruthlessness towards them since industrialisation, whether in the workplace (e.g. working to the point of exhaustion, starvation wages, dangerous working conditions) or in society, where (trade union) protest has been repeatedly suppressed over the centuries, not least by the police and the military. But even in the socially regulated and democratised welfare states, the struggle of workers with their subalternisation remained virulent, for example, in struggles for expanded possibilities of intervention and shaping in the workplace and society, be it in relation to wages, working hours or social security in old age and illness.

In other words, in contrast to the market-liberal assertion that capital and labour meet as equal contracting parties on the labour market, an examination of elite critique on the part of workers must focus on the fundamental socio-economic inequality of capitalist societies and, thus, on the structurally subaltern position of workers in companies and society. *Economically*, workers are those who have no (major) property, *historically and socially*, they are the disenfranchised and despised, the day labourers and vagabonds who can only sell their labour power to survive. Robert Castel (1995) has described in detail the path to social repositioning of this social group towards social recognition and material and immaterial participation in the Western welfare states. The compulsion to sell oneself and one's ability to work as a commodity continued, but it was relativised in the welfare state arrangement by the expansion of social rights and public infrastructure. However, social impoverishment and political disenfranchisement have been the order of the day again, and not just since the financial crisis (or currently the Corona crisis); the *return of the social question*, i.e. of extremely insecure employment, wages that do not secure one's livelihood and massive social vulnerability in general, has been an issue for years (Castel and Dörre, 2009; Schmalz and Brandon, 2019). In historically different forms, there is a basic social dependence of

DOI: 10.4324/9781003141150-8

workers in capitalist societies on decisions made above their own position based on the division of labour and hierarchy. Regarding the company, we can say in somewhat simplified terms that the *functional* division of labour manifests itself in a *social* hierarchisation within the company along the lines of manual and mental labour. The managers and other experts and their decisions are at the top and the manual workers, who are seen as non-experts, are at the bottom (Hürtgen and Wissel, 2018).[2] It is the company's experts who set the targets through production rates or the number of customer contacts and mouse clicks, as well as the design and functioning of machines or software programmes with which work is done. Entrepreneurial decision-makers also decide on working spaces (and their conditions), the concrete working position and activity in the company and, of course, whether people continue to be employed at all or whether there are "redundancies for operational reasons". This company social hierarchy is reproduced in wage labour-centred societies, often mediated as a hierarchised social space of possibility with unequal life resources and social relationships of recognition of work positions and occupations (Bourdieu, 1979; Solga, 2015).

This reminder of subalternity is important in order to take into account that a critique of elites in the workplace and society by workers always includes, on closer examination, a confrontation with their own (subaltern) position. What is at stake is the interpretation of a social relationship, a hierarchical social structure in which one is integrated and which one (critically) reproduces in a specific way.

Against this background, the thesis of my contribution is that the *way* in which workers locate themselves in this social relationship is central for further discussion and qualitative definition of elite critique. It is about how they understand their own social position and, above all, what kind of social agency they attach to this subaltern position. The following contribution shows that a progressive-intervening elite critique must be distinguished from a regressive-restrictive position. The former aims at the critique of capitalist hierarchisation and inequality, the latter *manifests* it. In order to elaborate these principally different forms of elite critique, the chapter combines empirical sociology of work with a concept of agency as developed in the historical-materialist paradigm of the Berlin school of Critical Psychology (Holzkamp, 1985; Tolman, 2013). The subject matter is, thus, presented conceptually and, at the same time, by means of exemplary interview statements and illustrations. The material cited comes both from my own qualitative research projects in the sociology of work and industry, which were carried out around the turn of the millennium, and from other studies in the sociology of work, which enquire about workers' understanding of society, and range from the 1950s to the present day. The findings refer geographically to different European countries (including Germany, France and Poland), each of which is characterised by a strong tradition of the workers' movement. The relationship of workers to superiors, managers and state elites in the material is an important component of the narratives surveyed, without this relationship itself having been elicited by an explicit question.

The chapter develops an ideal-typical typology of fundamentally different forms of elite critique by workers and then discusses their current crisis.

Subjective societal agency

The concept of *societal agency* comes from Critical Psychology and is fundamental to the rest of the discussion. Critical Psychology locates itself in (neo-)Marxist social theory and, thus, has a direct focus on social and economic inequality. However, it is equally critical of a theoretical detachment of the subject from society that is oriented towards "characteristics" and "attitudes" and, conversely, of the objectification of its actions as the direct result of relationships of domination. The concept of societal agency, on the other hand, aims at the connection of the (subalternised) subject to society. Societal agency means that the production of materially and immaterially *individually* relevant living conditions must be understood as conflictual participation in the production of *general social* living conditions (Holzkamp, 1992).

Individual ways of acting are part of the social whole and vice versa. Individual life is not only shaped *depending* on the hierarchical structures of society as a whole but necessarily also in permanent *confrontation* with them. In capitalist societies, this means the confrontation with economic and social relationships of domination, i.e. with the forms and mechanisms of objectification and subalternisation that the subjects themselves experience (and which in the economic sphere often appear as an unchangeable "practical constraint"). In order to bring together the connection between (one's own) subalternity and active agency, Critical Psychology distinguishes ideal-typically between restrictive and generalised societal agency. *Restrictive agency* interprets relationships of domination and "structures" as ultimately unchangeable and, therefore, seeks possibilities for shaping existence and life within them. *Generalised agency*, by contrast, sees social conditions as basically and necessarily changeable and seeks to intervene in and change them, essentially also in exchange and association with others. The improvement of one's own subjective space of possibility is here brought together with a generalised "structural" increase in the disposal of social conditions. In other words, the critical or approving reproduction of their social relationships takes place on the part of the subalternised essentially as a confrontation with the *scope of their social and societal agency*: Is it necessary – in the perspective of the subject – to arrange oneself in order not to endanger the existing spaces of action, or is it possible to ask for potentials of changes in the enabling relationships that extend beyond the individual, a consideration that is necessarily only conceivable in supra-individual associations. It is crucial here to understand that this distinction is an analytical one and overlaps in practice in many cases, and secondly, that the distinction is not a moral one but is *justified* from the point of view of the subject, who reflects the structures of society in his or her daily practical actions. Orientations of action are, therefore, not attitudes inherent in the individual, but – as emphasised above – the result of subjective confrontation with the (objectifying) structures of the society in which he or she is active and which he or she (also) reproduces.

In the following, I want to bring together these fundamental considerations on the societal agency of (subalternised) subjects with sociological studies of labour from different periods. I want to show that elite critique differs fundamentally according to the way in which workers constitute themselves as agents in their confrontation with economically and socially hierarchical social relationships, i.e. how they draw their own role in terms of practical action and how they situate it to guide their actions (cf. on this: Hürtgen and Voswinkel, 2014, pp. 23ff., Hürtgen and Voswinkel, 2017). To this end, it is helpful to analyse patterns of interpretation and orientation in everyday life and the world of work rather than trade union and party-political programmes and pamphlets. It then also becomes clear that a (radical) critique of the economic and political elites and of the socio-economic functioning of capitalist societies that sounds very similar at first glance can have thrusts which are diametrically opposed. In particular, not every "anti-capitalist" critique is to be seen as a progressive transgression and questioning of current distortions – a circumstance that remains hidden in the orthodox Marxist question of the "state of consciousness" of workers (Herkommer et al., 1979; Kudera et al., 1979). From the perspective of Critical Psychology, on the other hand, it can be shown that the critique of elites and hierarchies by workers must be ideally differentiated into an orientation aimed at progressive-emancipatory change of the grievances, on the one hand, and a regressive-restrictive position that ultimately affirms the social conditions *despite the sometimes massive criticism*, on the other. What is decisive is the way of social self-positioning, i.e. what kind of action one develops mentally and practically vis-à-vis the social conditions.

Restrictive elite critique: there's nothing you can do about "those up there" anyway

I begin with the restrictive form of elite critique. This has always been part of working-class culture, so, it is by no means a new phenomenon. A – sometimes massive – critique of social conditions in connection with a self-constitution that explicitly excludes one's own intervening social action is typical of this orientation. One sees oneself as a "little man" and "little woman", far below, at the mercy of the arbitrary decisions of "those up there". In sociology, this orientation towards action is associated with the "traditionless working-class milieu" or the underclass (Jonna and Bellamy Foster, 2016; Vester et al., 1993) and the underlying dichotomous world view is emphasised. According to this view, the world is divided into a rigid, unchanging top and bottom: "The worker is always the poor man who just has to work. The money, as they say, is made by the others" (Toolmaker, quoted in Popitz et al., 1957, p. 177).

What is decisive is that this hierarchy is depicted as worthy of criticism but, unfortunately, absolutely unchangeable. Workers with this orientation can describe the injustices they experience directly and indirectly very vividly and in detail and denounce them as unjust – but they almost always come to the conclusion that these injustices cannot be changed, that they have to submit. The researchers in the classic study "The Social Image of the Worker" by Heinrich Popitz, Hans Paul Bahrdt and Hanno Kersting also ask questions about the then newly introduced company and enterprise co-determination. A steelworker explains:

> We all feel cheated. The worst thing is that you can't work out our wages. [...] There are always big disappointments. The piecework system is completely opaque. Each tonne is calculated differently. [...] But we don't have much power. If someone protests and says that the work is too hard and the pay too low, they are quickly thrown out of the factory. It doesn't matter how long the person has been in the factory, how well he has worked so far. [...] We are all dependent on them. Of course, it is clear that nobody likes to be told what to do. I don't like it either when someone looks into my wallet and makes demands. In practice, that's all we do when we demand co-determination. [Co-determination] could be quite good if it went through. But you can't wait for that. In any case, I don't believe in it. The worker remains a worker, he remains dependent, he is never independent. He can only do one thing: his duty, earn as much as possible and leave in good time. That is all. Even co-determination won't change that.
>
> (Popitz et al., 1957, pp.244f)

This perspective can be summarised with Critical Psychology as the subjective-social constitution of restrictive societal action orientation (Holzkamp, 2013): The conditions appear miserable, but they always have been, that is the natural course of the world, you have to manage somehow. One's own (imagined) ability to act socially is aimed solely at oneself, at one's private environment, one's own person, one's family. This has to be achieved as skilfully and unscathed as possible in the face of social adversity. It requires not only a high degree of willingness to adapt, the renunciation of rebellion and opposition, but especially the willingness to be socially ruthless, which is what the circumstances in this orientation demand.

In a project carried out together with Stephan Voswinkel (Hürtgen and Voswinkel, 2014, 2017), we interviewed the chairman of the works council of a bank that was undergoing a permanent restructuring and downsizing process. The works council chairman lamented this development in detail and also expressed sympathy several times for the colleagues who had been dismissed, but, at the same time, he used his knowledge advantage and his social contacts "upwards" for years almost exclusively to change to another company department area in time before the upcoming closures and, thus, be spared from dismissals.

This way of acting is restrictive in the sense that it ultimately affirms domination and, thus, from the outset also excludes the expansion of one's own possibilities for action with others. This affirmation of domination and its consequences is explicitly not normative consent but is articulated as a deep insight into the conditions. The protagonists of this orientation know about the (intellectual) possibility that societal structures could be collectively changeable and emphasise all the more how dangerous such a *belief* in social change is. In the project just cited, we interviewed a secretary who repeatedly said that those who believe they do not have to adapt end up "in the gutter" (Hürtgen and Voswinkel, 2017).

As (cognitive) reproduction and interpretation of social contexts, imagined capacities for action are not random attitudes and opinions. Restrictive elite critique reflects and condenses experiences of powerlessness that have been experienced and/or handed down historically, as they have been typical particularly for workers over centuries. It articulates the social experience accumulated over generations of being the plaything of inscrutable economic developments and the arbitrary and massive exercise of political power. The only way to deal with one's own fate here is individually, by coming to terms with the rulers. What is more, the ability to skilfully get by, explicitly also the renunciation of resistance and protest, is, therefore, definitely associated with pride, namely, in having achieved something in one's life despite one's subaltern position as a "small worker", having secured an income, brought up one's family and mastered the imponderables of society.

Two remarks are important at this point regarding further discussion and theoretical consolidation:

Firstly, there is a particular affinity to xenophobic and racist positioning or, more generally, to the authoritarian character (Adorno et al., 1950) in this orientation, because social security and belonging are conceived in principle via the best possible correspondence and performance of adaptation to the requirements and rules formulated by "those up there". Accordingly, the restrictive agency remains structurally insecure and, thus, fearful because the basic dependence on arbitrary decisions is not addressed. Dependence and arbitrariness remain, even if one hopes to get away with it because of one's particular willingness to adapt.[3] Fear and anxiety in this authoritarian arrangement are transferred in the form of aggression, xenophobia and racism to those social groups and people who are already marked as "outsiders" in the prevailing social and ideological structures and are now said to be unable to adapt or to do so only insufficiently (Hürtgen, 2020a; Räthzel, 2002): Be it the "lazy Greeks" in the economic crisis of 2008 and what follows, the "social parasites" of one's own country who lie down in the "social hammock" and want to be paid for it, or the migrants who do not want to adapt to "our" culture, are not hard-working, etc.

The second remark concerns the centrality of the concept of labour for the workers' critique of elites, which has already become clear with these references: One's own (wage labour) performance in the restrictive orientation is understood as the best possible fulfilment of requirements, performed with diligence and care. It is, thus, *industriousness* or *servitude*, i.e. the orientation of one's own capacity to the requirements "from above", from the elites (Hürtgen, 2008, pp.128ff.). In this orientation, one is proud of one's ability to submit to the requirements and *not* to hold up social or other demands. This is what ultimately pays off.

This is what an East German interviewee in an earlier research project of mine relates of the time after the fall of the Wall and the "*Wende* (change)". He became a works council member in a newly built factory, where, however, the pay was initially very low and the working hours very flexible – while many employees of the old factory that was closed were put on paid short-time work:

> At the beginning it was depressing to see, in the first one to one and a half years, that our colleagues [in the new plant, SH] in two shifts always had fewer net earnings than my former colleagues in the [old plant] who went on *Kurzarbeit Null.*[4] A certain social peace was simply bought with money. [Many] former colleagues all said [at the time]: I'm not stupid to work for so little money! They went to the swimming pool the two summers we worked here and mocked us. Some of them came in and showed us their income statements, what they earned. They were happy about our anger. But they are still out today! [...] So in the end, the laughter was on our side.
>
> (Hürtgen, 2008, p.216)

Progressive-interventionist elite critique: the worker creates the values!

Only when one realises that the restrictive elite critique, directed towards the dichotomous 'down here, up there' model, represents the reflection and reproduction of a frequently subjectively experienced structural context does the second form of elite critique on the part of the workers presented here become comprehensively clear in its scope. In this progressive-intervening elite critique, the workers constitute themselves as an acting-intervening subject that can refer to transindividual norms and ideas of justice and/or social and political rights in order to address social grievances. Workers of this orientation have normatively legitimised *claims* on elites and superiors, to which they can refer either cognitively or, if the balance of power is right, also practically (Hürtgen and Voswinkel, 2014, 2017). The representatives of the restrictive orientation could only think, i.e. hope, for social improvements in a *particularistic* way (for themselves). The progressive-interventionist position also claims social improvements for itself but, simultaneously, goes beyond this. It invokes general rules of social coexistence and working together, norms of justice that should also apply to the decisions of the elites. Critique and self-constitution as an intervening subject takes place here on the basis of normative *orders of justification* (Boltanski and Chiapello, 1999; Boltanski and Thévenot, 1991; Dubet, 2009) or a *moral economy*, a concept that Thompson (1971) developed in historically concrete analysis in relation to actors (workers and others) in a subaltern social position. The ruling elites are always part of the morally-normatively imagined set of rules, so that their actions can be regulated to be less arbitrary, less unpredictable. A critique of elites in the progressive-interventionist perspective, thus, always includes an imagined or, depending on the assessment of resources, also practically realised *debate and struggle* about what applies or should apply in the company, in both close and long-distance social relationships. Current demands for justice, which are typically articulated by workers in companies, are demands that, for example, the targets in the company must be made more "human" (again), so that permanent hustle and time constraints stop, that the wage must be "right" for everyone (again) and not make them poor and that re-establishing pension security is overdue (Hürtgen and Voswinkel, 2014). But also on a small scale, in everyday (work) life, it is typically workers with this intervening orientation who spontaneously intervene, articulate dissent and stand up for and with colleagues "against the top". This is work, not the military, said one employee who had shown solidarity with a colleague who had been "put down" by her supervisor and was now being intimidated herself.

> What I can't have at all are threats, I don't have to threaten anyone. We are not in the military or anything like that [...] we are all equal people. Maybe one has a different position and can say something to the other [i.e., give commands], but it can't be that one is better than the other. Rather, we have different positions, but we are still human beings, each for themselves.
>
> (Hürtgen and Voswinkel, 2014, p.167)

This quotation shows clearly how the functional hierarchy between announcing and executing ("one person has a different position and can say something to the other") is embedded in a norm of the equality of all people that is applied as an overarching and general norm. This norm prohibits, for example, encroaching behaviour and, thus, provides a frame of reference that makes critique possible because it can also *oblige* those "at the top" to behave accordingly. Economic and political leadership elites are not simply "abolished" here, but the hierarchy itself must be subject to rules in this conception, it must not be arbitrary rule ("military or something").

This progressive-interventionist elite critique can be described with Critical Psychology as a *socially expanded orientation of action*. It is not a matter of particularist advantage but of intervening in overarching social structures. In union with others and with reference to something that transcends myself, the goal is to

> gain disposition of my respective individually relevant conditions of life. [...] Overcoming suffering, overcoming fear, satisfying the quality of human life can [...] only be achieved [...] through the disposal [...] of the conditions on which my possibility of life and development depends.[5]
>
> (Holzkamp, 1985, p.114)

The constitution of a specific concept of work, a specific idea of one's own (wage) work, is also fundamental in this orientation. In contrast to the restrictive orientation, wage labour is fundamentally seen here as something that is useful to society and contributes to it significantly. In this context, requirements on the part of management and superiors are also fulfilled, but the actual standard by which the duty of care, flexibility and one's own commitment are aligned is the *meaningfulness* of the self-created product or one's own work steps *for others* (Linhart, 2009). According to this, daily work, even if it is unqualified, exhausting or boring, is valuable for others, for society – it is *useful* to both (Nies, 2015). This perspective manifests itself in the common formulation that can still be found today: "The worker creates the values". One's own (wage) work is thought of socially here, it is "a fundamental prerequisite for society's existence" (Popitz et al., 1957, p.238).[6] In fact, the normative idea of wage labour as socially meaningful is a fundamental basis for not only making claims for socio-political improvements but also for conceiving of themselves, i.e. the workers and the labour movement, as those who intervene progressively in society as a whole (Hürtgen, 2021). According to this, it is they themselves, the workers and their interest groups, who, if they are strong enough, ensure that the elites, who are ultimately presented as reactionary, are controlled and the pursuit of profit is curbed, militarisation and wars are prevented, and human progress is achieved (Popitz et al., 1957, pp.201ff.; Hürtgen, 2008, pp.191ff., 2021).

This progressive-interventionist critique of the elite, related to wage labour, is also susceptible to stereotyping and exclusionary positions. It is based conceptually on an idea of work that, on the one hand, opens up a historically new political space for action but, on the other hand, runs the risk of devaluing and excluding those who supposedly do *not work* or do *not work well* in this form of wage labour, which is always oriented towards capitalist productivity: the unemployed, precariously employed, (female) domestic work (Hürtgen, 2017). Xenophobia and authoritarianism towards those who do not or not really usefully work are, thus, also structurally inherent in this orientation.

However, on the side of workers with a progressive-interventionist perspective, there are also normative resources which have been developed that can at least contain and partly even overcome the particularisation contained in productivism: the above-mentioned, action-effective justice principle of *humanity*, which proclaims a kind of universalistic basic right to human dignity, integrity, social belonging and political agency, regardless of the performance rendered.

Brutalisation and retreat into the private sphere: economic constraints as the Achilles' heel of progressively expanded capacity for action

Regarding the final consideration, it is central that *both* forms of elite critique, the restrictive-particularist and the intervening-generalising, must be analysed as currently deeply crisis-ridden. The background for this is undoubtedly the material neoliberal structural changes and the way they affect the working and living conditions of workers. The keywords here are the *recommodification* of wage labour in the form of a dismantling of socio-political protection and through direct authoritarian workfare measures to take up any kind of work, largely ignoring its conditions; furthermore, the far-reaching *precarisation* of work, which includes not only insecure working and employment conditions but also particularly (non-living) low wages for large sections of the working population; and thirdly, a *dismantling of public infrastructure*, with the consequence of crisis-ridden provisions for reproduction, for example, in the area of housing, healthcare, transport or child care.

It is not only these socio-political cuts as such that are important for an understanding of the double crisis of elite critique by workers (both restrictive and progressive). Instead, two normative-institutional shifts that permeate workers' orientations critically are also central, especially in the countries that are generally considered to be European welfare states.

The first shift refers to neoliberalisation as comprehensive social economisation. The business world's logic of competition is generalised into other, institutional and political forms of socialisation (*Vergesellschaftung*) and, in particular, the principle of social difference as a cost and locational advantage is elevated to a political demand and measuring device (Hürtgen, 2020b). The result is a socially generalised primacy of permanent competitive optimising of society, company and location down to the individual worker. The aim at all levels is to increase one's own competitiveness and position oneself optimally against others and, in case of doubt, to prevail. The logic of globalised competition itself becomes a presocial "out-there phenomenon" (Peck, 2002); it appears as a given, as a factual constraint, as a natural condition of existence to which other social, ecological and political concerns must be subordinated.

The second normative-institutional shift follows on from this and concerns particularly the relatively better-off workers, those who are not directly precarious or working poor: On the one hand, it is precisely for them that the institutions of their representation are still intact. Labour law, collective bargaining policy and codified negotiation processes are regarded in the European core countries as an indubitable part of social regulation and compromise; trade unions are a recognised part of social life and the premise of social integration through wage labour stemming from the Fordist era is normatively upheld (one should be able to live from one's work, etc.). On the other hand, "underneath" the institutionalised normality is the fact that even the better-off workers are experiencing a fundamental deterioration of their working conditions. The latter are characterised by far-reaching flexibilisation, extensification and intensification of work, by cost-related scarce personnel resources and the management of constantly short-term changes in work processes in the manner of permanent extra effort due to permanent restructuring and competitive strategic optimisation of business operations. The results are social cuts, social insecurity and social exhaustion of "vitality" (Jürgens, 2006) even among the institutionally "strong" and represented workforces (Hürtgen, 2020b). The normative-institutional subordination of socio-political concerns to the primacy of permanent competitive optimisation manifests itself in a massive social dissonance between the public-institutional socio-political integration and representation of workers, on the one hand, and their own, systematically deviating experience of a now only supposed, long since no longer normal normality on the other (Hürtgen and Voswinkel, 2014, 2017).

Against this background, both forms of elite critique, presented here in an ideal-typical way, enter a crisis, which can be described as the danger of brutalisation and the privatisation of elite critique.

Regarding the restrictive critique of the elite, the social cuts and changes outlined can be interpreted as a revocation of the compromise of subordination. The restrictive arrangement with which one endeavoured to achieve a certain calculable social security and ability to act in one's own life is acutely shaken. Adaptation to and affirmation of the objectifying-hierarchical conditions (and, thus, one's own subalternisation) offer no immediately visible prospect of relative social stability. "Adaptation [...] can hardly offer the security of not getting into an existential emergency" (Kalpaka and Räthzel, 1994, p.44). One reaction to this is a twofold, interrelated change: Xenophobic-racist articulations *in the workplaces* are becoming more offensive and extensive, not least because they are now also articulated offensively by (some) works councils and trade union representatives (Sauer et al., 2018). An aggressively heightened imagination of one's own ability to act as a threat to and destruction of the elites currently to be found has prevailed *in the social-public space* (demonstrations, rallies). Examples of this are physical attacks on representatives of the media (who are explicitly attacked as part of the establishment) or symbols of physical punishment and annihilation of politicians.[7] Unlike "on the street", this brutalisation of elite critique is not dominant in the workplace; as a place of necessary co-operation, this is also not to be expected here. However, due to the existing personal and discursive overlaps between the workplace and the street, corresponding aggressively brutalised orientations, which are by no means new as such, experience a socio-social reinforcement and affirmation (Sauer et al., 2018; cf. also Zoll, 1981; Bahl, 2014, pp.233ff.; Hürtgen, 2020a). In terms of its substance, it becomes apparent that even a brutalised critique of elites does not start from an intervening capacity to act in the sense of a social relationship to be changed. The critique of one's own socio-political vulnerability and hierarchical dependence on (elite) decisions is carried out rather through symbolic destruction, which, in case of doubt, is oriented towards an *exchange* of elites when it imagines new right-wing populist leaders as the "real" representatives.

Secondly, and above all, the progressive elite critique has also entered a deep crisis with the economisation of institutions and society. The orientation developed here of a fundamental ability to shape society becomes questionable because social and political concerns must constantly move within the framework of increasing competitiveness and are, as it were, undermined by this premise – while, at the same time, the normative importance of social integration, partnership-based participation, and so on continues to be upheld. Even more, the generalisation of the principle of economic competition means the particularisation (deregulation) of labour rights and collective bargaining policy, so that social disputes are conducted at the local or workplace level, sometimes with partial success, but without an overarching, socio-political rejection of the socio-political rollback. Rather, trade unions are now themselves involved in the fragmented negotiation processes and promote them in the mode of *concession bargaining* (Burawoy, 1985, pp.150ff.).

Against this background, it is hardly surprising that labour research has been noting extensive subjective experiences of powerlessness in the workplace for a long time. Especially those who refer to an interventionist tradition of the workers' movement, or who, for other reasons, develop a progressive, expanded orientation of action, experience a massive subjective failure of this orientation and their critique of the elite. This widespread experience of powerlessness has been well researched regarding works councils, shop stewards and trade union activists (Bergmann, 2001; Bergmann et al., 2002; Hürtgen, 2008; Menz et al., 2011) and it also applies to ordinary workers (Bahl, 2014; Hürtgen and Voswinkel, 2014). The ideas of a fateful and objective position in the company and society, rejected in the progressive tradition of the workers' movement, are coming up against fundamental limits. The ability to shape the social is broken by the primacy of ever new competitive optimisation (including the corresponding withdrawal of social and political rights) and the overriding need to secure one's own company, location, job, etc. The norms and institutions that are supposed to act as a reference for (collective) critique and counter-policies in this process are not effective. The result is the widespread experience of powerlessness, coupled with an *addressless rage* (Menz et al., 2011): The massive critique of the conditions that are seen as in urgent need of change has no or only very limited effectiveness in its concrete normative-institutional articulation. Ideally formulated (i.e. ignoring multiple overlaps), one's own powerlessness is not acted out symbolically-aggressively on certain representatives of the elite. Instead, the anger remains without an addressee, because the focus in this orientation is on the social relationships that need to be changed and not on the elites as such. A typical consequence of this orientation is, therefore, a retreat into the private sphere, the resigned abandonment of (earlier) ideas of intervening and shaping action (Hürtgen and Voswinkel, 2014). It remains for further analysis to follow on from this and to show that progressive-intervening elite critique must necessarily develop a new, progressive perspective of the economic sphere.

Translated from German by Philipp Saunders

Notes

1 I speak of "workers" in the following in a very broad sense. This refers to all those people who do not have any major assets and, therefore, have to live "from their hands" (and their heads) and have only limited authority in the hierarchical structure of the workplace. The term "workers" in this chapter is not meant in an ouvrierist way, it includes not only male workers in blue coats but also (minor) employees, service workers, (pseudo-self-employed), time workers, etc.

2 At this point, it cannot be discussed that even so-called executive or "unqualified" activities actually involve a high degree of professional expertise and decision-making competence. In any case, the hierarchical division into decisive mental work and executive manual work is a social construction, corresponding to the division into experts and/or non-experts (cf. Hürtgen and Wissel 2018).
3 "Insofar as I try to achieve freedom within the framework of the existing relationships of domination, I am in a certain sense negating the freedom itself, because the freedom is granted by the rulers and can be withdrawn at any time" (Holzkamp, 1985, p. 17).
4 "*Kurzarbeit Null*" (short-time work zero) was a form of subsidy for company restructuring measures in which workers receive short-time allowance but, in fact, undergo qualification measures (typically in transfer companies founded for this purpose). The subsidy instrument was used on a massive scale after the so-called reunification and in the course of the economic deindustrialisation in East Germany but – as the interviewee also points out – often represented a transitional stage to unemployment.
5 Intervening generalised action is, in this respect, related to the question of solidarity; see Billmann and Held (2013).
6 While the authors in their study from the 1950s stated that contributing social work is essentially imagined as muscle-related, physically strenuous (manual) work and that there is a deep mistrust of the "white collars" in the offices, whether they work at all, the picture is now more differentiated. Nowadays, the interviewees (works councillors and company trade unionists from various European countries) emphasised the high level of work input and the importance of the activities in the offices (Hürtgen, 2008) – even if the classic prejudice structure of manual workers against the "Wasserköpfen" (slang for excessive bureaucracy) in the offices still exists.
7 Examples from Germany are posters showing a gallows for Angela Merkel (during the right-wing populist protests against her migration policy) or current posters and pictures in the so-called social networks, professionally produced on a mass scale during the so-called Corona protests, where well-known doctors, virologists and politicians are locked up as supposed representatives of a "Corona dictatorship" and shown in prison clothes.

Bibliography

Adorno, T., Frenkel-Brunswik, W.E., Levinson, D. and Sanford, N., 1950. *The Authoritarian Personality*. New York: Harper & Row.

Bahl, F., 2014. *Lebensmodelle in der Dienstleistungsgesellschaft*. Hamburg: Hamburger Editionen.

Bergmann, J., 2001. Krisenerfahrungen und Zukunftsängste. In: H. Wagner, ed. *Interventionen wider den Zeitgeist. Für eine emanzipatorische Gewerkschaftspolitik im 21. Jahrhundert*. Hamburg: VSA Verlag, pp. 72–85.

Bergmann, J., Bürckmann, E. and Dabrowski, H., 2002. *Krisen und Krisenerfahrungen. Einschätzungen und Deutungen von Betriebsräten und Vertrauensleuten*. Hamburg: VSA Verlag.

Billmann, L. and Held, J., eds., 2013. *Solidarität in der Krise*. Wiesbaden: Springer VS.

Boltanski, L. and Chiapello, E., 1999. *Le nouvel esprit du capitalisme*. Paris: Gallimard.

Boltanski, L. and Thévenot, L., 1991. *De la justification: Les économies de la grandeu*. Paris: Gallimard.

Bourdieu, P., 1979. *La distinction: Critique sociale du jugement*. Paris: Minuit.

Burawoy, M., 1985. *The Politics of Production*. London: Verso.

Castel, R., 1995. *Les métamorphoses de la question sociale: Une Chronique du Salariat*. Paris: Folio

Castel, R. and Dörre, K., eds., 2009. *Prekarität, Abstieg, Ausgrenzung. Die soziale Frage am Beginn des 21. Jahrhunderts*. Frankfurt am Main/New York: Campus.

Dubet, F., 2009. *Injustice at Work*. New York: Routledge.

Herkommer, S., Bischoff, J., Lohauß, P., Maldaner, K.H , and Steinfeld, F., 1979. Organisationsgrad und Bewusstsein. *Gewerkschaftliche Monatshefte*, 30(11), pp. 709–720.

Holzkamp, K., 1985. Gerundkonzepte der Kritischen Psychologie. Reprint 1987. In: AG Gewerkschaftliche Schulung und Lehrerfortbildung, eds. *Wi(e)der die Anpassung. Texte der Kritischen Psychologie zu Schule und Erziehung*. Soltau: Verlag Schulze, pp. 13–19. https://www.kritische-psychologie.de/files/kh1985a.pdf [Accessed 31 December 2020].

Holzkamp, K., 1992. On doing psychology critically. *Theory and Psychology*, 2(2), pp. 193–204.

Holzkamp, K., 2013. The development of critical psychology as a subject science. In: E. Schraube and U. Osterkamp, eds. *Psychology from the Standpoint of the Subject. Selected Writings of Klaus Holzkamp*. Basingstoke/New York: Palgrave/McMillan, pp. 28–45.

Hürtgen, S., 2008. *Transnationales Co-Management. Betriebliche Politik in der globalen Konkurrenz*. Münster: Westfälisches Dampfboot.

Hürtgen, S., 2017. Verwerfungen in der moralischen Ökonomie des Wohlfahrtsstaates: Ausgrenzungen prekär Beschäftigter und die Problematik sozialer und politischer Bürgerschaft. *Arbeits- und Industriesoziologische Studien*, 10(1), pp. 23–38. http://www.ais-studien.de/home/veroeffentlichungen-17/april.html [Accessed 31 December 2020].

Hürtgen, S., 2020a. Labour-process-related racism in transnational European production: fragmenting work meets xenophobic culturalisation among workers. *Global Labour Journal*, 11(1), pp. 18–33.

Hürtgen, S., 2020b. Precarization of work and employment in the light of competitive Europeanization and the fragmented and flexible regime of European production. *Capital & Class*, online first, February 2020. https://journals.sagepub.com/doi/full/10.1177/0309816819900123 [Accessed 31 December 2020].

Hürtgen, S., 2021. Meaningful work and social citizenship. In: A. Kupfer, ed. *Work Appropriation and Aocial Inequality*. Wilmington: Vernon Press (forthcoming).

Hürtgen, S. and Voswinkel, S., 2014. *Nichtnormale Normalität? Anspruchslogiken aus der Arbeitnehmermitte*. Berlin: Sigma.

Hürtgen, S. and Voswinkel, S., 2017. Non-normal normality? Claims on work and life in a contingent world of work. *International Journal of Action Research*, 13(2), pp. 112–128.

Hürtgen, S. and Wissel, J., 2018. Subalternity and the social division of labor. Sociology of work meets materialist state theory. In: K. Dörre, N. Mayer-Ahuja, D. Sauer and V. Wittke, eds. *Capitalism and Labor. Towards Critical Perspectives*. Frankfurt am Main/New York: Campus, pp. 143–156.

Jonna, R.J. and Bellamy Foster, J., 2016. Marx's theory of working-class precariousness—and its relevance today. *Alternate Routes: A Journal of Critical Social Research*, 27, pp. 21–45.

Jürgens, K., 2006. *Arbeits- und Lebenskraft. Reproduktion als eigensinnige Grenzziehung*. Wiesbaden: Springer.

Kudera, W., Mangold, W., Ruff, K., Schmidt, R. and Wentzke, T., 1979. *Gesellschaftliches und politisches Bewusstsein von Arbeitern. Eine empirische Untersuchung*. Frankfurt am Main: Europäische Verlagsanstalt.

Kalpaka, A. and Räthzel, N. (Eds). 1994. *Die Schwierigkeit, nicht rassistisch zu sein*. Köln: Dreisam.

Linhart, D., 2009. *Travailler sans les autres?* Paris: Du Seuil

Menz, W., Nies, S., Sauer, D. and Detje, R., 2011. The German miracle? Interests and orientations for action during the crisis – the perspectives of those affected. *Transform! Europe* [online] 10 May. https://www.transform-network.net/en/publications/yearbook/overview/article/journal-082011/the-german-miracle-interests-and-orientations-for-action-during-the-crisis-the-perspectives-of-th/ [Accessed 6 December 2020].

Nies, S., 2015. *Nützlichkeit und Nutzen von Arbeit. Beschäftigte im Konflikt zwischen Unternehmenszielen und eigenen Ansprüchen*. Baden-Baden: Sigma bei Nomos.

Peck, J., 2002. Political economies of scale. Fast policy, interscalar relations and neoliberal workfare. *Economic Geography*, 78(3), pp. 331–360.

Popitz, H., Bahrdt, H.P. and Kersting, H., 1957. *Das Gesellschaftsbild des Arbeiters. Soziologische Untersuchungen in der Hüttenindustrie*. Tübingen: Mohr Siebeck.

Räthzel, N., 2002. Developments in the theories of racism. In: The Evens Foundation, ed. *Europe's New Racism. Causes, Manifestations, and Solutions*. New York/Oxford: Berghahn.

Sauer, D., Stöger, U., Bischoff, J., Detje, R. and Müller, B., 2018. *Rechtspopulismus und Gewerkschaften. Eine arbeitsweltliche Spurensuche*. Hamburg: VSA Verlag.

Schmalz, S. and Brandon, S., eds., 2019. *Confronting Crisis and Precariousness. Organised Labour and Social Unrest in the European Union*. London/New York: Roman and Littlefield.

Solga, H., 2015. The social investment state and the myth of meritocracy. In: A. Gallas, H. Herr, F. Hoffer and C. Scherrer, eds. *Combating Inequality: The Global North and South*. London: Routledge, pp. 199–211

Thompson, E.P., 1971. The moral economy of the English crowd in the 18th century. *Past & Present*, 50, pp. 76–136.

Tolman, C., 2013. *Psychology, Society and Subjectivity: An Introduction to German Critical Psychology*. London: Routledge

Vester, M., von Oertzen, P., Geiling, H., Hermann, T. and Müller, D., 1993. *Soziale Milieus im gesellschaftlichen Strukturwandel – Zwischen Integration und Ausgrenzung*. Köln: Bund Verlag.

Zoll, R., ed., 1981. *Die Arbeitslosen, die könnt ich alle erschießen. Arbeiterbewusstsein in der Wirtschaftskrise*. Köln: Bund Verlag.

7 "Social rage" against the oligarchs

Justice, Jews and dreams of unity in current Russia

Olga Reznikova

Against the elites

Criticism of the political and economic elite and the appeal to the "little man" from the "common people" (*prostoi narod*) are important components of many protests against the government and against grievances in Russia.[1] Whether implicitly or explicitly, with a positive or negative connotation, they figure in the vast majority of movements, initiatives and appeals that operate as "resistance", "protest" or "opposition" in Putin's Russia. The criticism of the political elites and the oligarchs can be linked to various demands: For example, the release of political prisoners is called for, and criticism is levelled at the tax system, yet aspirations of a free market are also articulated, and social inequalities in the country are discussed. At the same time, the defeats in the battle against the elites are often linked to the passivity and backwardness of the majority of "common people", and hope is placed in them soon "waking up". The acts of mobilisation that begin to respond to such hopes are usually associated with strong emotions that are often perceived by both the actors involved and observers alike as rage or anger. In the period between 2015 and 2018, following the political and economic crises, a wave of such protests can be observed that address a variety of social issues (see Reznikova and Ege, 2019; Бизюков, 2018; Щелин, 2017).

This type of protest by "common" or "ordinary" people, their rage and the hopes that accompany this form of resistance are also part of the international debate. Like for many left-wing protest researchers in Russia (e.g. Erpyleva, 2019; Медведев and Журавлев, 2020), it is the prospects of such protests that are first and foremost emphasised, for example, in the analyses of the "yellow vest" protests in France (Clément, 2020; Susser, 2021). The theoretical examination of left-wing populism (Laclau, 2005; Rancière, 2017) also enquires about the positive outlooks of the new configuration of "the people" and their rage during the protests (for a critique of this, see Narotzky, 2019). In this chapter, I will approach this topic from a different perspective. Against the background of the *conjunctural analysis* approach (see Clarke in this volume), and using an emphatic notion of emancipation that is based on the critical theories of the Frankfurt School (Adorno and Horkheimer, 2010) and examining the Russian debate about protests and strategies of the left,

DOI: 10.4324/9781003141150-9

I enquire about the nature of the historical and social constellation in Russia and the references the various actors make to one another in which this rage against the elites develops and receives its relational meaning. In the second part of the chapter, I then work out, in particular, what role anti-Semitism plays in this type of protest. In doing so, I analyse thought patterns and constellations within Russian[2] protests (2014–2019) – some of which I attended and followed in solidarity as a field researcher. In some respects, however, they may also be similar to other contexts.[3] I am thus concerned not least with covert dynamics and with the dark side of such protests, without thereby wishing to suggest – as I finally argue – that this has to be the central and decisive question when it comes to declaring solidarity or a lack of solidarity with the protests.

Protests in Russia in the constellations of *vlast, narod* and the intelligentsia

For those looking at Russia from the outside in, the question of the extent to which protest there can be classified as democratic, left-wing or, rather, nationalistic seems to be of relevance. Human rights activists, democrats and political groups from the USA and Western Europe ask themselves to what extent those leading the protests adhere to ethical standards and commit to tolerance and equality.[4] Yet, the temptation to divide up the Russian protests and movements into a left, a liberal and a right wing often misses the essential point, because these ideological points of reference do not constitute the central relevant axes of alignment for the protests' agenda and their strategy for present-day Russia, and this is also true of the organised left there. Neither anti-racism or the fight against Russian Great-Power Chauvinism nor anti-capitalism or the demand for social justice nor feminism and the fight against domestic violence are what currently drive most of the Russian left onto the streets. The authoritarian government, with its harsh acts of repression, murders (as well as death threats and attempted murders) and aggressive right-wing rhetoric at the level of foreign and domestic policy, shift most of the forces to a large extent towards an "anti-Putin" line of confrontation. Among other things – and this is the thesis that I would like to present here – this leads to the reactivation and updating, or reconfiguration, of an old thought pattern in which protest (or a movement, resistance or opposition) is categorised within a triangle between "*vlast*" (government, the political elite or even Putin himself), "*narod*" (the majority of the population that tends to be underprivileged and ethnically Russian) and the "collective West" – that is, above all, EU Europe, the UK and the USA. This is not just true of the left; the current relevance of engaging with this thought pattern pertains to an even greater extent to the liberal and right-wing liberal forces. However, it is articulated differently: *vlast*, *narod* and intelligentsia are figured in distinct ways, set in various different relations to one another and used in a variety of manners for mobilisation. Taking a closer look at the role of such a thought pattern and the constellations found within protests makes it possible to shift the focus onto the implications, the potential and also the limits of protest movements.

A methodological approach such as this does not lead directly to answers to the burning questions about what the right strategy and tactics are for such movements, or about how to show solidarity with them. Steering attention away from the search for a "right strategy" for protest (something which most intellectuals in Russia currently participate in, for good reason) and towards the analysis of these and similar thought patterns is possible in my academic work because I primarily view protests as cultural phenomena: phenomena that are products of the balance of power, cultural processing and the relationship structures in society. This cultural–analytical approach has the effect that the protests are not explained on their own terms or by the demands, strategies and criticisms articulated within them, but are rather classified in the context of the current *conjuncture*. In view of how they are articulated, the relations between *vlast* and *narod* can be sorted on the basis of two axes: populism/anti-populism and anti-Western/pro-Western. Within the broad anti-Putin coalition, which, in January/February 2021, is striving even more than in previous years for an anti-Putin consensus, this thought pattern and these axes come to be of crucial importance for the formation of alliances and the choice of strategy.[5]

In this chapter, I concentrate on the populism/anti-populism axis and try to use the example of "social protests" (and "social rage") and their relationships to the other forms of protest to elucidate the above-mentioned thought pattern and to set it in relation to the analysis of anti-Semitism. The theses of this article are based on participatory observation and interviews with actors taking part in protests, as well as on the analysis of media reports and social media discussions about the protests in the period between 2015 and 2018.

The Kemerovo tragedy and social protest

On 25 March 25, 2018 in Kemerovo, a medium-sized city in eastern Russia, 60 people died in a fire in the "Winter Cherry" shopping centre, 41 of whom were children. This tragedy mobilised the people of Kemerovo to protest. In particular, the victims' family members took to the streets to express their shock and anger about the events. People in many other parts of Russia also declared their solidarity with the relatives, and soon thousands took to the streets and squares of numerous cities not only to mourn together but also to protest against corruption and the government. The last audio recordings of the victims, which quickly spread on social media, touched and concerned people: Children called their parents from within the shopping centre, calling for help, or saying goodbye before they died.[6]

The reason for the high number of victims, according to the criticism from relatives, protesters and many journalists, is a combination of many factors, including, above all, deficiencies in compliance with fire protection and safety precautions and corruption on the part of the owners, the state inspectorate for fire protection and the mayor (e.g. Baza, 2020).[7] In addition, the way the shopping centre was constructed made it impossible for visitors to find the exit quickly. The owner of the shopping centre has therefore been accused, in the aftermath, of having prioritised greater profit over the safety of consumers. During the demonstrations, this deadly construction and the lack of fire protection were again seen as effects of the corrupted regional government and of corruption across the entire country.[8]

Within the landscape of Russian protests, this mobilisation against corruption in March 2018 (similar to the partially simultaneous mobilisation against the landfill site in Volokolamsk) was referred to as a "social protest". The wide use of the term "*социальный протест*" [social protest] is relatively new, although the phenomenon that the term describes has a long history. For social protest is about the mobilisation of people not previously involved in protest, who mostly take a stand more or less spontaneously for their own interests, and mostly in the form of morally justified anger. In the case of Kemerovo, the protests were labelled as such because the victims' relatives (who were central figures in the protest) called for those responsible for the deaths of their children to be prosecuted.[9] Yet, the previous workers' protests where they campaigned to improve their working conditions, the protests against piles of rubbish near the protesters' places of residence or, for example, the protests against construction projects in the protesters' own district are also emically and etically referred to as "social protests" (especially in the opposition media and in other movements). Like in Volokolamsk and Krasnodar, the relatives of the fire victims turn to President Putin in hope. These appeals are heavily criticised by the representatives of the "political protests", and hope is expressed that the protests in Kemerovo will soon "wake up" and "politicise themselves".[10] Many different groups and individuals who consider themselves part of the anti-Putin movement are increasingly positioning themselves as "political protest" as opposed to "social protest".

The debate about the terms "politics" and "politicisation" underwent an interesting development within Russian protests during the period of the protests in Kemerovo. The protests from 2011 to 2013 that had arisen out of the movement for free elections (which, in turn, had its origins in disgruntled election observers) were not motivated and mobilised by social or economic issues, but by the desire to maintain "dignity". Up until 2015, the people who were newly mobilised at that time also for the most part explicitly described themselves as non-political and used the rhetoric of wanting to take a stand "for what is good". Only through contact with and demarcation from protests such as those in Kemerovo or workers' protests (which campaigned for their own particular interests and often sought help from the president) did a new term of self-reference develop: From then on, the movement that opposed election fraud, undemocratic reforms, centralisation of political power, imperial claims in Russian politics or political persecution called itself "political protest" and accused the "social protests" of depoliticisation, because they were not directed against Putin.[11] For their part, the protests that, since 2015, had been increasingly emerging spontaneously around burning social issues, and which primarily made social or economic demands, mostly disassociated themselves from the opposition figures and from an agenda that was critical of the government, because they expected that to lead to a disadvantage for them in the dialogue with the *vlast*.[12]

Social protest in the context of populism and anti-populism

Even if this clear division into "political" and "social" protests is new, it has its antecedent in Russian and Soviet movements and formations of alliances. In this tradition, the question of a common agenda between the so-called intelligentsia (the educated middle class) and the "*narod*" (everyone else, but, above all, the non-privileged, and, at the same time, predominantly Russians [ethnic: russkii]) plays a central role for the opposition to the "*vlast*" (the term can be translated as regime, power, government or authority).[13] The notion of "social protest" thus reflects the homogenised view that the intelligentsia have of the people they call the "*narod*". The term "political protest", on the other hand, alludes to an understanding of politics in which a universal need of all people in Russia is described from the liberal view of political and economic development.

The literary scholar Rossen Djagalov (Джагалов, 2011) argues that the balance of power in the pre-revolutionary Russian Empire, where the intelligentsia considered itself to be in an alliance with the exploited and oppressed *narod*, was increasingly replaced in the Soviet, and especially the post-Soviet, era by anti-populism on the part of the intelligentsia. Anti-populism is first and foremost characterised by the strongest demarcation of the intellectuals from "the masses", with the writers, poets and artists, etc., also using dehumanising vocabulary towards the underprivileged and workers.

> The gradual withdrawal of the traditional intelligentsia from the historical bloc with 'the people' has led to the populist rhetoric and the claim to "*народность*" [here: proximity to the people] being taken over by a fringe group of the intelligentsia – the nationalists. It is this voluntary abandonment of the traditional social functions of the intelligentsia that can turn them into what Georgy Knabe termed a "page turner".
>
> (Джагалов, 2011)

My empirical, mostly ethnographic research in the 2010s also shows that anti-populism plays a central role in the political protests. As also documented by Djagalov on the basis of the literature, my conversation partners from the political protests impute a complicity with Putin's government to the "common people" and associate this with the almost innate character traits of the uneducated masses, who are doomed to "remain slaves". In one interview, my partner in conversation (a Moscow lawyer in her early 40s who took part in the anti-Putin protests) argues that the "Russian people" have a "slave gene", by which she means that the underprivileged majority of Russians do not want freedom and, therefore, do not defend themselves against the authoritarian government.

However, the theory of a close or even causal connection between the anti-populism of the intelligentsia and the successes of the right-wing movements in the late 2000s and early 2010s does not ultimately convince me. For neither among the intelligentsia of tsarist Russia nor in Soviet dissident culture (in which Djagalov sees the cause of the problem) was the question of alliances fully resolved. The left-wing dissidents (such as Valery Ronkin from the group "Kolokol" or philosopher Mikhail Molostvov, historian Lev Krasnopevzev and many others) mostly represented more populist positions – if that category can even be projected into the past at all. Yet, some right-wing dissidents who viewed themselves in the tradition of the pre-revolutionary pochvennichestvo[14] were also populists according to the current understanding of the word and further developed a ethno-nationalist understanding of *narod* in their texts (such as anti-elitist anti-Semitism, e.g., as elaborated by Razuvalova in 2015 for right-wing dissident literature).[15] Furthermore, it does not always make sense to trace the processes in society as a whole (such as the growing right-wing mood in the 2000s or the low level of participation in the protests in the first half of the 2010s) back to the attitude of the intelligentsia. Here, despite his justified criticism, Djagalov echoes the intelligentsia's exaggerated opinion of itself as always being of crucial importance for society.

The perception of the "rage of the common man" or the "rage of the people" thus veers to a particular extent within the opposition faction of Russian public life, and also on the part of the government rhetoric, between two poles, which can be schematically divided into populist and anti-populist. However, the assignment of concrete movements and speaker positions to these poles remains difficult, and, of course, there are also numerous nuances and overlaps. This ambiguity is discussed in such a charged manner not least because, also historically, this axis does not (or does not always) coincide with the right–left axis, but was and is always connected with it. On the academic and journalistic left, a number of international authors, such as Oliver Marchart (2017),[16] argues against an unnuanced view of populism and against its blanket rejection. I furthermore advocate for a nuanced view of anti-populism, because it too has different political faces and lines of argument. For example, it can be derived from an anti-democratic, (neo-)liberal attitude towards the personal responsibility the underprivileged bear for their plight (as in Russia in the example cited above concerning the slave people, though also from a politician like Grigory Yavlinsky), but also from the justified criticism of Russian Great-Power Chauvinism[17] and the rejection of ethno-nationalist ideology. On the theoretical and ideological level, a distinction must also be made between an anti-populist neo-liberal shift of responsibility onto the individual, on the one hand, and the emphasis on the possibility of the subjects' emancipatory agency, on the other, which is also often anti-populist in its effects (cf. Salzborn, 2019). The categorical distancing of the left from anti-populism can therefore result in having to back away from the emancipatory substance of left-wing utopias.

Terms such as "social protest", as a means of demarcation from "political protest", even if they originate with the workers, neighbours and desperate parents themselves, thus tend to describe the moral rage of the *narod* from a homogenised perspective of the intelligentsia, which is caught in the dichotomy of populism and anti-populism. Yet, at the same time, the term "social protest" actually describes a form of mobilisation of those people who perceive their own situation as an emergency and develop moral rage towards (mostly local or economic) elites, figuring them as opponents. Despite serious differences in their demands, the workers' protests of the truck drivers (2015–2016), the protests against rubbish heaps (in Volokolamsk from March to April 2018 and in Schies from 2018 to 2020) or the local protests in the big cities against development in the parks (especially from 2015 to 2019) also have a lot in common in terms of how they are articulated:

- The protesters see themselves as *ordinary men and women* from the "*prostoi narod*" (common people). This self-designation actively distances itself from the opposition liberal intelligentsia, on the one hand, and from the *vlast* and any party, on the other.[18] The collective commonness, ordinariness and normality can be articulated in different ways depending on the subject of the protest, yet one of the following three leitmotifs is always present: 1) Protesters see themselves as "ordinary people" because they are collectively members of a professional group (that tends to be poorly paid). This was the case, for example, with the farmers' protests or the truck drivers' strike, but also with the doctors' protests. 2) Protesters link their "ordinariness" to their long-term local connection to a neighbourhood or place of residence (protests against urban development, against waste incineration plants); or 3) protesters underline their "normality" based on their marital status (it is characteristic of all social protests for people to position themselves as "the mother of two children" or as "a family man"; in the case of Kemerovo, however, this articulation of the "common person" is pushed to the fore, with the result that solidarity with the relatives of the victims is often also linked to one's own marital status[19]).
- The social protests mostly arise *spontaneously* or are portrayed as spontaneous in retrospect. Organisation and the ability to plan contradict the idea of social protests in a certain sense, so that if a trade union, for example, emerges from a protest, its actions are usually neither referred to by the participants themselves nor by outside observers as "social protests".
- The protests are directed *against the elite*, especially against the local elite or against the oligarchs, whereby the president is seen as protection against them (at least at the beginning of the protests). In Kemerovo, it is the oligarchs and the local political elite who figure as the elite, among the truck drivers, it was an oligarchic family, and, in Volokolamsk, it was the city of Moscow, the abstract oligarchy and the local elite. In all of these protests, the activists issue a plea to the president to protect them from despotism,

because they are powerless against the powerful elite. In Kemerovo, for example, the president was supposed to monitor the investigation into the causes of the fire, because, without him, the corrupted structures would falsify the results of the investigation.

- The social protests are extremely emotionally charged. The participants explain their spontaneous mobilisation by means of their *rage*[20] and perceive themselves as "enraged". Depending on which leitmotif prevails in the construction of "ordinariness", figures emerge who reflect their own and external perceptions of the particular form of social rage in question (e.g. "enraged long-distance truck drivers" during the strike, "angry residents" during the local and ecological protests or "enraged fathers" in Kemerovo).

This similarity between the protests in terms of the manner in which they are articulated cannot be explained in and of itself, but is rather to be classified by means of the societal relations within the protest landscape and, beyond that, in the context of the prevailing balance of power: above all, the authoritarian form of government, a (neo-)patrimonial capitalism and weak horizontal structures. Furthermore, not only the political opposition, but also the pro-government media initially contrasted the social protests – as protests by "normal people" – with the anti-Putin protesters.[21] On the other hand, the liberals were portrayed by the authoritarian-populist government rhetoric as "alienated from the people", Western-oriented and elitist. In the case of some social protesters (most notably in Kemerovo, but also in Volokolamsk), the government (local, regional and, in Kemerovo, also at the federal level) repeatedly pretended to want to enter into a dialogue. However, this rhetoric quickly petered out again, without the demands being addressed, as soon as the heated mood had calmed down.

The growing role of social protests with an anti-corruption agenda and the rage that stems from a feeling of injustice have many interesting and important aspects. Above all, they are often an opportunity for workers, pensioners and residents of small and large cities to demonstrate their agency in the course of mobilisation. This is shown by both the interviews that I conducted and those of other researchers (cf., e.g. Erpyleva, 2019). In one conversation about the moments of mobilisation, an activist from a workers' protest said:

> [Back then, before the protest,] I thought like many of our colleagues do today. That we can't do anything, that it doesn't matter what you do anyway, you can't change anything. Only now do I understand: if we do something, we can, of course, change something [...] And we can actually change everything! And this idea has actually changed for me during the period.[22]
>
> [since the mobilisation]

The outcome, make-up and course of these and similar experiences of collective agency are, in principle, open after mobilisation. Collective narratives and rhetoric can also change significantly over time, depending on the experiences the protesters have and how they classify them. In the following, however, I would like to address only one aspect of the social protests that is particularly relevant during the first spontaneous and emotionally charged mobilisation – anti-Semitism. It correlates more with the experience of powerlessness than with that of agency.[23] It is not my intention here to say that this is the most significant element of social protest. In my estimation, the social rage and the loud (whether to a greater or lesser degree) demand for justice that could be observed in this new wave of social protests from 2015 to 2018 are the most notable practice of the current protests, including in relation to the emancipatory substance of the movements. Moreover, open anti-Semitism is not a unique feature of social protests, because it is also often found in other "political" forms of protest (and, indeed, in right-wing and liberal as well as left-wing political protest). However, I am interested in a specific articulation of the anti-Semitic worldview within social protest that is directed against the elites, the oligarchs and corruption, and which emanates from the people who were not politicised before the outrage expressed during the protest.

Anti-Semitism and slave rhetoric

Let us now return to March 2018. In Kemerovo, panic struck after the fire, and rumours spread that there were many more victims – not 60 but 400 people.[24] After the demonstrations, the crowds of people went to the mortuary to find the allegedly "hidden victims". On social media, in Telegram and WhatsApp channels reporting on the "real situation" in Kemerovo, there were a lot of messages about the "lies of the elite", who were hiding the victims of the fire. The criticism of the corruption, the horrifying news and the regret over the incapacitation of the people were overlaid with conspiracy myths about the cause of the fire and the extent of the tragedy. After two days, there was a rally – "*народный сход*" [people's assembly] – in Kemerovo, at which "angry city dwellers" and "angry parents" were supposed to learn the truth. The gathering in front of the government building only gradually dissipated when the delegation gained access to inspect the mortuary. However, that same day, and the next day, there were calls to look for the dead children in the synagogue.[25]

Many other social protests showed solidarity with the rage of the Kemerovo residents, some referring to the conspiracy myths in a positive manner. On 30 March 2018, the following thoughts about the fire in Kemerovo circulated in a Telegram group organised by the activists of a workers' protest in solidarity with the other social protests:

> The tragedy in "Winter Cherry" happened shortly before the Jewish festival of Passover. The tragedy in the "Lame Horse" club in Perm[26] occurred shortly before the Jewish festival of Hanukkah. The "Saratov Airlines" flight with 71 people on board crashed near Moscow just before Purim. These were all sacral sacrifices by God's chosen people.[27]

Another activist of the group replied to this message in the same chat:

> Yes, we have to learn to count, one plus one – if we don't want to be slaves anymore. Who is behind all this: behind the tragedy in Kemerovo, behind the war and behind the stealing and robbing from the Russian population?!

Even if the other empirical material shows that this position was not shared by all activists in this group, it should be noted that such messages appeared regularly and were not prohibited, refuted or sanctioned by the group. In such "solidary" bridge-building as this between one's own concerns and those of the relatives of the fire victims, the anti-Semitic worldview coalesces with the idea of being able to fight the cause of the evil together and united. For, the reactions to the Kemerovo fire on the part of other social protests (for instance, the vigils held in solidarity in other cities) were also mostly not about the specific demands of the victims' relatives (e.g. access to the investigation documents). Rather, they are characterised by the rhetoric of "uniting", of a collective "waking up", and the like.

> In parallel to this dynamic, which mobilised many social protests to engage in campaigns of solidarity with Kemerovo, the activists of the "political protests" discussed the question of whether the *narod* and Igor Vostrikov himself, who was seen as the personification of the protest in Kemerovo, would one day see the "true cause" of all problems: namely, Putin. It was about the question of why the people from the *narod* were demanding protection from the president[28] and when they would finally "wake up", and the *vlast*, fall.

The Levada Centre (Левада-центр 2018), an independent institute for social research, as well as various Jewish organisations, spoke after the fire in Kemerovo about the new surge in anti-Semitism using old anti-Jewish stereotypes. The President of the Federation of Jewish Communities of Russia, Alexander Boroda, reported that after the fire in Kemerovo, anti-Semitic agitation increased rapidly, with the revival of various myths about "Jewish power", "Jewish blood lust" and, above all, "ritual murder": "People are now afraid to go to the synagogue. People do not understand where this might lead; they are even afraid of pogroms" (Лехаим, 2018).

A similar protest dynamic can be observed in the protests against the piles of rubbish in Volokolamsk. There, too, it seemed important for the protesters to hold a social protest and to demarcate it from "political protests".[29] Again, during this protest against environmental pollution in a small town and against the rubbish that gets delivered from nearby Moscow, people came to the demonstrations with "Putin, help us" posters. And here, too, some of the demonstrators and some of the supporters blamed the Jews for the mass poisoning of the residents (especially the children) by means of the gas from the rubbish.

The anti-Semitism researcher Viktor Shnierelman (Шнирельман, 2019) establishes a connection between the revitalisation of anti-Semitic "myths" after the Kemerovo tragedy and the increase in right-wing sentiments in society at large. In his earlier texts, Shnierelman (Шнирельман, 2017) also links the frequent anti-Semitic forms of expression to the growing importance of the Orthodox Church in Russia.[30] But what does anti-Semitism have to do with "social rage"?

It is striking that in this articulation of rage, the focus is not on grief, but on powerlessness and self-identification as slaves. The recourse to this slave metaphor can have two main functions in present-day Russian protests: On the one hand, it can be used to express a differentiation made on the part of the intelligentsia from the uneducated and passive *narod*, often with a reference to the Soviet, and even the tsarist, era. On the other hand, actors in the social protests use the same term to articulate their own powerlessness (often projected into their own past: with the repeated allusion to having themselves been "still slaves" in former times). On the basis of my ethnographic research during the strike of the long-distance truck drivers in 2015/2016, it is possible to illustrate this metaphor and trace its connection to anti-Semitic resentment. Over the course of the mobilisation for labour rights and against the perceived injustice of the change in the law, the label of *raby* (slaves) came up more and more among the truck drivers as a means for referring to themselves, colleagues and other workers. The narrative about the necessity of waking up from the state of being a slave took on greater significance alongside the conception of themselves as representatives of the "working class" (primarily understood as people who perform physical labour).

The actors' view of the connection between mobilisation and waking up is made clear in the following quotation:[31]

PIOTR: "Do you know, I didn't think about it before. I only worked. I never thought about it, about the question of what has happened in our country. I myself was a typical *vatnik*.[32] To be honest, it did not matter for me [before joining the protest]. It is embarrassing now [...]. But then, we had a lot of rage [...] I didn't know what I wanted at first. I only had this rage against injustice. And then, one thing led to another and we began to analyse things [during the protests] [...] Why do they get richer and richer, while we get poorer and poorer?

OR: Who are "they" actually? Umm, what do you think now, who are they?

PIOTR: Hmm, maybe I can reformulate it in this way: Why do the oligarchs have so much money, and why do we have so little? Why is it not allowed for us to protest for our rights? [...] And then a new question emerged. Why does Medvedev[33] have so many Jews around him? Is he also one of them? [...] Our fight [...] opened my eyes. I understood how the network had developed, and what kind of connections were in charge of my country. I understood who controlled everything here. I understood, step by step, why the Soviet Union crashed, why and by whom we were robbed

> [...] We all are *raby* [slaves]. Slaves who work for their owners. Slaves don't have rights; slaves have to pay for everything. They only work for free [...]

OR: And the owners are the Jews?

PIOTR: [...] *zhidy* [slur against Jews]!

Piotr uses two derogatory terms here for his past self: *vatnik* (Russians manipulated by Putin) and *rab* (slave). The first term was borrowed from the liberal opposition movement. *Vatnik* is used to insult people like Piotr, who are generally older, generally not privileged and also uneducated, with feelings of nostalgia for the Soviet era, and who tend to be loyal to Putin. The other term, slave, is also increasingly used by the newly mobilised workers to judge their past selves (as well as their non-protesting colleagues). With this in mind, it should be emphasised that both terms are products of mobilisation and of the relating of one's own point of view to the point of view of the liberal protests against Putin.

Piotr, but also some other social protesters, describe their mobilisation using the following metaphors: The social rage helped them to recognise the "manipulation", "to wake up from their sleep" and "to do something to combat their powerlessness". The powerlessness that spread after the fire in Kemerovo and the anti-Semitic responses to it are, it seems, a magnifying glass of the short-circuit that is becoming more prevalent in social protests. This is, at the same time, integrated, both in the social and the political protests, into the rhetoric of the *объединения* (union/community of all) against those who are evil.

"We have to unite"

Alexander and Margarete Mitscherlich (1977), in their psychoanalytic study of "specifically German" behaviour "typical of that time" (ibid., 17), posed the question of how their patients dealt with their own complicity in National Socialism. The anti-Semites, on the one hand, showed an infantile identification with the government and, on the other, an inability to grieve. For the anti-Semites, grieving is replaced by melancholy with a narcissistic object choice. Here is their argument, following Sigmund Freud (1982):

> "In these attempts to shake off guilt, remarkably little thought is given to the victims – regardless of whether they are one's own or those of the other side. This reveals the extent of the energy that needs to be expended to deny what is, in truth, a by no means so clear-cut predicament of the past [...] When analysing the psychological events that constitute grief, we find the pain of the loss of a being with whom the mourner was connected in a deeper emotional relationship between fellow humans. Something was lost with the mourned object that was a valuable part of our experienced environment. However, there is a pathological increase in grief, melancholy [...] In grief, I feel impoverished, but not degraded in my self-worth. The melancholic, however, has this latter experience. He

suffers an extraordinary diminishment of his sense of self, a tremendous impoverishment of the ego.

(Mitscherlich and Mitscherlich 1977, 36–37)

For the research into anti-Semitism, these and similar analyses of the deep psychological structures of anti-Semites are often crucial in order to understand the false projections that accompany anti-Semitic affect (Horkheimer and Adorno 2010, 196ff). The absence of the reflexive moment when experiencing the injustice within the violent structure of bourgeois society drives the anti-Semites to collective madness, which strives to create a just world through the extermination of the Jews as a pathic way of dealing with their own wound. The anti-Semite is incapable of grief, which presupposes love and happiness, whereas melancholy, by contrast, can give fresh impetus to the manic.

With the aid of this hypothesis, which is based on psychoanalysis and critical theories, one can understand the relations around the "social protests" in Russia – in an obviously completely different constellation than for Horkheimer/Adorno and the Mitscherlichs – as follows: The Kemerovo residents initially did not mourn the actual victims of the fire but instead flocked to the mortuary to look for the imagined dead children and quickly hit upon the idea of besieging the synagogue as well. The workers, who experience an absence of agency in their struggle for their own rights and interests, shared this melancholic rage in order to imagine a "community of the oppressed" in the fight against those who are evil. And the intelligentsia was annoyed that that *vatniki* people did not form an alliance with it to oppose the *vlast*, and thereby "finally experiences the truth". The rhetoric of unification [объединение, единство], which can empirically be found in all of these groups, plays a decisive role in this. The concept of "*prostoi narod*" is admittedly used in various different ways in the protests, sometimes in an ethno-nationalist sense, sometimes in a class-specific one and, in some instances, also as a term for the entire population of the Russian Federation. Yet, in all of these cases, the *narod* is something that is not unified and that needs to be unified. In this sense, there is a connection between the populist dream of the intelligentsia about a common agenda with the *narod* (or the anti-populist repression of this dream) and the dream of the social protesters to unite as the *narod* in their (partly anti-Semitic) fight against "those up at the top".

The reasons why people view the real balance of power through such a lens in which powerlessness and the fantasy of omnipotence merge (cf. Erdheim, 1988, 371–435, Mitscherlich and Mitscherlich 1977), and why they articulate a rhetoric of unification, lie not only in the deep socio-psychological structure of the actors. It is not this alone that is the problem of the described dynamics in the protests, but also the actually existing balance of power in society, in which the actors have to operate. There are currently fewer and fewer opportunities to organise. Any opposition – whether populist or anti-populist, whether it holds a romanticised or demonising view of the *narod* or sees itself as a part of it – is currently at risk of severe repression. In addition, the daily life of the "apolitical" people in and outside of Russia (especially in Belarus and Ukraine, but also beyond) is also severely impacted by the tough authoritarian line of the Russian government, which uses political murders, neo-imperial wars and aggressive right-wing rhetoric as a tool to achieve its goals. The powerlessness is therefore also partly a reaction to the tangible political situation. Nevertheless, anti-Semitism in the protests, which can be analysed as a consequence of this conflation, should not be ignored or played down.

Just as Detlev Claussen (1989), following Horkheimer and Adorno (2010), understood anti-Semitism as the limit of the Enlightenment, it is perhaps possible in the anti-Semitism of the social protests to identify a constituent limit to the emancipatory substance of the unification of the enraged against the elite. Historically, this is not a new topic.[34] In the current-day situation, however, in which people are looking for a new explanation for the obvious injustice in their lives, the subject of the regressive and anti-Semitic content of the union of the oppressed takes on a new relevance. Unsuccessful liberation and the failure of emancipation constitute a part of the polarisation and formation of consciousness in popular movements, including the worker's movement, both historically and in the present day. If the analysis of inequalities only distinguishes between the majority of the people and a small, privileged minority, without considering any anti-emancipatory elements that form a part of this constellation and develop within it, then it contributes to this problem.

Conclusion

The anti-Semitic elements described are not necessarily intended to diminish hopes for progressive change and a "revolution against Putin". With this article, I do not intend to criticise the protest movements in Russia or to claim that anti-Semitism and conspiracy myths are the central problem of the protest movement. Nor do I presume to propose possible solutions for the daily challenges of the protests. The cultural science view compels this modesty. It is thus not my aim to influence the current mobilisation, nor to question the forms of the protests. However, by using the analysis from my empirical research to consider a thought pattern, I hope to make the anti-Semitic implications of the "rage of the common people against the elites" accessible for collective reflection, which is an important part of solidarity. How exactly this reflection could be used strategically and tactically for the protest movement is beyond the scope of this article. I am convinced, however, that the revolution cannot just be about recognising the balance of power and fighting for hegemony, but must also be about conceptual work, which, in turn, deals with real events and their ambivalences. In order for the revolution not to degenerate into a regressive revolt, it also needs a utopia that contains reflection on that revolution and its conditions.

Translated from German by Josephine Draper

Notes

1 This chapter was completed in 2021, before the full-on Russian attack on Ukraine.
2 In Russian, a distinction is made between ethnic affiliation (*russkii*) and affiliation with the politico-geographical entity or nation state (*rossiiskii*). Since there is no established translation of this distinction in English, "Russian" is used in this text without elaboration, in both adjectival and noun form, for "rossiiskii", but when the sense of ethnic affiliation is intended, this is noted in parentheses as follows: (ethnic: *russkii*).
3 Research was supported by a grant from the German Research Council (DFG), FOR2101, 371/1-1 and -2.
4 On this, cf. the discussion about Alexei Navalny and Amnesty International (e.g., Gessen, 2021) at the time this chapter was written.
5 Leonid Gozman (Гозман, 2021), for example, traces the need to unite all forces – left, liberal and right – around Navalny to the argument that the "necessary condition for protest to develop is the feeling of the common group identity that is emerging today". A collective notion of togetherness, Gozman's reflection continues, arises from the confrontation between "the bad" and "the good". This view is plausible in the context of Russian politics. However, the question I ask myself here is what social, political and overall societal effects can result from a depoliticised group identity such as this.
6 See media reports from the time, for example, RenTV (2018), Удовченко (2018), Ильченко (2018), TOK (2018).
7 Not only did such an assessment by Alexei Navalny (Навальный 2018) spread widely among the public; the other critical opposition members and some victims also primarily connected the fire to corruption (see Info24, 2018; Антикоррупционный портал, 2018).
8 The anti-government and anti-corruption responses to such catastrophes are not specifically Russian. Similar, partly spontaneous reactions with acute criticism of state and local corruption were also central to the mass protests in Bucharest after the fire tragedy in 2015 and resulted in broad scrutiny of the established political structures in Romania (Habit 2021, 121,126). Unlike in Russia, however, in Romania, the more exact connection between corruption and the high number of victims of the fire in the club was investigated and confirmed. In Russia, on the other hand, the relatives and those showing solidarity with them complained that the state of corruption was not being combatted and continued to feel powerless, even after the trial of the employees of the shopping centre who were responsible for fire protection (cf. Baza, 2020; Ананьев, 2020; Востриков, 2019). I would like to thank Daniel Habit for pointing out the similarity between the protests in Bucharest and Kemerovo.
9 Igor Vostrikov, who lost his wife, sister and three children in the "Winter Cherry" fire, quickly became a symbol of this protest. He published videos on his social media accounts with information about the course of the investigation, appeals for mobilisation and a call to Putin to monitor the investigation.
10 On social media, those in opposition expressed particular criticism of Igor Vostrikov personally for his insufficiently critical attitude towards the president.
11 For an alternative perspective, see Клеман et al. (2010).
12 Characteristic and typical of this were, for example, the reactions to Alexei Navalny's plan to come to Volokolamsk during the protests, which the opposition politician regarded as an important movement. The protesters, campaigning against a landfill site, firmly rejected the idea, stating: "We don't need that radical here", "We're not here to solve political issues", "Now he's going to come and scream, "Down with Putin". We don't need Navalny here" (see video from Antilop, 2018). The striking truck drivers also reacted in a largely similar manner during the first few weeks of

the truck driver protest (2015) when Navalny's fellow campaigners and other Moscow opposition members went to the drivers and offered their support.

13 This juxtaposition between anti-populism and populism has an interesting historical dimension in Russia. The question of the *narod* had already been the crux of the political strategy of the intelligentsia at the end of the 19th/early 20th century, both among the left-wing *narodnichestvo* and the right-wing *slavianofilstvo* or *pochvennichestvo*. Later on, the question of the relationship to the *narod* also crops up among the dissidents, with the left-wing intelligentsia or left-wing dissidents mostly representing more populist, but pro-Western and pro-modernist positions, while some right-wing dissidents in the *pochvennichestvo* tradition had elements of ethno-nationalist populism, anti-Westernism and the glorification of the Russian village in their texts. On the topics of the heritage of traditions, change and continuities in the dissident literary texts, see Ann Razuvalova (Разувалова, 2015); on the role of anti-Semitism in the updating of the *slavianofilstvo* or *pochvennichestvo* movements among the post-Soviet intelligentsia, see, e.g., Rossman (2002).

14 Literally: soil-ism, an ideological current within *slavianofilstvo* that launched the idea of the *narod*'s return to the soil and which glorifies the Russian village and peasantry.

15 See also footnote 11.

16 For a treatment without any explicit reference to anti-populism, see also the classic Mouffe, 2005; Stavrakakis et al. 2018; D'Eramo (2013) offers a historical analysis of the shift in the terms "populism" and "the people" before and after the Cold War (with a strong thesis against the "antipopular despotism" of the oligarchs); with regard to the situation in Russia, Ilya Matveev, Ilya Budraitskis, Kirill Medvedev and Oleg Zhuravlev, in particular, represent a similar line of argument (see, e.g., Budraitskis et al., 2017; Budraitskis and Matveev, 2021; Медведев and Журавлев, 2020).

17 The contemporary texts in this regard (e.g., by Arkadij Babchenko, Alexander Skobov or Boris Stomahin) are mostly quite polemical and provocative, but, in my opinion, provide important impetus for reflection within the protest movement.

18 With the result that, if a protester is a member of a party or a member of parliament, he/she underlines that he/she is taking part in the protests not in this capacity, but rather as an "ordinary person".

19 For example, when the reason for one's solidarity is explained as follows: "We are all mothers and fathers. I also have two children. I cannot be silent." (solidarity initiative in Vologda, March 2018).

20 One conversation partner also describes his emotions to me as "social rage" (in the sense that the rage arose because of injustice).

21 This aspect correlates with the assessment of the British situation by John Clarke (2010, 2013), where regressive populists use the figure of "ordinary people" as an "object of desire" of the government.

22 From a taped group discussion. The research data I collected includes transcripts of interviews and discussions, field notes, screenshots from social media and similar types of materials that are not distinguished in detail for the purpose of this chapter.

23 For a theoretical and sociological derivation of this, see also Hürtgen in this volume.

24 According to media reports, the origin of the rumours about such a large number of victims was a targeted campaign by a blogger who wanted to spread panic; however, it was quickly latched onto by other bloggers and Kemerovo residents (for more information, see, e.g., TOK, 2019).

25 The following summons was issued by a former nationalist hieromonk of the Orthodox Church, Antony Shlaihov (https://vk.com/id285978785): "Burned

children are a great opportunity for the bastards to celebrate Passover, but this is also an opportunity to hide the kidnappings. Kemerovo, you should start a search. Look for the children! Organise the Russian troops! Check every synagogue in Russia to see if there are Russian children there!" When further disseminated by Orthodox and nationally focussed groups, the appeal was quickly shortened to "Organise troops! Look for the children! Search every synagogue!" The actual address of the Jewish Centre in Kemerovo was often published in these appeals (sometimes with the note "It's a 5-minute walk from Winter Cherry to the synagogue" (e.g., https://vk.com/id245435503). Many of these posts in the various blogs bear the headline: "Kemerovo is the Russian Holocaust", though later, this wording also makes it into the official media (e.g., Корсакова 2018). In the WhatsApp and Telegram groups of the various protest groups, the call to search for the kidnapped children in the synagogues then appears in the form of text, picture memes or stickers.

26 A fire in this club in Perm in December 2009 killed 156 people.

27 This is an almost verbatim quote from Artem Nikiforov, a pagan anti-Semite from Kemerovo who shot one of the anti-Semitic videos that were spread during those days. Over the course of the chat, other quotations and videos were also shared by Nikiforov and other public anti-Semites, for example, by Anton Šlyachov, an Orthodox Christian fundamentalist and Holocaust denier who calls the Kemerovo tragedy a "sacrificial offering", or by the esotericist Andrei Perez, using similar vocabulary.

28 The victims' families write open letters and record video appeals to the president, asking him, among other things, to protect their city and punish the local government, the fire department and the management of the shopping centre.

29 For an example, see footnote 10.

30 The role of the Orthodox activists and even officials of the Russian Orthodox Church in spreading anti-Semitic videos and messages after the fire in Kemerovo can serve as additional evidence for Shnierelman's analysis.

31 The contextual and ethnographic classification of these dialogues can be found in Reznikova, 2020 (chapter 4).

32 A disparaging term for a person who supports Putin's policies and is not interested in politics. It is derived from the name for a quilted jacket, a cheap garment of the Soviet era.

33 Prime Minister Dmitry Medvedev is meant.

34 Charters Wynn (1992), for example, shows how closely the labour movement that in its struggle against employers and the tsarist government in around 1905 increasingly saw itself as working class was linked to anti-Semitic pogroms.

Bibliography

Budraitskis, I. and Matveev, I., 2021. Putin's majority? *Sidecar* [online] 13 February. Available at https://newleftreview.org/sidecar/posts/putins-majority [Accessed 1 March 2021].

Budraitskis, I., Matveev I. and Guillory S., 2017. Not just an artifact. Jacobin [online] 8 January. Available at https://www.jacobinmag.com/2017/08/russa-alexey-navalny-anticorruption-movement-left [Accessed 1 March 2021].

Clarke, J., 2010. Enrolling ordinary people: governmental strategies and the avoidance of politics? *Citizenship Studies*, 14 (6), pp.637–50.

Clarke, J., 2013. In search of ordinary people: the problematic politics of popular participation. *Communication, Culture & Critique*, 6, pp. 208–26.

Claussen, D., 1989. Grenzen der Aufklärung. Antisemitismus als Destruktionskraft. *Widerspruch: Beiträge zur sozialistischen Politik*, 9 (18), pp.20–31.

Clément, K., 2020. "On va enfin faire redescendre tout ça sur terre!": penser une critique sociale ordinaire populaire de bon sens. *Numéros*, 1 [online]. Available at https://revues.mshparisnord.fr/chcp/index.php?id=115&lang=pt [Accessed 1 March 2021].

D'Eramo, M., 2013. Populism and the new oligarchy. *New Left Review*, 82 [online]. Available at https://newleftreview.org/issues/ii82/articles/marco-d-eramo-populism-and-the-new-oligarchy [Accessed 1 March 2021].

Erdheim, M., 1988. *Die gesellschaftliche Produktion von Unbewußtheit. Eine Einführung in den ethnopsychoanalytischen Prozeß*. Frankfurt/M: Suhrkamp.

Erpyleva, S., 2019. *The New Local Activism in Russia: Biography, Event, and Culture* (dissertation thesis) [online]. Available at https://helda.helsinki.fi/handle/10138/301023 [Accessed 1 March 2021].

Freud, S., 1982[1917]. *Trauer und Melancholie*. Berlin: Volk und Welt.

Habit, D., 2021. "The good, the bad and the ugly": bucharest's urban core as a moral playground. In: M. Ege and J. Moser, eds. *Urban Ethics. Conflicts over the Good and Proper Life in Cities*. Milton Park/New York: Routledge, pp.112–199.

Horkheimer M., Adorno, Th. W., 2010 [1944]. *Dialektik der Aufklärung: Philosophische Fragmente*. Frankfurt/M: Fischer.

Laclau, E., 2005. *On Populist Reason*. London: Verso.

Marchart, O., 2017. Liberaler Antipopulismus. Ein Ausdruck von Postpolitik, Bundeszentrale für politische Bildung [online] 27 October. Available at https://www.bpb.de/apuz/258497/liberaler-antipopulismus-ein-ausdruck-von-postpolitik [Accessed 1 March 2021].

Mouffe, Ch., 2005. *On the Political*. London: Routledge.

Mitscherlich, A., Mitscherlich, M., 1977. *Die Unfähigkeit zu trauern : Grundlagen kollektiven Verhaltens*. München: Piper.

Narotzky, S., 2019. Populism's claims: the struggle between privilege and equality. In: B. Kapferer and D. Theodossopoulos, eds. *Democracy's paradox: Populism and its Contemporary Crisis*. Oxford: Berghahn, pp.97–121

Rancière, J., 2017. *En quell temps vivons-nous? Conversation avec Eric Hazan*. Paris: La Fabrique èditions.

Reznikova, O., 2020. *Wütende Fernfahrer: Geschichte eines sozialen Protests und seiner Suche nach Gerechtigkeit und Politik in Russland* ([Angry truckers: The history of a social protest and its search for justice and politics in Russia] unpublished manuscript of dissertation thesis). Göttingen.

Reznikova, O, and Ege, M. 2019. Ethische Proteste oder moralische Ökonomien? Arbeiter- und Mittelschichtsproteste in Moskau. In: K. Braun, J. Moser, and C.-M. Dietrich, eds., *Wirtschaften. Kulturwissenschaftliche Perspektiven*, Marburg: Philips-Universität, pp.343–54.

Rossman, V., 2002. *Russian Intellectual Antisemitism in the Post-Communist Era*. Jerusalem: Sassoon International Center for the Study of Antisemitism.

Salzborn, S., 2019. Alltagskulturelle Selbstdisziplinierung: Über die Macht der Ohnmacht in Fernsehen. In: W. Bergem, P. Diehl and H.J. Lietzmann, eds. *Politische Kulturforschung reloaded: Neue Theorien, Methoden und Ergebnisse*. Bielefeld: transcript.

Stavrakakis, Y., Katsambekis, G., Kioupkiolis, A., Nikisianis, N., Siomos, Th., 2018. Populism, anti-populism and crisis. *Contemporary Political Theory*, 17, pp.4–27.

Susser, Ida, 2021. "They are stealing the state": commoning and the Gilets Jaunes. In: M. Ege and J. Moser, eds. *Urban Ethics. Conflicts over the Good and Proper Life in Cities*. Milton Park/New York: Routledge, pp.277–294.

Wynn, Ch., 1992. *Workers, Strikes, and Pogroms: The Donbass-Dnepr Bend in Late Imperial Russia, 1870–1905*, Princeton: Legacy Library.

Бизюков, П., 2018. Трудовые протесты в России в 2008–2017 гг. *Центр социальнотрудовых прав* [online]. Available at http://trudprava.ru/expert/analytics/protestanalyt/2015 [Accessed 1 March 2021].

Джагалов, Р., 2011. Антипопулизм постсоциалистической интеллигенции. *Непрекосновенный Запас*, 75 (1), pp.476–486.

Клеман, К., Мирясова, О. and Демидов А, 2010. *От обывателей к активистам. Зарождающиеся социальные движения в современной России*. Москва: Три квадрата.

Левада-центр, 2018. Мониторинг ксенофобских настроений: июль 2018 года. [online] 27 August. Available at https://www.levada.ru/2018/08/27/monitoring-ksenofobskih-nastroenij/ [Accessed 1 March 2021].

Медведев, К., Журавлев, О., 2020. Новый патриотизм — новая оппозиция? Наблюдения слева. *Colta* [online] 10 February. Available at https://www.colta.ru/articles/society/23550-kirill-medvedev-i-oleg-zhuravlev-o-progressivnom-patriotizme [Accessed 1 March 2021].

Разувалова, А., 2015. *Писатели-«деревенщики»: литература и консервативная идеология 1970-х годов*. Москва: Новое литературное обозрение.

Шнирельман, В., 2017. *Колено Даново: Эсхатология и антисемитизм в современной России*. Москва: ББИ.

Шнирельман, В., 2019. Осознания того, какое именно зло было побеждено, в обществе нет. *Коммерсантъ* [online] 8 May. Available at https://www.kommersant.ru/doc/3965631#id1529096 [Accessed 1 March 2021].

Щелин, П., 2017. Российский опыт политического протеста? Прошлое и настоящее. *Хвиля*, 13 October 10 [online]. Was available at https://hvylya.net/special-projects/ukrainian-institutefuture/rossiyskiy-opyit-politicheskogo-protesta-proshloe-i-nastoyashhee.html [Accessed 02.10.2020, no longer available, copy in research archive OR].

Sources

Antilop, 2018. Реакция людей на приезд Навального в Волоколамск [online]. Available at https://t.me/akitilop/4454 [Accessed 1 March 2021].

Baza, 2020. Полное интервью Александра Ананьева. *Зимняя вишня* [online] 26 March. Available at https://www.youtube.com/watch?v=gOIjdjIc3u4&feature=youtu.be [Accessed 1 March 2021].

Gessen, M., 2021. Why won't Amnesty international call Alexey Navalny a prisoner of conscience? *The New Yorker* [online] 24 February. Available at https://www.newyorker.com/news/our-columnists/why-wont-amnesty-international-call-alexey-navalny-a-prisoner-of-conscience [Accessed 1 March 2021].

Info24, 2018. Потерял семью и решил бороться с коррупцией. *Кто такой Игорь Востриков* [online] 11 May. Available at https://info24.ru/news/poteryal-semyu-i-reshil-borotsya-s-korruptsiey-i-propagandoy-kto-takoy-igor-vostrikov.html [Accessed 1 March 2021].

RenTV, 2018. Папочка, я умираю: Последние звонки детей, погибших в. *Зимней вишне* [online] 28 March. Available at https://www.youtube.com/watch?v=gERPjhJRXcs [Accessed 1 March 2021].

TOK, 2018. Передай маме, что я ее любила [online] 26 March. Available at https://www.youtube.com/watch?v=ul8PdhO_wXA [Accessed 1 March 2021].

ТОК, 2019. По следам огня. Фильм о трагедии в. *Зимней вишне* [online] 24 March Available at https://www.youtube.com/watch?v=V9ia193QYzY&feature=youtu.be&has_verified=1 [Accessed 1 March 2021].

Ананьев, А., 2020, Письмо Путину В.В [online] 18 March, Available at https://vk.com/doc47180414_542911565?hash=2369b63108c434d97e&dl=b1e09af04f34f3c5bf [Accessed 1 March 2021].

Антикоррупционный портал, 2018. МЧС, прокуратура и коррупция. Игорь Востриков рассказал о расследовании пожара в. *Зимней вишне* [online] 10 May. Available at http://anticorr.media/mchs-prokuratura-i-korrupciya-igor-vostrikov-rasskazal-o-rassledovanii-pozhara-v-zimnej-vishne/ [Accessed 1 March 2021].

Востриков, И., 2019. Черная пропаганда. *Как с ней бороться* [online] 8 February. Available at https://www.youtube.com/watch?v=TYQKxTDK7LI&feature=youtu.be [Accessed 1 March 2021].

Гозман, Л., 2021. Люмос максима. Или почему власти боятся фонариков. *Новая газета* [online] 13 February. Available at https://novayagazeta.ru/articles/2021/02/13/89204-lyumos-maksima?fbclid=IwAR1_jwYmdvC6hsfsDiMM-qgMLvXny5cXWfyegpo2CCrdX801DmBBhqHn41A [Accessed 1 March 2021].

Ильченко, Н., 2018. «Прощай, мы, скорее всего, погибнем…»: последние звонки, которые успели сделать родственникам из горящего кинотеатра в. *Зимней вишне* [online] 27 March. Available at https://www.krsk.kp.ru/daily/26810/3846850/ [Accessed 1 March 2021].

Корсакова, Т., 2018. Новый русский холокост, Комсомольская правда [online] 2 April. Available at https://www.kp.ru/daily/26813/3849863/ [Accessed 1 March 2021].

Лехаим, 2018. Российские евреи требуют разобраться с антисемитскими вбросами в связи с трагедией в Кемерово [online] 30 March. Available at https://lechaim.ru/news/rossijskie-evrei-trebuyut-razobratsya-s-antisemitskimi-vbrosami-v-svyazi-s-tragediej-v-kemerovo/ [Accessed 1 March 2021].

Навальный, А., 2018. О чем сегодня говорим: Кемерово напугало Путина. *Екатеринбург, все на митинг* [online] 29 March. Available at https://navalny.com/p/5832/ [Accessed 1 March 2021].

Удовченко, А., 2018. Последние слова погибших в Кемерово были о любви. *Комсомольская правда* [online] 2 April. Available at https://www.kp.ru/daily/26813.5/3849397/ [Accessed 1 March 2021].

"AGAINST THE ELITES!"

SPATIAL AND TEMPORAL DIFFERENTIATIONS

PART III

BRIGITTA SCHMIDT-LAUBER, ALEXANDRA SCHWELL, SANAM ROOHI

8 Countryside versus city?

Anti-urban populism, Heimat discourse and rurban assemblages in Austria

Brigitta Schmidt-Lauber

Many indicators point towards a dissolving of the dichotomy between the city and the countryside in current post-industrial societies. This concerns cultural semantics, economics, infrastructure and society more broadly. It has also manifested in changes to the balance of political power. The outcome of national elections in various European states, dynamics in the United States of America (Hochschild, 2017; Maxwell, 2019; McKee, 2008) and endeavours to leave the European Union (EU), such as Brexit, have given rise to an awareness that educational background has been joined by geography as a key determining factor in political debate and positioning. The simplified version posits the "right-wing countryside" against the "left-wing city" (Andersson et al., 2009; Burschel, 2010; Emanuele, 2018; Gimpel and Karnes, 2006; Ivaldi and Gombin, 2015). Cities are associated with political, economic and – particularly – cultural elites in almost all cases. This article takes a view from and of Austria, a small state in central Europe, where this issue has been the object of particularly intense focus in recent years.[1] I will begin by sketching the differences between the city and the countryside in Austrian politics and cultural history, before moving on to an analysis of the concrete ways in which the countryside and the city have become symbolically and politically charged dichotomous categories in recent Austrian election campaigns. Following this, I will report on research projects that focus on lived realities in towns of different sizes and rural areas. Taking the findings from these projects, I move away from the assumption of a city-countryside dichotomy and, instead, posit the thesis of so-called rurban assemblages, which encompass the different relationships and ties people have to "city" and "countryside", depending on their life circumstances. These are expressed through everyday imaginations, practices and attributions of the spaces that people frequent.

Countryside versus city in Austrian politics

Austria is a country in which there are marked differences between extensive rural areas, on the one hand, and a few larger towns and the dominant capital Vienna, on the other, in contrast to countries with a greater number of larger cities, for example, Germany, and where, furthermore, the prevailing structure

DOI: 10.4324/9781003141150-11

is strongly centralist. The metropolis of Vienna functions as a centre for the entire country in a number of ways and stands uncontested as Austria's foremost city for various different forms of capital, for example, cultural, political and historical. A total of 1.9 million of a total population of 8.86 million (in 2019) live in the federal capital of Vienna (Mohr, 2021). In 2019,

> close to 4.7 million [were living] [...] in urban areas and close to 4.2 million in rural areas. This means that the urban population, with 52.8 per cent, represents a slightly higher proportion than the rural population with 47.2 per cent.
>
> (Falter, 2019)

The dichotomy is reflected in political rhetoric and has deep roots in society, politics and history. Regarding politics, the two main traditional parties in Austria – ÖVP (Austrian People's Party) and SPÖ (Social Democratic Party of Austria) – are associated with different spaces and stand symbolically for either the countryside or the city. The ÖVP represents conservative bourgeois values and has traditionally recruited its loyal clientele from among farmers, business and voters affiliated with the Roman Catholic Church. The social democratic SPÖ, by contrast, is considered the earliest "left-wing" political power in Austria, with close ties to the urban workforce, trade unions and, specifically, the metropolis of Vienna. Currently, since the era of "Red Vienna" (1919–1934), during which numerous social reforms in housing, education and social policy were implemented (in particular the construction of social housing with affordable apartments), the "Red City" stands in symbolic opposition to the "Black Countryside" on the political map of Austria. This distinction between "Red Vienna"[2] and the "Black Countryside" (the colour black traditionally symbolises the conservatives and Catholicism) continues to be negotiated and (re)produced on many levels to this day.

Austria, as elsewhere in Europe, has observed a shift to the right over recent years, resulting in considerable popularity among voters of the right-wing populist FPÖ (Freedom Party of Austria) and a right-wing conservative ÖVP under Sebastian Kurz, rebranded as "The New People's Party" with turquoise replacing black as the party colour. The symbolic attributions of the city and the countryside also play their role here. The FPÖ has a longer history of electoral successes than similar parties in many other countries. After some stability in the post-war party system, FPÖ and ÖVP had fallen into disrepute due to corruption scandals and widespread clientelism on the local, regional and national level. The FPÖ managed to participate in government for the first time in 1983 and, under former party leader Jörg Haider – at that time, the right-wing populist governor of Carinthia, an Alpine province in southern Austria characterised by agriculture and tourism – it enjoyed widespread popularity among the population as a right-wing populist, nationalist and Eurosceptic party, before party divisions led to a split and the founding of the BZÖ (Bündnis Zukunft Österreich/Alliance for the Future of Austria) under Haider. Nowadays, the FPÖ – following fresh scandals and splits caused by the Ibiza scandal[3] – continues to represent right-wing, conservative and nationalist values and repeatedly makes the headlines with its proximity to right-wing extremism.

While the right-wing populist FPÖ has been enjoying electoral success above all since the millennium (Weidinger, 2016) and has widened its appeal to voters in (small and medium-sized) urban areas and working-class voters who used to vote socialist (Flecker et al., 2018; Flecker and Kirschenhofer, 2007; Oesch, 2008), the ÖVP can continue to rely on a traditional and particularly strong level of support in the countryside, above all, in agricultural milieus (Dworczak, 2006; Pelinka, 2002).[4]

Trends in public petitions over recent years also confirm growing geographical frictions along party-political lines: Urban voters in Austria in the *Frauenvolksbegehren* – a petition to demand gender equality – and in the national and EU elections in 2017 and 2019 were more likely to express a preference for green, left-wing or social democratic parties and their projects, while the proportion of votes for the right-wing conservative and populist parties, the ÖVP and the FPÖ, was higher in the countryside and generally depended on the size of the voting district (Gavenda and Resul, 2016).

The Heimat *concept in Austria*

It makes sense in political overviews such as these to speak of frictions between the city and the countryside. These are articulated in multiple ways, increasingly along a line of conflict between the federal state/Vienna (the federal government sits in Vienna) and the federal provinces. Established dichotomous images and stereotypes remain entrenched. These frictions are fuelled by political parties through electoral campaigns that use vivid words and images to stir up a cluster of stereotypes and a set of ideological arguments. It is possible to observe a pronounced differentiation between "us" and "the others" in the right-wing populist FPÖ, especially regarding migrants and asylum seekers but also urban intellectuals. This distinction culminates in an emotionalised appeal to ideological concepts such as *Heimat* (the homeland) and *Volk* (the people).

These terms have carried a particular political, namely, exclusionary – even racist – meaning in the German language and the intellectual and social history of Austria and Germany, and they remain loaded terms. The anti-urbanism encountered in this imagery, one directed against intellectual city dwellers, academics, artists and other milieus – and against Vienna in particular – finds its historical complement in a powerful anti-Semitism, which was especially pronounced in the countryside (Botz et al., 2002; Wistrich, 2002).

The election campaign for the office of Federal President of Austria in 2016 showed this all too clearly. The FPÖ in particular drew on clichéd images of an idyllic and untouched, yet, also endangered countryside and used these to evoke a dynamic of social division in which the term *Heimat* played a key role. The party laid ideological claim to the term in a style consistent with a right-wing populist campaign: *Heimat* was framed through the countryside – green, usually mountainous landscapes – and as a familiar social environment of down-to-earth people like "us", an environment conceived as under threat. "Preserve tradition, nurture custom, protect identity" ran a slogan on one election poster. Men in lederhosen, a handshake to signal social connection and commitment, the radiant faces of top politicians and sun-drenched meadows suggested peace, quiet, familiarity and closeness. These images were complemented by magazine covers depicting armed and masked figures in threatening poses, which referenced alleged daily police operations at Styrian refugee hostels. In this election, the strategy paid off.

The ideologised concept of *Heimat* developed particularly in the nineteenth century, during the time of the multinational Austro-Hungarian Empire, and was manifested in various contexts. In Austria, a *Heimatschutzbewegung* – a movement to protect "homeland" culture and tradition – emerged around the turn of the twentieth century predominantly in "small-town settings and in the ranks of educated, middle-class provincial dignitaries" (Nikitsch, 1995, p.24). From there, however, the movement was soon able to declare "the former international city of Vienna a bastion of native custom" (Nikitsch, 1995, p.24) as well. The 1930s saw contradictory meanings ascribed to the city-countryside relationship in Austria. The metropolis of Vienna, which, like the monarchy, stood for multiculturality and internationality, was, to some extent, a barrier to the pan-German ideas of National Socialism. Alpine regions within Austria, on the other hand, were eminently suited to ideas of annexation to the German Reich – of *Anschluss* – and these ideas soon sought to connote Vienna as *völkisch* and anti-elitist as well. At the same time, Vienna, with its associated symbols and meanings, also offered Dollfuß and the Austro-fascists, who opposed annexation, opportunities for distancing themselves from Germany. In brief, the

> image of the countryside [...] in twentieth-century Austria underwent several shifts, connected, on the one hand, to its reduction after 1918 to a state territory largely defined by mountains, and, on the other hand, to changes in its political, economic and cultural needs. As the Danube Monarchy was superseded by the Alpine Republic, new places, views and compositions gained in significance.
>
> (Johler et al., 1995, p.188, translated by the author)

Hence, today's evocative – and successfully marketed – images of what is perceived as an "authentic" Alpine *Heimat* emerged especially after the two world wars, which had separated the country "from its traditional hinterland" (Johler, 1995, p.37) of the former k. u. k. (kaiserlich und königlich) Empire – just as generally any claimed "tradition [...] is an invention of bourgeois thinking and industrial values" (Johler, 1995, p.18). In this context, the Alps became an ever stronger symbol of Austria and, after the Second World War, skiing rose to become *the* national sport, producing the country's popular heroes; ski-lifts were open over the 2020/2021 Christmas holidays in Austria even during the coronavirus pandemic in spite of a lockdown. Considered thus, *Heimat* is always a question of perspective and an expression of a particular social situation.[5] While the concept of *Heimat* is fought over mainly in socio-political debates, it is also experiencing a renaissance in academic discourse (Egger, 2014; Tauschek, 2005), which clearly illustrates the entanglement of social processes and academic analysis and knowledge production, of social history and the history of social science and the humanities.

This invocation of *Volk* and *Heimat* gives rise to the image of a (flexibly defined) hostile antagonist who threatens the imagined community of those who are ostensibly the same. Solidarity is created against "the others" who might be "foreigners", in one instance, intellectual urban dwellers, Islam or "the East", in others. Elite-bashing and racist depictions go hand in hand in the discourse of the populist right. Linking the image of *Heimat* to both *völkisch*-racist and anti-elitist connotations has appealed to many people in Austria. While the FPÖ candidate Norbert Hofer, whose political base is in the Burgenland region, narrowly missed being elected as the new federal president in favour of the independent candidate and former Green Party politician Alexander Van der Bellen in a second-round vote, the FPÖ became the third strongest party in the general election in 2017 and, until the so-called Ibiza scandal in 2019, was part of a coalition government with the ÖVP. Yet, with Van der Bellen, a different type of figure entered the political arena. A university professor of economics and public finance, he embodied the liberal Viennese intellectual who, in addition, presented himself as an outsider with a migration background. He was born in Vienna in 1944 and grew up in Tyrol, after his parents had moved to the German Reich as Baltic Germans in 1941 and came to Austria via the Würzburg resettlement camp. As a "father of the nation" who stresses democratic and humanitarian values, he campaigned for acceptance for refugee families and for climate protection and equal recognition for same-sex marriage, thus, positioning himself against the values of the FPÖ.

The presidential election campaign demonstrated the active construction and production of social boundaries within Austrian society. This was done by blending different references to the countryside, to "the people" and to *Heimat*, i.e. by selecting and utilising powerful, entrenched imagery and attributions, and a convincing rhetoric (Lehner, 2019, p. 46). Hofer sought to distance himself from economic, artistic and intellectual elites, thereby suggesting his particular closeness to and solidarity with "the people". Part of this strategy was to address voters with the personal, informal "Du" form for "you" (as opposed to the more formal "Sie") and to use language such as "our homeland" (*unsere Heimat*) to evoke community. A story in the news magazine *Profil* ran:

> So it came as no surprise that the Freedom Party candidate Norbert Hofer, in response to a remark by a TV journalist that several artists were calling on people to vote for Alexander Van der Bellen but none were supporting Hofer, took this as something to boast about. "He has high society, the people are for me", said Hofer, with more than a little self-assurance
>
> (Zöchling, 2016; "*Der hat die Hautevolee, bei mir sind die Menschen*", translated by the author).

Presenting a love of *Heimat* – framed as a love of the countryside, the ordinary, the people and nature – is an electoral strategy that is gaining in influence, especially in right-wing populist programmes, where it makes use of conspicuously aggressive language, as in the example above. An insistent rhetoric of exclusion and social decay that framed migration as a danger and a threat and declared immigration in general to be a problem for society (Rheindorf and Wodak, 2018) was blended with the figure of anti-urbanism, which was based on ideas about "true country life" and social closeness among "peers" and a promise of "authenticity". It is this imaginary, the historically rooted dichotomy between the city and the countryside, together with the simultaneous existence of densely intertwined social relationships between these spaces, that provides, so my thesis runs, particularly fertile soil for the ideologisation of *Heimat* as countryside in Austria.

The anti-elitism in evidence here is, of course, one of the basic features of populism (Priester, 2010, p.4; Weckwerth, 2013) and is also characteristic of the FPÖ's overall style, even though the party cultivates close contacts with self-proclaimed elites, for example, right-wing university fraternities and their alumni organisations. The latter, however, have not been detrimental the FPÖ's fostering of the "myth of the common man". Socio-economic elites in Germany and Austria are indeed becoming increasingly distanced from the broad mass of society, as sociologist Michael Hartmann shows in his research. However, while Hartmann addresses economic elites, FPÖ elite-bashing is aimed predominantly at cultural and intellectual elites (Hartmann, 2018; Kontrast, 2018). As part of the coalition government, so Hartmann finds, the FPÖ was in fact more likely than other parties to contribute to growing social exclusivity (Kontrast, 2018).

However, in recent times, *Heimat* has not remained the sole preserve of right-wing parties. On the contrary, opposing presidential candidate Alexander Van der Bellen also employed the theme of *Heimat*, albeit with a very different interpretation and content. The independent candidate attempted to pitch an alternative, democratic-pluralist concept of *Heimat* and give the term more open political connotations. He was depicted on posters standing in front of a Tyrolean mountain landscape with slogans such as "Those who love our homeland do not divide it", which urged people to embrace a humanistic and democratic idea of solidarity (Die Presse, 2016), or with his dog, turned towards the camera while leaning casually on a fence. This combination of picturesque natural views, a casual pose and socially inclusive slogans does not evoke dichotomy or exclusion like the appeals to *Heimat* by the FPÖ and, instead, promotes compassion and integration over social division. The depiction of Van der Bellen as a nature- and animal-loving Austrian in touch with the *Heimat* complements his image as an "intellectual" and is intended to render him "relatable" in the country at large. The candidate was also shown in statesmanlike poses in venerable historical buildings, but, significantly, "urbanity" was not used as a motif in Van der Bellen's campaign. Countryside and nature stand for a "homeland" to which people everywhere in Austria are entitled.

The effectiveness of these posters and their reception cannot be investigated further here. However, the visual language fell upon voters whose eyes were predisposed to certain ways of looking and it followed well-practiced viewing conventions. It quoted from the social imaginary and, with that, drew on a repertoire of stereotypes. Thus, the images used in this campaign, especially those of the "countryside", also implicitly or explicitly reproduced the dichotomous pair of concepts: "city" versus "countryside". They summoned up established clichés in which the countryside emerges as a space of tradition, harmony and idyllic life, in turn, suggesting authenticity, belonging (together) and an intact world. They connoted ordinary people, down-to-earth types, as opposed to the intellectuals in the city or, indeed, "foreigners" or "others" who were not visualised in the context of the Alpine *Heimat*. This was contrasted, at least implicitly, with the city and its embodiment of plurality, density, anonymity and individuality, which was often framed as a threatening and alienating space and discussed in connection with social problems caused by migration (Hill and Yildiz, 2018; Yildiz and Hill, 2015).

That the loaded symbolism and contrasts placed on geographical classifications and dichotomies (countryside = the people, city = elite/intellectuals) continue to play a significant role in socio-political discourse, for example, in the country's media, was again in evidence in 2020. Following the Ibiza scandal and the dissolution of the ÖVP-FPÖ coalition government, the former leader of the FPÖ, Heinz-Christian Strache, stood as a candidate in Vienna's city council elections – even though his personal residence was outside Vienna, in Klosterneuburg, the third-largest town in Lower Austria. His personal choice of residence meant that he had no legitimate claim to the office of mayor in the eyes of media reporters. This also drew on the established antagonism between Vienna and the rest of the country, between Vienna and the "countryside". In turn, a joke made the rounds among the progressive Viennese particularly concerning Federal Chancellor Sebastian Kurz, a Viennese, and his claim to be a "Meidlinger from the Waldviertel", i.e. the ÖVP politician's claim to unite the city (Meidling, a district of Vienna) and the countryside (Waldviertel, a very rural region) and, thus, his attempt to service old party-political obligations towards the countryside.

Relativising the city-countryside dichotomy: on the necessity of a differentiated social analysis of everyday life

There is no question that a dichotomy between the city and the countryside and the continual invocation of this difference in discourse are well established in social and political rhetoric, as well as finding expression in voting behaviour. As appealing and convincing as such a diagnosis may be on these levels and regarding meaning structures, there are many arguments for going beyond simplified representations and taking a more differentiated look at this relationship in actual practice. The semantic dichotomy reduces the heterogeneity of lifestyles and conceals the fact that towns and cities can vary greatly and "countryside" can also be constituted in different ways and mean different things. A social and cultural analysis of everyday life and its underlying ethics and possibilities can offer more rounded findings here about lived realities in different spaces and the connotations they carry. Building on the notion of "rurbanity," I propose the concept of "rurban assemblages" to recognise the very different ways in which the "city" and the "country" are interwoven in everyday practice. The experiences and everyday routines of Austria's inhabitants contradict the city-countryside dichotomy; spaces are lived and connected in ways that are much more multilayered.[6]

The existence of very close links between the "city" and the "countryside" is evident on many levels in Austria. Many businesses draw workers into the city from rural areas. Rural exodus is still an ongoing issue, and, especially single women move from the countryside to the city (Weber and Fischer, 2012). A considerable proportion of the workforce in Vienna comes in from other provinces. People from rural regions bring new ideas into the city and live their routines and ties in "the city". Different spaces are intertwined in individual and family biographies; many Viennese people have links to other regions through family and, above all, very many Viennese are themselves incomers from other provinces. Children of all social milieus are regularly sent to relatives in the countryside during the long summer holidays or to spend weekends, summers or other long holidays with one parent in a rural setting. The seasonal rhythms of relocation give even city dwellers access to and experience of the countryside, be it through commuting between work and home, visiting relatives or second homes. The practice of escaping the city for the cool of the countryside in summer known as *Sommerfrische* – originally an aristocratic and bourgeois form of relocating to the countryside in summer, which has spread successively from one class to another, even shaping the rhythms of migrant workers – is still in evidence in Austria and is currently being heralded as having something of a "comeback" and actively promoted (Brandenburg et al., 2018; Schmidt-Lauber, 2014, 2019). This is just one example of the close intertwining of city and countryside. Similar to so-called elites, urban and rural dwellers often cannot be confined to one particular space in their real lives.

The trend towards second home ownership, which is an important lifestyle option for the city's more well off intellectual, academic and artistic clientele in addition to their urban living and work spaces, is another example that argues against the dichotomy between city and countryside. Veritable colonies have formed in the Waldviertel and Weinviertel regions of Lower Austria, especially since the 1990s and 2000s (Statistik Austria, 2019), and ever more "refugees from the city" are to be found there who participate in local life and organise social and cultural events. The trend of moving to affluent commuter belts around cities – the so-called *Speckgürtel*, lit. "bacon belt" – will also continue, according to forecasts by the "Regional Prognosis 2010 to 2030". This heralds not only the culturalisation of rural regions, as posited by Andreas Reckwitz (2018), i.e. the economisation and marketing of "rural culture", but also a ruralisation of imaginaries and utopias among urban milieus (Springer, 2014). The Lower Austrian village of Drosendorf boasts

KuKUK (*Kunst Kultur Umwelt Kommunikation* – "Art Culture Environment Communication"), Haugsdorf in the Weinviertel has a society for *Kunst und Wein* ("Art and Wine") and other villages organise film, discussion or musical evenings. A collector is putting together an ethnological museum in the Waldviertel on the Czech border, and many other examples could be given that show a culturalisation of rural regions closely connected to urban resources and stemming from a variety of motivations. This has diversified the groups of people seeking to realise an idealised image of the "countryside" in many rural regions, especially those close to larger towns. Some romanticisation notwithstanding, the new part-time residents renovate old buildings and, on weekends and in summer at least, bring new life to the village streets and shops. Initiatives in food or agricultural policy that seek to create a link between rural and urban residents, life-worlds and products are also becoming more frequent, as shown by the boom in food cooperatives in cities or by agricultural producers supplying city dwellers directly ("organic vegetable boxes", e.g. Adamah BioHof, n.d.).

These kinds of constellations can also lead to unexpected coalitions of interest arising from the entanglement between the "city" and the "countryside". Long-standing residents in and new arrivals to the village of Drosendorf in the Waldviertel region on the Czech border participated in efforts to modernise the main square. An artist, resident in the village since the 1990s, contributed an installation. It was criticised for being out of keeping with the local community – this criticism came, above all, from recent arrivals, as well as local farmers (Kalchhauser, 2018). However, coalitions of interests can also form along other parameters, strengthening established dichotomies between "countryside" and "city". In a research project on the implementation of rural development programmes in Germany, Oliver Müller, Ove Sutter and Sina Wohlgemuth observed conflicts between old residents and newcomers over a green space in the centre of the village, which had been used previously as a parking lot and was to be converted into a biotope with EU funds, and which led to friction (Müller et al., 2019). Sutter elaborates on that conflict in a forthcoming paper. He noted that a group of newly arrived citizens who had moved from the city and the countryside saw the project as a successful contribution towards more sustainability and, thus, took part in the development. However, the long-established citizens met the meadow with suspicion because of, among others, the "wild growth" ("*unordentliche[r] Bewuch[s]*") in front of their doors (Sutter, 2021, translated by the author and Ove Sutter). Similar examples could be given from Austria as well.

In fact, aesthetic differences also manifest themselves materially: Old houses along the main road in some villages in the Weinviertel near the Czech border are increasingly being bought up by university-educated and artistic Viennese who lovingly restore them, thereby preserving the village's "traditional" appearance that is the basis of conservative *Heimat* aesthetics. Meanwhile, working people whose primary residence is in the village often do not want to forego the comforts of a newbuild and prefer to settle in uniform, prefabricated houses on the edge of the village. A 55-year-old man born in the district commented on the social shifts caused by people from outside buying houses in his village thus: "Better the Viennese than dilapidated houses" (Fieldnote BSL, 25 August 2018). Different arrangements of commuting – to work in the city on a daily basis, to a second home in the countryside at weekends – overlap and shape everyday life in both places.

On the other hand, urban planning in towns and cities is increasingly focused on qualities more usually associated with rural areas, such as social and spatial manageability. The "city of short journeys" has become an important ideal in urban planning, conveying a sense of defined neighbourhood and social closeness, citizens' participation in public space and proximity to nature, for example, in projects such as urban gardening (Müller, 2011). The weekly market in Wels, a medium-size city, was declared a "Du Zone", in which everyone would be addressed with the informal "Du" for "you" – an example of social closeness that is usually performed in rural regions and towns (Wolfmayr, 2019, p.217f). These are just a few examples that argue against a dichotomous attribution of an intellectual, cosmopolitan, elitist city, on the one hand, and down-to-earth countryside, on the other.

Still, places confer unequal levels of symbolic capital on their residents – and this is relevant for people's perception of their place in society. But this cannot simply be mapped onto the city-countryside dichotomy or a (de)valuation of specific places. We noticed during a research project based at the Department of European Ethnology at the University of Vienna on negotiations of everyday life in so-called medium-sized towns in Germany and Austria that living spaces took on different connotations and normative meanings in different circumstances and milieus. Evaluations of a place depended greatly on the individual resident and their stage in life. In the medium-sized towns studied, those who complained most about the town and expressed an aim of moving to the city if at all possible were overwhelmingly young adults, i.e. people who had yet to establish careers or start a family. They described their own town as a deficient space that combined all the disadvantages of the city and the countryside, and they aspired to relocate to the metropolis of Vienna or to Berlin, the currently most desirable destinations for a creative and youthful clientele (Eckert, Schmidt-Lauber and Wolfmayr, 2019; Wolfmayr, 2019). Indeed, the city as a place of personal growth and autonomy plays a central role in and constitutes an inherent part of a "normal biography", especially from the perspective of middle-class milieus. In many cases, a "good life" for today's young people includes experiences of and an aptitude for the city, which, in turn, can fuel a discourse that defends and justifies the rural.

The values associated with different spaces can evoke moments of shame in the case of "still" living in the countryside or smaller towns. They correspond, to some extent, to structural inequalities. Many regions of Europe are experiencing a sustained rural exodus, which has given rise to EU funding streams, such as the LEADER programme for strengthening local communities. This also exists in Austria and, as part of the multistage "Programme for

Rural Development in Austria 2014–2020", it seeks to develop future strategies for rural areas. One hallmark of the Austrian city-countryside relationship is a historically determined centralism, in which a symbolically all-powerful Vienna acts against the interests of the other provinces – in short, "the countryside". Economically small, agricultural (family) businesses predominate there, which reinforces the contrast. Due to these economic structures, but also the geographical proximity, the situation of rural areas and their residents in Austria differs from that in many other countries, particularly those with large landowners in remote areas.

Finally, transformations are also emerging in other respects: We can observe the beginnings of a more general societal shift away from social (class) and political (party-political, voting behaviour) allegiance towards biography and lifestyle as guiding factors, with life goals and ways of living determined by age and interest; for example, when young entrepreneurs deliberately decide to base themselves in rural areas and implement sustainable forms of production, as in the Waldviertel region. This should by no means suggest a levelling of socio-structural factors in social development or the full-scale individualisation of life paths. Economic and educational factors, as well as gender, are still significant parameters determining the way in which we each live.

As a result of my urban-rural research and socio-political observations, I propose a change of perspective: It makes sense not only to follow obvious polarisations between city and countryside but, at the same time and above all, to use examples of concrete living arrangements to take a more differentiated look at everyday lives in our societies. The forms and histories of these entanglements must be traced more closely in order to reveal the lived practices and social realities that produce new assemblages of city and countryside, i.e. not an either/or but a combination of "routines and ties to both city and countryside".

Notes

1 I would like to thank Manuel Liebig and Maren Sacherer for their pointers and discussions on the topics dealt with in this article and Joanna White for her translation into English resp. linguistic revision.

2 "Red Vienna" designates the capital for the period from 1918 to 1934 when the Social Democratic Workers Party of German Austria repeatedly won an absolute majority in elections and initiated significant changes in municipal politics and planning, especially in the areas of housing – through large-scale social housing programmes – and in social, health and education policy. Within Austria, these politics has had a lasting impact on the city and how it sees itself, which can still be felt in everyday life and public awareness today.

3 This was sparked by the release of a secretly filmed video, in which two high-ranking FPÖ politicians express an openness to corruption. Contacts with the SPÖ, for example, continue to play a major role in the city when it comes to filling important positions or realizing public projects, and the reform programs in social and housing policy also continue to have an impact to the present day, such as in the "Wiener Gemeindebau", a block of municipal social housing that has left a sustainable mark on Vienna's architecture and everyday culture.

4 Gender and age are also relevant factors in voting behaviour, as an analysis of voting patterns in Austria carried out by the Institut für Strategieanalyse has shown:

> In the 2017 parliamentary elections, men voted with above-average frequency for the FPÖ, women gave their vote more often to the SPÖ and the Greens, for the ÖVP the differences were minimal. [...] In terms of age, SPÖ and ÖVP enjoyed greater approval among older voters. [...] Overall, the FPÖ was stronger among the under-60s than among the generation 60-plus.
> (SORA/ISA, 2017)

5 Positive references to *Heimat* have an important flipside in "Anti-*Heimat* literature" and concomitant structures of feeling in Austrian literature and social observations and attitudes more broadly. As Robert Menasse puts it:

> It is certainly no accident that the so-called Anti-Heimat literature emerged in Austria and created an internationally entirely unique, new literary genre: For Austria is Anti-Heimat par excellence. But the Anti-Heimat literature is not only a distinctly Austrian genre, it is essentially the most significant, dominant form of literature in the Second Republic: The authors who shaped and developed this form and genre constitute an all but complete who's who of modern Austrian literature.
> (Menasse, 1993, pp.101-102; translation by the editors)

6 The following section is based on interviews and observations gathered in 2018 for an exploratory project on "rurban assemblages" at the Department of European Ethnology, University of Vienna, and on a completed research project on middle-size cities in Austria and Germany.

Bibliography

Adamah BioHof, n.d. Genuss direkt nach Hause und ins Büro. Adamah.at [online]. https://www.adamah.at/BioKistl [Accessed 25 February 2021].

Andersson, K., Eklund, E., Lehtola, M. and Salmi, P., eds., 2009. *Beyond the Rural-urban Divide: Cross-continental Perspectives on the Differentiated Countryside and Its Regulation*. Bingley: Emerald Group Publishing Limited.

Botz, G., Oxaal, I., Pollak, M. and Scholz, N., eds., 2002. *Eine zerstörte Kultur. Jüdisches Leben und Antisemitismus in Wien seit dem 19. Jahrhundert*, 2nd ed. Vienna: Czernin.

Brandenburg, C., Czachs, C., Jiricka-Pürrer, A., Liebl, U., Juschten, M., Unbehaun, W., Prutsch, A., Offenzeller, M. and Weber, F., 2018. *Refresh! Revival der Sommerfrische. Aus der städtischen Hitze in die Sommerfrische*. Rahmendokument, Wien: Inspiration für stadtnahe Destinationen [online]. https://sommerfrische-neu.boku.ac.at/pdf/Rahmendokument.pdf [Accessed 25 February 2021].

Burschel, F., ed., 2010. *Stadt – Land – Rechts. Brauner Alltag in der deutschen Provinz*. Berlin: Dietz.

Die Presse, 2016. Neue Van-der-Bellen-Plakate: "Wer Heimat liebt, spaltet sie nicht". *Die Presse* [online] 26 April. https://www.diepresse.com/4975729/neue-van-der-bellen-plakate-wer-heimat-liebt-spaltet-sie-nicht [Accessed 25 February 2021].

Dworczak, H., 2006. Modernisierter Rechtsextremismus und Rechtspopulismus am Beispiel Österreichs. In: P. Bathke and S. Spindler, eds. *Neoliberalismus und Rechtsextremismus in Europa. Zusammenhänge – Widersprüche – Gegenstrategien*. Berlin: Dietz, pp.84–87.

Ebermann, T., 2019. *Linke Heimatliebe. Eine Entwurzelung*. Hamburg: KVV "konkret".

Eckert, A., Schmidt-Lauber, B. and Wolfmayr, G., 2019. *Aushandlungen städtischer Größe. Mittelstadt leben, erzählen, vermarkten.* (= Ethnographie des Alltags. Schriften des Instituts für Europäische Ethnologie Wien, 6). Vienna: Böhlau.

Egger, S., 2014. *Heimat: wie wir unseren Sehnsuchtsort immer wieder neu erfinden.* Munich: Riemann Verlag.

Emanuele, V., 2018. The hidden cleavage of the French election: Macron, Le Pen and the urban-rural conflict. In: L. De Sio and A. Paparo, eds. *The Year of Challengers? Issues, Public Opinion, and Elections in Western Europe in 2017.* Rome: CISE, pp.91–95.

Falter, 2019. Wie groß ist die Landflucht in Österreich wirklich, und bekommen Frauen auf dem Land mehr Kinder? *Falter.at* [online] 27 February. https://www.falter.at/zeitung/20190227/wie-gross-ist-die-landflucht-in-oesterreich-wirklich-und-bekommen-frauen-auf-dem-land-mehr-kinder/dfcead717e?ver=a [Accessed 25 February 2021].

Flecker, J., Altreiter, C. and Schindler, S., 2018. Erfolg des Rechtspopulismus durch exkludierende Solidarität? Das Beispiel Österreich. In: K. Becker, K. Dörre and P. Reif-Spirek, eds. *Arbeiterbewegung von rechts? Ungleichheit, Verteilungskämpfe, populistische Revolte.* Frankfurt; New York: Campus, pp.245–256.

Flecker, J. and Kirschenhofer, S., 2007. *Die populistische Lücke. Umbrüche in der Arbeitswelt und Aufstieg des Rechtspopulismus am Beispiel Österreichs.* Berlin: Edition Sigma.

Gavenda, M. and Resul, U., 2016. The 2016 Austrian presidential election: a tale of three divides. *Regional & Federal Studies*, 26(3), pp.419–432.

Gimpel, J.G. and Karnes, K.A., 2006. The rural side of the urban-rural gap. *Political Science and Politics*, 39(3), pp.467–472.

Hartmann, M., 2018. *Die Abgehobenen. Wie die Eliten die Demokratie gefährden.* Frankfurt am Main: Campus.

Hill, M. and Yildiz, E., eds., 2018. *Postmigrantische Visionen. Erfahrungen – Ideen – Reflexionen.* Bielefeld: transcript.

Hochschild, A.R., 2017. *Fremd in ihrem Land. Eine Reise ins Herz der amerikanischen Rechten.* Frankfurt am Main; New York: Campus.

Ivaldi, G. and Gombin, J., 2015. The Front National and the new politics of the rural in France. In: D. Strijker, G. Voerman and I.J. Terluin, eds. *Rural Protest Groups and Populist Political Parties.* Wageningen; Netherlands: Wageningen Academic Publishers, pp.243–264.

Johler, R., 1995. Das Österreichische. Vom Schönen in Natur, Volk und Geschichte. In: R. Johler, H. Nikitsch, and B. Tschofen, eds. *Schönes Österreich. Heimatschutz zwischen Ästhetik und Ideologie.* Begleitbuch zur Sonderausstellung "Schönes Österreich", 26 October 1995 to 25 February 1996. Vienna: Österreichisches Museum für Volkskunde, pp.31–39.

Johler, R., Nikitsch, H., and Tschofen, B. 1995. Österreichische Landschaft. In: R. Johler, H. Nikitsch, and B. Tschofen, eds. *Schönes Österreich. Heimatschutz zwischen Ästhetik und Ideologie.* Begleitbuch zur Sonderausstellung "Schönes Österreich", 26 October 1995 to 25 February 1996. Vienna: Österreichisches Museum für Volkskunde, pp.188–189.

Kalchhauser, M., 2018. Ausstellung in Drosendorf. Bruch Spur Zeichen Künstlerin mit "Zwischenbericht". *NÖN.at* [online] 10 March. https://www.noen.at/horn/ausstellung-in-drosendorf-bruch-spur-zeichen-kuenstlerin-mit-zwischenbericht-interview-bruch-spur-zeichen-sabine-mueller-funk-81534008 [Accessed 25 February 2021].

Kontrast, 2018. Eliten-Forscher Hartmann: Eine kleine abgehobene Gruppe bringt die Demokratie in Gefahr. *Kontrast.at* [online] 22 October. https://kontrast.at/elite-soziale-ungleichheit-hartmann/ [Accessed 25 February 2021].

Lehner, S., 2019. Rechtspopulistische Rhetorik revisited am Beispiel der FPÖ-Wahlkämpfe in den Jahren 2015 und 2016. *Linguistik Online*, 94 (1/19) [online] 29 March. https://bop.unibe.ch/linguistik-online/article/view/5433 [Accessed 26 February 2021].

Maxwell, R., 2019. Why are urban and rural areas so politically divided? *The Washington Post* [online] 5 March. https://www.washingtonpost.com/politics/2019/03/05/why-are-urban-rural-areas-so-politically-divided/ [Accessed 25 February 2021].

McKee, S.C., 2008. Rural voters and the polarization of American presidential elections. *Political Science and Politics*, 41(1), pp.101–108.

Menasse, R., 1993. *Das Land ohne Eigenschaften. Essay zur österreichischen Identität.* Vienna: Sonderzahl.

Mohr, M., 2021. Bevölkerung von Österreich von 2011 bis 2021. *Statista* [online] 11 February. https://de.statista.com/statistik/daten/studie/19292/umfrage/gesamtbevoelkerung-in-oesterreich/ [Accessed 25 February 2021].

Müller, C., ed., 2011. *Urban Gardening. Über die Rückkehr der Gärten in die Stadt*. 2nd ed. Munich: Oekom Verlag GmbH.

Müller, O., Sutter, O. and Wohlgemuth, S., 2019. Translating the bottom-up frame. Everyday negotiations of the European Union's rural development programme LEADER. *Anthropological Journal of European Cultures*, 28(2), pp. 45–65.

Nikitsch, H., 1995. Heimatschutz in Österreich. In: R. Johler, H. Nikitsch, and B. Tschofen, eds. *Schönes Österreich. Heimatschutz zwischen Ästhetik und Ideologie*. Begleitbuch zur Sonderausstellung "Schönes Österreich", 26 October 1995 to 25 February 1996. Vienna: Österreichisches Museum für Volkskunde, pp. 19–29.

Oesch, D., 2008. Explaining workers' support for right-wing populist parties in Western Europe: evidence from Austria, Belgium, France, Norway, and Switzerland. *International Political Science Review/Revue Internationale de Science Politique*, 29(3), pp.349–373.

Pelinka, A., 2002. Die FPÖ in der vergleichenden Parteienforschung. Zur typologischen Einordnung der Freiheitlichen Partei Österreichs. *Österreichische Zeitschrift für Politikwissenschaft*, 31(3), pp.281–290.

Priester, K., 2010. Wesensmerkmale des Populismus. *Aus Politik und Zeitgeschichte B*, 5–6, pp.3–9.

Reckwitz, A., 2018. Die Gesellschaft der Singularitäten: Zur Kulturalisierung des Sozialen. In: H. Busche, Th. Heinze, F. Hillebrandt and F. Schäfer, eds. *Kultur – Interdisziplinäre Zugänge*. Wiesbaden: Springer VS, pp.45–62.

Rheindorf, M. and Wodak, R., 2018. Borders, fences, and limits – protecting Austria from refugees: metadiscursive negotiation of meaning in the current refugee crisis. *Journal of Immigrant & Refugee Studies*, 16(1–2), pp.15–38.

Runciman, D., 2016. A win for "proper people"? Brexit as a rejection of the networked world. *Juncture IPPR*, 23(1), pp.4–7.

Schmidt-Lauber, B., ed., 2014. *Sommer_frische. Bilder. Orte. Praktiken*. Vienna: Institut für Europäische Ethnologie der Universität Wien.

Schmidt-Lauber, B., 2019. Sommerfrische: eine österreichische Institution. In: S. Eggmann, S. Kolbe and J. Winkler, eds. *Wohin geht die Reise?* Basel, pp. 307–317. *Geruch der Zeit* [online]. https://www.geruchderzeit.org/schmidt-lauber/ [Accessed 26 February 2021].

SORA/ISA, 2017. Wahlanalyse Nationalratswahl 2017. *SORA and Institut für Strategieanalysen* [online]. https://www.sora.at/fileadmin/downloads/wahlen/2017_NRW_Wahlanalyse.pdf [Accessed 25 February 2021].

Springer, G., 2014. Speckgürtel: Wo besonders viele hinwollen. *DerStandard* [online] 3 September. https://www.derstandard.at/story/2000005055307/speckguertel-wo-besonders-viele-hinwollen [Accessed 25 February 2021].

Statistik Austria, 2019. Wanderungsstatistik. *Statistik Austria* [online] 6 July. https://www.statistik.at/web_de/statistiken/menschen_und_gesellschaft/bevoelkerung/wanderungen/wanderungen_innerhalb_oesterreichs_binnenwanderungen/023066.html [Accessed 25 February 2021].

Sutter, O., 2021. Erzählen, Wissen, Hegemonie. Zur narrativen Formierung epistemischer Sozialitäten. In: P. Hinrichs, M. Röthl, and M. Seifert, eds. *Theoretische Reflexionen. Perspektiven der Europäischen Ethnologie*. Berlin: REIMER.

Tauschek, M., 2005. Zur Relevanz des Begriffs Heimat in einer mobilen Gesellschaft. *Kieler Blätter zur Volkskunde*, 37, pp.63–85.

Weber, G. and Fischer, T., 2012. Gehen oder Bleiben? Die Motive des Wanderungs- und Bleibeverhaltens junger Frauen im ländlichen Raum der Steiermark und die daraus resultierenden Handlungsoptionen. In: U. Bechmann and F. Christian, eds. *Mobilitäten*. Graz: Montagsakademie, pp.199–214.

Weckwerth, C., 2013. Es gibt einen europaweiten Trend zum Anti-Elitarismus. *Die Zeit* [online] 26 February. https://www.zeit.de/politik/ausland/2013-02/interview-populismus-hartleb [Accessed 25 June 2019].

Weidinger, B., 2016. The far right in Austria. Small on the streets, big in parliament. In: M. Fielitz and L.L. Laloire, eds. *Trouble on the Far Right. Contemporary Right-Wing Strategies and Practices in Europe*. Bielefeld: Transcript Verlag, pp.43–48.

Wistrich, R.S., 2002. Sozialdemokratie, Antisemitismus und die Wiener Juden. In: G. Botz, I. Oxaal, M. Pollak and N. Scholz, eds. *Eine zerstörte Kultur. Jüdisches Leben und Antisemitismus in Wien seit dem 19. Jahrhundert*. 2nd ed. Vienna: Czernin, pp.169–180.

Wolfmayr, G., 2019. Lebensort Wels. Alltägliche Aushandlungen von Ort, Größe und Maßstab in einer symbolisch schrumpfenden Stadt. (= Ethnographie des Alltags. Schriften des Instituts für Europäische Ethnologie Wien, 5). Vienna.

Yildiz, E. and Hill, M., eds., 2015. *Nach der Migration. Postmigrantische Perspektiven jenseits der Parallelgesellschaft*. Bielefeld: Transcript Verlag.

Zöchling, C., 2016. Norbert Hofer und das Eliten-Bashing: Die Vertreibung der Vernunft. *Profil* [online] 27 May. https://www.profil.at/shortlist/oesterreich/norbert-hofer-eliten-bashing-vertreibung-der-vernunft-6382028 [Accessed 25 June 2019].

9 Invoking urgency

Emotional politics and two kinds of anti-elitism

Alexandra Schwell

Introduction

"There is no planet B!" – "We need to act now!" In 2019, climate activists were waving posters with these and similar slogans at Fridays for Future demonstrations all over the globe. It is not only students and pupils who warn of the consequences of climate change. Global warming is, a US government report states, "already deadly serious and without urgent, dramatic change, it will be catastrophic" (BBC News, 2018). The call for action to change the pace of current processes and avoid a coming catastrophe is a powerful practice of emotional politics in climate change discourse. Greta Thunberg is an icon of the protest movement against global warming. A vulnerable yet eloquent teenager, she conveys an authentic sense of urgency that affects her audience and calls them to action while attacking the political elites. Climate change protests go hand in hand with movements such as "March for Science", which advocates scientific expertise and science-informed public policies worldwide.[1] In addition, experts agree that a growing ecological urgency on a global scale exists and that consciousness-raising and educative initiatives are needed to increase public acceptance of climate action (Falk, 2009, p.52).

In January 2016, Frauke Petry, the then leader of the German far-right party "Alternative für Deutschland", said in a newspaper interview that police officers must prevent illegal border crossings by refugees, and "if necessary, also make use of the firearm". No police officer wants to shoot a refugee, she continued, "I don't want that either. But the use of armed force is a last resort". This scenario must be prevented, she said, by slowing down the influx of refugees through agreements with Austria and controls at the European Union's external borders (Mack and Serif, 2016). Right-wing extremists were not alone in evoking fear of refugees during the "refugee crisis". Politicians from different corners of the political spectrum trumped each other in calling for extreme measures to prevent further influx, deport those who had already arrived or enforce strict separation from the German population. Instead of presenting a well-thought-out political strategy to address the current challenges, they tried to convey a sense of urgency and determination to the electorate, despite better knowledge (Greven, 2016).

DOI: 10.4324/9781003141150-12

Urgency seems to lie at the heart of both the climate change issue and right-wing populist politics. In both domains, it functions as a driving force, as a discursive and emotional practice employed to underline the issue's relevance and timeliness. To invoke urgency and generate a sense of urgency is meant to evoke this same feeling in other actors – the feeling that "something needs to be done, now!" As such, urgency is a crucial element of processes of securitisation (Buzan, Wæver and de Wilde, 1998). Securitisation must address emotions and feelings, such as fear, rage, anger or helplessness, to be successful (Schwell, 2015). These feelings convey a sense of urgency, of time running out and a diminishing opportunity for agency. In the concept of urgency, time and emergency are condensed and reinforce each other; they are enacted performatively, disciplining actors and having an affective effect. To invoke urgency prompts actors to react, to be affected. At the same time, the capacity to affect is a potentially powerful political device. As such, I consider urgency an integral part of emotional politics, following Sara Ahmed's claim that "emotions 'matter' for politics; emotions show us how power shapes the very surface of bodies as well as worlds. So in a way, we do 'feel our way'" (2014, p.12). Emotion is both a force and resource in political life and in mediated politics (Papacharissi, 2015; Wahl-Jorgensen, 2019). The invocation of urgency is a key component of many anti-elite discourses today in that it portrays democratic politics and elected politicians as distant and detached from "the people", both on the left and the right wing of the political spectrum, as unable to deal with crisis threat scenarios and, worse, as unwilling and malicious.

In this contribution, I seek to trace the concept of urgency and "follow the concept" (Marcus, 1995) of urgency in the two very different domains of climate change policies and right-wing populism. In this way, I link different and often geographically and socially very distant areas, thereby defining and identifying basic characteristics of urgency that it maintains in all these different areas. Thereby, I show how the concept of urgency changes and is instrumentalised in the process of translation. The goal of this chapter is twofold: to understand how urgency functions as a tool of mobilisation in general and to explore its anti-democratic potential in particular. In doing so, I do not seek to provide a comprehensive comparison of these two domains but illuminate the respective uses of urgency and its effects. Obviously, I do not argue that the political demands of these movements are equivalent in normative terms. I will return to this at the end of the chapter. Nevertheless, the similarities in their use of the concept of urgency are remarkable and worth spelling out. The chapter proceeds as follows: As a first step, I trace the notion of urgency through various social arenas and identify its key features: the centrality of the valued good, urgency's temporal dimension, and urgency as an element of emergency. Together, these features highlight the importance of the concept of urgency for emotional politics. Focusing on climate change protests and policies, I subsequently explore the roles knowledge and "truth" play in the creation and perpetuation of the valued good as a reference object of urgency. In a third step, I show that the temporality of politics is pivotal for how emergencies are experienced and political consent is generated. Fourthly, using the example of the Austrian Emergency Decree draft proposal, I argue that urgency is an integral part of recent populist politics by presenting imagined threats as real and imminent. This example not only serves to shed further light on the mechanisms of urgency, but it also allows to delineate the fine but remarkable line between urgency and the concept of the state of exception that is often employed in analyses of the "current moment". Finally, I argue that populist politicians use a sense of urgency as a self-referential mobilisation tool insofar as it does not refer to something else but creates its own purpose. While climate activists protest the inaction of ruling elites, populist politics position elites as opponents of a "people" they themselves created in the first place.

Defining urgency

Urgency plays an integral role in many social domains in late modern societies, both implicitly and explicitly. Categorizations of urgency determine the separation of patients in the emergency room; degrees of urgency are defined by specific parameters regarding heartbeat, respiration, blood pressure, and the like (Wohlgemuth, 2009; Schmitz-Luhn, 2013).[2] In the field of law, exigent circumstances create the urgency to act. This type of urgency is in itself strongly regulated by law, such as entering without a warrant to avert imminent danger, prevent the destruction of evidence or preserve a greater good. Temporary restraining orders are issued only in the face of an imminent emergency.

The business world has adopted the language of crisis and urgency as well. In the last 10 to 20 years, business management literature has discovered urgency as a "central success factor in management". Business guru and best-selling author John Kotter (2008) calls for managers to create a "burning platform" in their companies, "because it's only when people are convinced that change needs to happen, and happen soon, that things will start to move" (Sidhu 2012). A good manager succeeds in implementing a "sense of urgency" as a guiding principle, where permanent speed and restless dedication to a company's success are paramount.

Even though the fields of law, health, business, and political and social movements may seem worlds apart in practice, they all adapt and use the notion of urgency to separate, prioritise, mobilise, and affect. However, they exhibit different uses of the concept of urgency. Both the judicial and the medical systems use claims of urgency to prioritise, shortcut procedures, and even legitimise breaking laws and regulations for the sake of a greater good. Their rule-breaking, however, again follows clearly defined rules. The business literature uses the sense of urgency to set actors in motion, but this mobilisation is self-referential. In defining this kind of urgency as a self-referential practice, I borrow from the Copenhagen School's concept of securitisation: "'Security' is thus a self-referential practice, because it is in this practice that the issue becomes a security issue – not necessarily because a real existential threat exists but because the issue is presented as such a threat" (Buzan, Wæver and de Wilde, 1998, p.24). Similarly, a sense of urgency that creates its own *raison d'être* is a means in itself, as it is sustained by constant reference to itself. The way urgency is used, thus, highlights crucial differences but also significant similarities, three of which are discussed in more detail below: the centrality of the valued good, the temporal dimension of urgency, and urgency as an element of emergency.

Urgency and valued good

Urgency is relational. The invocation of urgency places the subject in relation to an object or a good that acquires a specific value through this positioning or brings the object into existence in the first place through its naming ("the people", "the planet", "health"). Its value and relevance are social constructions. No object or good *a priori* calls for or evokes a sense of urgency. While some objects are framed as valuable by social convention and tradition, other valued referent objects are entirely idiosyncratic and subjective.

Similarly, the late 19th-century German economist Andreas Voigt noted that "the urgency of a need or urgency to satisfy a need plays a significant role in the subjective evaluation of the value of a good" (1891, p.372; my translation). It follows from Voigt's argument, firstly, that urgency is not inscribed in the value of a good; on the contrary, it is essential to the definition of that value. At the same time, as Voigt writes, this value is not an inherent property but is determined by the subjective need associated with the good. Since value depends on the urgency of the need, urgency is relational, potentially changeable, and dependent on context and other power factors. Secondly, Voigt's emphasis on the importance of need points to the affective potential of urgency. The identification of a need implies an urge to fill a void and satisfy a desire. It is the relationship between the object and the need that generates or evokes an emotional response.

Urgency and temporality

No matter where the term urgency appears, it always implies that something must be done and that it must be done quickly. Urgency suggests a call to action with a strong temporal emphasis. Two temporalities characterise urgency: Firstly, invoking urgency places a subject on a timeline between a pre-emergency past and a future. Geographer Ben Anderson claims that this time "is the time of an omnipresent Present: there is no time except the time of *now* that requires some form of urgent action" (Anderson, 2017, p.8; emphasis in original).

The acting subject imagines a future in the present, and this future scenario calls for action. The threatening future looms over the present. Urgency plays a crucial role in any action deemed necessary to influence or prevent a future scenario from occurring in the present. Thus, by introducing urgency, a temporal link is established between the present and the future. The goal is to achieve a transformation of that same future to either restore a pre-emergency state or initiate a changed, new future. The relationship between the future and urgency will be discussed in more detail below.

The second temporality associated with urgency is speed and acceleration. Anderson suggests that the sense of urgency "involves the presence of (or construction of a sense of) an on-rushing future that severs the present from the past and compresses the time for decision and action" (Anderson, 2017, p.8). When something is urgent, time runs out faster and faster, and the looming future scenario intensifies the call for urgent action.

Urgency and emergency

We have seen that urgency is a temporal notion that interrupts the cycle and pace of everyday life. The event that brings about this interruption is an emergency. For Anderson, urgency is a critical element of emergency. An event is considered an emergency

> if urgent, time-limited action is deemed necessary to forestall, stop or otherwise affect some kind of undesired future. Central to the uses of the term emergency is, then, a sense that something valued (life, health, security) is at risk and, importantly, a sense that there is a limited time within which to curtail irreparable harm or damage to whatever it is that has been valued.
>
> (Anderson, 2017, p.3)

Voigt's earlier definition of urgency rings familiar, as something labelled valuable is threatened and urgent, requiring immediate action that does not allow for delay. For Anderson, whether something qualifies as an emergency is again highly subjective and depends on whether a valued good is perceived as threatened. This broad definition of an emergency is crucial.

So far, we have identified the key characteristics and properties of the concept of urgency: the valued good, temporality, and the concept of emergency. Urgency links the emergency in the present with a looming and on-rushing future scenario. The sense of urgency serves to mobilise, affect, and initiate action that refers to a valued and desired good. It is not enough to call something an emergency; to set things in motion and initiate change, a speaker must also create a sense of urgency and seek consent. Urgency, then, is a means of mobilisation. Each characteristic is discussed in detail below to capture the different facets of the concept of urgency. The next section examines the role of urgency in defining the valued good and debating the existence and nature of the emergency in question.

Urgency and the valued good: climate change truths and their consequences

As there is no ontologically defined urgency that affects everyone equally, the scenarios conjured up by reference to urgency are potentially volatile, contested, and contingent upon subjective evaluation. Such evaluation is based on information, dispositions, historical legacies and other contextual factors. As a result, the sense of urgency as a mobilisation tool is fuelled by competing discursive "truths". Following Foucault, "truth" is linked to power: "'Truth' is linked in a circular relation with systems of power which produce and sustain it, and to effects of power which it induces and which extend it. A 'regime' of truth" (Foucault, 1980, p.133). Therefore, "truth" is a battleground not so much regarding objective facts but to "the ensemble of rules according to which the true and the false are separated and specific effects of power attached to the true" (Foucault, 1980, p.132). Foucault works with two concepts of truth; the validity of scientific truth does not simply follow from external social factors (Kappeler, 2008, p.267). The more facts are considered negotiable, trust in facts erodes or is challenged by "alternative facts" or belief in a supreme being, as exemplified by climate change scepticism.

The overwhelming majority of scientists worldwide agrees that climate change is real and has tangible and observable impacts. At the same time, climate change sceptics cultivate a deep mistrust of scientific evidence. This resonates with other forms of distrust or even enmity towards "knowledge elites" in powerful institutions. In the US, the 2014 PRRI/AAR National Survey on Religion, Values, and Climate Change found that in the contemporary US, climate change is a matter of "belief" rather than of knowledge or fact, dividing the American population into "believers, sympathizers, and skeptics" (Jones, Cox and Navarro-Rivera, 2014, p.2). In addition, the survey shows that belief or non-belief in climate change correlates strongly with right-wing populist and far-right attitudes. Tea party followers are much more likely to doubt the existence of climate change (53 %) than Democrats (13 %) (also see Skocpol and Williamson, 2012; Hochschild, 2016). In relation to the question of urgent action, this means that an alternative interpretation of facts, or even "alternative facts", results in actors experiencing a different present in which the sense of urgency cannot be generated because the foundation (the sense that a valued asset must be saved) is not there. Distrust in or ignorance of facts means non-recognition of the emergency ("global warming"). When climate change deniers fail to acknowledge the emergency, the linear time of past-present-future flows on undisturbed. In the same vein, Davies argues, "there is a perceived 'anti-science' dimension to the rhetoric of many populists, which undermines the public credibility of issues such as climate change and the extreme urgency of addressing them" (2020, p.647).

Judging from the scientific evidence, global warming is a fact and has practical and measurable impacts. If it qualifies as an emergency, it has great policy implications, albeit with considerable regional and political differences. Depending on where and when necessary action on climate change is discussed, the values at risk and the future scenarios differ significantly. Anthropologist Silja Klepp (2018) analysed climate change and migration on the island of Kiribati in the Central Pacific. The islands of Kiribati are facing their total disappearance due to rising sea levels. Klepp focuses on migration policies and emerging legal regimes to adapt to climate change. The urgency for Kiribati to develop an exit strategy for its people to address the incumbent changes, i.e. the islands' drowning, underlies the problem. How can a resettlement take place? Who will host the climate refugees? What will their legal status be? How will the evacuations proceed? In the face of a catastrophic future scenario, priorities must be set urgently. Klepp explains how the Kiribati government under President Anote Tong has developed the "Migration with Dignity" policy, which aims to ensure that Kiribati citizens are welcomed as migrants in potential host countries such as New Zealand, by raising the level and quality of education. The government also purchased land in Fiji to serve as housing for future climate migrants. The "Migration with Dignity" strategy provides us with an interesting example of urgency. It is an attempt to prioritise political action in a way that moves the issue – the total loss of the homeland – from crisis to risk mode, from insecurity to control, to manage the emergency from the perspective of a rational policy based on scientific expertise.

However, the same emergency does not necessarily lead to identical interpretation by different actors. In Kiribati, a policy shift occurred that illustrates that the governance of an emergency and the type of actions that are taken in the remaining time depend on political attitudes and historical, political, cultural, economic, and social legacies. The new president, Taneti Maamau, who was elected in 2016, refuses to follow the path of his predecessor and stopped the "Migration with Dignity" programme. Maamau does not believe that climate change is human-made:

> Instead, he believes only divine will can unmake Kiribati and has "put aside [Tong's] misleading and pessimistic scenario of a sinking/deserted nation." [...] As he told Kiribati's parliament, Kiribati citizens must "try to isolate [themselves] from the belief that Kiribati will be drowned [, as] the ultimate decision is God's".
>
> (Ray, 2019)

Maamau is not a climate change denier, but he believes that only God can prevent the islands from drowning and that all residents should stay. In other words, both Maamau and his predecessor acknowledge the same emergency but evaluate it in an entirely different light and interpret it in a different framework. The new parameters identified as relevant to the current emergency alter the future scenario envisioned and limit the options for action to change the future. By delegating decision-making to a higher power and, thus, shifting the question from rational to religious, from evidence-based to affective politics, the new president not only dismisses academic expertise. He also disempowers Kiribati citizens by relieving them of their responsibility to take decisions based on scientific expertise and to act in their own best interests. Citizens of a democratic society are turned into subjects that form part of a religiously motivated, homogenised "people". The sense of urgency is not meant to empower citizens to demand change but, on the contrary, to leave urgent decisions to a superior power, or to the populist who claims to act on its behalf.

Right-wing populist and far-right scenarios in the countries of the Global North share concerns about climate change, sometimes even curiously coexisting with a denial of anthropogenic climate change. The German "Alternative für Deutschland" (AfD) and the Austrian Freedom Party (FPÖ) are among the most notable organised climate change sceptics in their countries (Schaller and Carius, 2019). However, their concern is focused on a different but related reference object. While the people in Kiribati fear for their homeland, far-right activists in the Global North are stoking fears of "climate refugees" seeking sanctuary in the Global North (Chaturvedi and Doyle, 2010; Bettini, 2013; Baldwin, Methmann and Rothe, 2014). Climate change, it is suggested, destabilises the climate refugees' host countries, leading inevitably to a collapse of Global North societies. The temporary US withdrawal from the Paris Climate Agreement under the Trump administration epitomises a shift in priorities and a realignment of urgency driven by economic and national-populist motives. Here and elsewhere, the Trump administration mobilised a sense of urgency to underscore the need to "Make America Great Again" and subordinate earlier pressing needs to this goal.

Two points emerge from these examples: Firstly, urgency and emergency are contingent. Even when different actors in different parts of the world recognise the same emergency, the subjectively valued good at risk, the urgency to address the threat, and the underlying motivation to respond may differ significantly, as the reactions of Kiribati politicians illustrate. A plethora of factors influences how an event becomes an emergency and how it is framed. There is no necessary link between urgency and the referent object of urgency, the endangered good, but this link is deliberately created and a product of hegemonic practices. The Kiribati case shows that it is misleading to draw a clear boundary between objective and socially constructed emergencies, as the way each emergency is interpreted and treated reflects inequalities and power struggles.

Secondly, there is a clear divide between science-based climate activists and policymakers, on the one hand, and populist politicians and climate change sceptics, on the other, regarding the role of elites, facts and expertise. The former claim to act on behalf of science- and evidence-based rational policy and criticise political elites for not taking facts and their consequences seriously. Urgency is used to mobilise both supporters and elites to work to save the planet. The latter tend to reject or even dismiss scientific expertise and draw a clear line between the ruling (or other, such as academic) elites and "the people", i.e. this ominous entity they created in the first place. A sense of urgency is generated to reinforce the divide between "the people" and the rest.

Urgency and the temporality of the in-between

The invocation of urgency locates the present on a timeline between a pre-emergency past and the future. The future either signifies a return to the past, whether a pre-emergency past or the restoration of an imagined past, or the present evolves into a future that ranges from catastrophic or apocalyptic (and also cathartic) to a new and better future. As argued by Arjun Appadurai (2013), the future is not a natural but a cultural fact. He distinguishes between imagination, aspiration and anticipation: (1) Imagination is a collective practice, a social energy that has the potential to change the order of things. The past provides us with a basis from which we experience the present and draft the future. (2) Aspiration means to strive for something. For Appadurai, aspiration is the political instead: counterpart of imagination. It embodies the politics of hope and includes ideas of a good life. However, Appadurai argues that the "capacity to aspire" is distributed unequally across the globe (Appadurai, 2013, p.289). In large parts of the Global South, for instance, bare life must be ensured before actors can begin to contemplate their future. Finally, (3) anticipation entails the evaluation and assessment of risk. This includes scientific prediction but also pure speculation or witchcraft and magic, all of which are ways to anticipate and possibly influence coming events. Each of these three concepts carries inherent insecurity and uncertainty, and all are deeply ambivalent.

Two points follow from Appadurai's concept: Firstly, imagining, aspiring and anticipating the future are crucial to political mobilisation. The future may appear as an imaginary pre-emergency state, or it may appear as a hopeful and purified future in which collective efforts and/or rational politics will have prevented the catastrophe. Secondly, insecurity, uncertainty, and ambivalence are integral aspects of a future that is not yet known but is approaching. The previous section explored the importance of data, "truths" and facts, alternative or otherwise, in gaining consent to recognise an emergency in the present. The future is characterised by a lack of knowledge that creates space for speculation, modelling, promise, and fantasies. Insecurity is deeply engrained in the process of negotiating the future.

Climate change again provides a vivid example, as unambiguous predictions are impossible. Even the most experienced experts will not claim to know what precisely will happen. There is a wide range of scenarios used to explore possible futures.[3] Even if the threatening phenomenon itself, climate change and global warming, is undisputed, there remains a wide margin of discretion for future scenarios and the urgency attributed to the referent object. The historian Philipp Blom calls this the "amplitude of insecurity":

> A shared, underlying fear unites both sides, a fear which may center on very different concerns and be expressed along very different lines of argument, but which always carries with it a brutal uncertainty, a monstrous possibility: what if the change is catastrophic and sudden, not slow and limited?
>
> (Blom, 2018, p.8)

Urgency and the in-between

The invocation of urgency creates a time span between the present emergency and the on-rushing future characterised by ambivalence and insecurity. This insecurity in the in-between, between the present and an uncertain but potentially disruptive future, resembles a state of liminality (Turner, 2001). To declare urgency is a decision to declare time expiring, to acknowledge both the acceleration of time and the looming possibility of the worst-case future scenario occurring. The invocation of urgency is a performative act that creates an in-between that is posited outside of normality. To declare urgency implies a loss of control, a space in-between. The in-between is a crisis in the sense of the word, a tipping point and development with an unclear outcome: "Crises are moments of potential change, but the nature of their resolution is not given" (Hall and Massey, 2010, p.57). A crisis requires urgent extraordinary action to change or influence the future. Everything, the entire future, depends on action in the present.

The in-between opens a space and a limited and ever-decreasing window of opportunity for action that can still make a difference. Anderson calls this the interval:

> The interval is an interruption to linear time: it defines a space-time for action in-between the onset of something new and the temporary stabilization of a changed present. To govern emergencies and through emergency is to enact and act within 'intervals'.
>
> (Anderson, 2017, p.9)

The insecurity of the in-between cannot be left ungoverned but calls for action. The sense that time is running out exacerbates fears and anxieties about the future. When a loss of control is declared, the goal must be to remedy the emergency by taking back control, by mastering the crisis, by knowing the future again. The promise of knowing and mastering the future serves to alleviate fears and convey security, stability, and authority.

But when does the present end and the future begin? What is the extent of the in-between, the interval within which action to alter the future is possible? The declaration of urgency implies a point of no return, a point in time until which things can still be changed. At this point, the present ends and the future begins and can no longer be altered. The "point of no return" rhetoric urges us to act before it will be too late, and change becomes irreversible. That sense of urgency has been a favourite motif of the environmental movement and anti-nuclear activists since their heyday in the 1980s. Today's climate change activists use similar strategies and rhetoric to their predecessors, such as the Friday for Future demonstrations that call adults and politicians to action. Slogans such as "It's 5 to 12" or "11th Hour" travel across the most diverse media and geographical locations, such as the Austrian anti-nuclear protest at Zwentendorf in 1978, or Hollywood, where Leonardo Di Caprio produced the film *The 11th Hour*. Here as there, urgency implies the idea of irreversibility, the ever-looming possibility of a final loss of the good, a transition to a catastrophic future. The absolutely irreversible is death, extinction. Nevertheless, despite its emphasis on crisis and potential misfortune, urgency discourse is ultimately profoundly optimistic. It foreshadows the looming disaster but, at the same time, projects the possibility of averting that very disaster. The future is alterable. In urgency lies hope.

The Emergency Decree: juxtaposing urgency and the state of exception

The following example uses a third feature of urgency, its close connection to an emergency, to illustrate how urgency functions as a tool for mobilisation. Juxtaposing the concepts of urgency and the state of exception, this section expands on the performative potential of urgency as a tool of securitisation. I argue that while the state of exception conveys a security promise, urgency exacerbates insecurity and uncertainty, explaining its significance for emotional politics.

The 2015 "refugee crisis" marked a heyday of urgency politics. Following the events in 2015, hundreds of thousands of refugees fleeing the wars in Syria, Afghanistan, and Iraq arrived in Austria from Hungary and other neighbouring countries (Rheindorf and Wodak, 2018). Most of the refugees did not intend to stay in Austria but planned to travel further north, mainly to Germany and Sweden. In 2016, Austria introduced a cap on asylum applications. In the first year, 2016, asylum applications were not to exceed 37,500, and in the following year, no more than 35,000 applications were to be accepted. It is important to note that limiting asylum applications meant a discontinuity in Austrian politics. Similar to Germany, Austria had, at least on paper, considered it a humanitarian obligation to grant asylum to people fleeing war and persecution, as a lesson from World War II. Since suffering and prosecution could not be measured in numbers, there was no fixed limit for asylum applications. However, the number of recognised asylum seekers in Austria did not necessarily relate to the number of applications. In the aftermath of the "refugee crisis", when the situation had already cooled down and new arrivals had decreased significantly, right-wing populist and far-right politicians saw an opportunity. They proposed introducing a limit on the number of asylum applications.

The proposal to adopt a bill on an Emergency Decree was discussed in parliament in the autumn of 2016. The proposal referred explicitly to an incumbent threat of a state of exception, which would affect all realms of social life. The Decree's proponents expected an increase in crime, and challenges to public administration, schools, housing and the labour market. The draft proposal foresaw an "enormous challenge to general security", a "total collapse of institutions", an increase in crime and radicalisation among prison inmates, a "massive overstressing", "supply shortfalls" and similar disaster scenarios (Bundesministerium für Inneres, 2016; Oswald, 2016; my translation). However, the number of applications had declined significantly since December 2015. Therefore, it was no surprise to parliamentarians at the time of the discussion in the autumn of 2016 that the upper limit would not be reached in that year.

The Austrian bishops, the Chamber of Lawyers, the trade union federation, the Ludwig Boltzmann Institute for Human Rights, and the United Nations

High Commission for Refugees harshly criticised the attempt to curtail the right to asylum and qualified it as breaking a taboo (Ludwig Boltzmann Institut für Menschenrechte, 2016). Nevertheless, Parliamentary Speaker Sobotka remained in favour of the Emergency Decree, arguing on national television that there was no point in buying a fire engine once the fire was already burning, i.e. when the emergency had already occurred. The Ministry of the Interior argued in the same vein in September 2016 that the Decree should enter into force before the magic number of 37,500 asylum applications was reached. By already allowing the rejection of asylum seekers at the national border, the number of applications could be kept low, a ministry spokesperson argued (Trescher, 2018). The Emergency Decree was, thus, meant to be a proactive policy to prevent the very emergency it so fearfully invoked.

The Emergency Decree did not pass, but a legal basis was created in Austrian asylum law. This basis allows for a tightening of regulations at a lower level, below the threshold of the national state of emergency, the details of which are beyond the scope of this chapter. For our purpose, it is vital how the state of exception is used in the debate on the Emergency Decree. In the wake of the "refugee crisis", right-wing populist political actors have identified a threat that is argued to have the potential to be existential in the future. The distinction is rooted in the different temporalities to which the Emergency Decree narrative appeals. The emergency that Sobotka and his colleagues invoke to justify a state of exception is not close or imminent; it is not at our doors, but it is a potential crisis. The state, this narrative insinuates, is not actually but potentially in a state of self-defence, in a state of emergency. A potential state of exception may not exist, but it could become an actual emergency at any moment.

Theoretically, Giorgio Agamben (1998) has elaborated on the implications of the concept of the state of exception for social analysis. His use of the idea dates back to the conservative fascist thinker Carl Schmitt, who defines sovereignty as the power to declare the state of exception and act outside the law. The sovereign can declare a loss of control while claiming at the same time that control can only be regained by way of a state of exception. The declaration of a state of exception is a performative act that transforms the present with far-reaching effects. It permits the introduction of extraordinary measures that allow for the suspension of rights and fundamental freedoms to restore safety and security. These measures are taken top-down with explicit reference to urgency.

The Emergency Decree illustrates the particular relationship between the state of exception and the concept of urgency and its performative effect. Firstly, the proclamation of the state of emergency and the invocation of urgency already involve a statement about sovereignty, by either restoring sovereignty or making a claim to it. The speech act performs a social reality and sends a message to the public. Secondly, in the discussion of the Emergency Decree, the state of exception is not (yet) an empirical reality but a future that is actively imagined and anticipated and aspired to by some. Thirdly, the emergency thereby acquires a quality as a real future object: The

Decree's proponents argue that if we act now, we can still avert the emergency and retain control, but to do so, we must act outside normal politics. If we wait too long and let time pass idly, we will be doomed, and our civilisation as we know it will collapse. This is one of many criticisms which the Austrian Judges' Association raised concerning the draft: "the regulation does not comply with the law because the terms 'internal security' and 'state of emergency' are interpreted incorrectly, and the law does not require the possibility of threat but an actual threat" (Richtervereinigung, 2016; my translation).

Against the backdrop of these events, I argue that the concepts of the state of exception and urgency are related but not identical. Urgency fills the in-between, a timespan between a moment when an emergency has been identified and the loss of control. When urgency is invoked, the present is no longer in a pre-emergency state, and the irreversible fantasy of the state of exception is present. The state of exception is supposed to deal with an existing or imagined emergency in the present and allows the sovereign to declare the problem solved in the present. Not only is the sovereign able to declare the state of emergency but he also demonstrates his sovereignty by ending it and handling the emergency.

The concept of urgency, on the other hand, is formulated in the present but projects a disastrous future, the extent of which can only be imagined since its actual dimensions remain in the dark. The crisis is threatening, yet, at the same time, unclear and hardly predictable. This inherent uncertainty of the in-between is why the idea of urgency is so powerful. In a state of exception, the public can resist the sovereign's actions, submit to them or actively support them, relying on the sovereign's capabilities, knowledge, and power. Paradoxically, the state of exception contains a promise of security. The concept of urgency, however, carries an inherent uncertainty that makes it much more difficult to place trust in political or other actors.

Insecurity and uncertainty make semantic shifts much more likely, as symbolic representations are increasingly prone to destabilisation in times of crisis. Such a semantic shift took place during the "summer of migration" in 2015, a shift with tangible consequences. Refugees who had just been welcomed with enthusiasm were quickly and collectively labelled a security threat. The events at Cologne central station on New Year's Eve 2015, when dozens of women were harassed and sexually assaulted by men, most of whom appeared to be of North African and Arab origin (Arendt, Brosius and Hauck, 2017), heightened tensions and galvanised public opinion against refugees in general. While this incident took place in Germany, Austrian populist and far-right politicians quickly embraced it, as German political and public discourse is continuously received and compared to Austria's domestic affairs.

Against the backdrop of the Cologne events, the Emergency Decree draft was formulated when the tide was already turning against refugees in both Germany and Austria. Public opinion had initially been ostensibly favourable and self-celebratory due to a new "welcome culture" that was supposed to show the friendly face of a post-nationalistic Germany. Interestingly, the self-celebrating attitude of 2015 was reminiscent of the so-called summer fairy tale of 2006, when Germany hosted the Football World Championship (Sonntag, 2007). Once more, it seemed, the country had managed to overcome racism, right-wing extremism and its Nazi past and was celebrating difference and diversity and, above all, itself. As the initial enthusiasm of this welcome culture waned, some people involved in the day-to-day work of taking in refugees complained that Merkel's "We can do it!" was followed by a lack of practical support for those active in cities, towns, and civil society initiatives. There are many reasons why the tide turned against refugees and, indeed, Cologne is a decisive factor, as is the tendency of tabloid media to scandalise and generalise crimes committed by refugees and migrants.

However, the Emergency Decree in Austria was not drafted and formulated in response to actual crimes and terrorist attacks. Instead, the Decree framed the unprecedented high number of refugees of 2015 as an emergency "new normal" threatening Austrian society. By externalising refugees and migrants, the draft uses a language of urgency that draws on a far-right register and well-established linguistic and visual narratives on the Other in general. The Emergency Decree does not exist in isolation but is part of a larger discursive and visual narrative that is not confined to Austria (Wodak, 2015; Bauman, 2016). In these xenophobic discourses, visual and linguistic metaphors that refer to migration and refugees possess processual and temporal qualities: The flood, the wave, the stream or the avalanche are metaphors that not only suggest an emergency but also performatively create urgency (Wright, 2002; Demos, 2013).

Linguistic metaphors are paralleled by the visual imagery of approaching masses on the highway towards Vienna or contrasting the untouched Austrian nature with their sheer overwhelming presence (Schwell, 2021). Images and imaginaries suggest a slowly but steadily approaching catastrophe, creating a collective anxiety that transcends time and space and shrinks the Global North into an imaginary space under siege. Visual and linguistic metaphors create and reinforce urgency by projecting a future catastrophe. Visual images have an appellative power that prompts the viewer to become emotionally involved (Schirra and Sachs-Hombach, 2013).

Creating an emergency and securitising an issue through linguistic, visual, and sensual practices is successful if a feeling is conveyed that "we" are doomed unless something is done quickly. The representation of refugees as a security risk and the imagination of a loss of sovereignty mutually reinforce each other and are linked through emotional practices, emphasising the importance of emotions-as-practice which "emerge in the doing of emotion" (Scheer, 2012, p.220) in urgent politics. The invocation and declaration of urgency to prevent or end a projected state of exception is a performative practice that makes a lasting impression on social actors and has far-reaching implications for democratic political culture.

Conclusion: two types of anti-elitism

This chapter set out to explore how urgency operates as a means of mobilisation. The invocation of urgency creates an in-between characterised by crisis, insecurity and a declared loss of control that calls for immediate action.

As we have seen, urgency is not an absolute category, not an independent variable, but results from a strategic practice reflecting existing societal interests and power. Having traced the concept of urgency in politics and activism against climate change and in right-wing populist politics, we note that a sense of urgency emphasises the importance and timeliness of an issue. It is an argument to prioritise issues or bypass the usual democratic process for the sake of a higher purpose. In both areas, criticism of powerful elites is central. In a broad sense, elites are accused of being detached, not paying attention, not caring enough or being too technocratic.

Nevertheless, there is a crucial difference between the two kinds of anti-elitism: While climate activists generate a sense of urgency to engage elites and remind them of their responsibility to solve the current emergency, populism and the far-right focus on an exclusivist imaginary "people"; they "create the homogeneous people in whose name they had been speaking all along" (Müller, 2016, p.49). Studies of neo-nationalism have shown that the idea of the pure and moral people is conceptualised in negation, as being threatened by two other groups:

> One group of 'them' is constructed, in terms of power, as being 'above us': the EU authorities in Brussels and their mysterious associates elsewhere. A second stratum of 'them' is perceived as being ranked, in terms of status, 'below us': local immigrants and other cultural and linguistic minorities living in the EU, plus their 'dangerous' associates in Africa, Asia and elsewhere.
>
> (Gingrich, 2006, p.199)

Right-wing populism does not attack the elites to remind them of their democratic responsibility to the entire population, and when it does, it does so with the aim of replacing and excluding them from the imagined homogenous "people". Rather, it positions both elites and experts ("above us") and refugees ("below us") as external and destructive to the people. In the same vein, Zulianello and Ceccobelli (2020) argue that while Greta Thunberg's performance and politics exhibit many features of populism, her message is more aptly described as technocratic ecocentrism:

> her critique of the political, economic and media elite is not justified by evoking a moral superiority of the people as in the case of the populists [...]; instead, it is grounded on the exaltation of science and of the scientific elite.
>
> (Zulianello and Ceccobelli, 2020, p.626)

In both cases, the invocation of urgency has a potentially anti-democratic effect as it is used to circumvent democratic procedures and legitimise the curtailment of political participation. However, there is another crucial difference: Climate activists maintain that they act on behalf of a goal which is claimed to be legitimate because it is based on legitimate knowledge. The value of the good (saving the planet; survival) is argued with reference to evidence-based facts that provide both the grounds for discussion and the search for a solution. In this respect, urgency as an instrument of mobilisation resembles the way it is ideally employed in the medical or judicial field.

For populists, on the contrary, urgency is a self-referential tool in that it creates and sustains the urgency it invokes. Similar to the business and management worlds, it is a gimmick, a rhetorical device and a tool for affective political practice. It is a strategy to "create a burning platform", to mobilise for the sake of mobilisation. The constant repetition of urgency creates a permanently looming crisis. It creates a desire for immediate fulfilment, but it cannot redeem it. The permanent sense of urgency projects a utopian future promise that it will not deliver. Urgency fosters uncertainty and fear because it creates the feeling that time is running out, time that we do not have for democratic processes, somebody better do something fast! In the same way, Anderson draws upon Elaine Scarry to show

> how 'claims of emergency' function through an affect of urgency that forestalls processes of deliberation and dissensus. Democratic procedures and habits become impediments to timely action, since 'the unspoken presumption is that either one can think or one can act, and given that it is absolutely mandatory that an action be performed, thinking must fall away'
> (Anderson, 2017, pp.8–9).

Urgency is inextricably linked to fear as an emotional practice. This fear is not just a fear of a future scenario but a more general fear of standstill, immobility, incapability, and failure to act. Its emphasis on timeliness and the need for speed makes urgency the opposite of deliberation; urgency is the ultimate affective politics. Urgency makes slowness seem weak or, worse, hostile. Here, contemplation almost implies an intentional will to harm. Democratic procedures with their discussions of pros and cons seem futile when time is running out. If "we must act now" and the time for discussion is over, then it is only a small step for the political opponent to become the traitor to and the enemy of the people, not only unable but unwilling to face crises and threat scenarios decisively in order to save "us". Urgency, then, is an indispensable tool for political mobilisation in times of emergency. The less a sense of urgency is directed at a legitimate good, the more likely it is to function as a self-referential practice with a twofold effect: A self-perpetuating cycle of fear and anxieties and an increasing distance from and rejection of normal politics, both of which lead to an erosion of trust in democratic structures and institutions.

Acknowledgements

Published with the support of the Faculty of Cultural Studies of the University of Klagenfurt. The author wishes to thank Monique Scheer and Guido Tiemann for valuable comments on earlier drafts of this contribution. I also thank colleagues at the anthropology department at LMU Munich and at the department of cultural analysis in Klagenfurt for stimulating discussions. They all contributed significantly to shaping this contribution.

Notes

1 See https://marchforscience.org/
2 Urgency in the medical field lies beyond the scope of this chapter. Suffice it to mention that urgency plays an integral role in the medical field, from the triage system to emergency rooms to varying degrees of urgency in relation to different health insurances. Here, the blurring of objective criteria, on the one hand, and economic and cultural capital, on the other, becomes particularly obvious. Recently, the COVID-19 pandemic familiarised the wider public with the concept of triage beyond the emergency room. Triage became synonymous with a war-like worst-case scenario where doctors separate those who get treatment from those who do not qualify. In the pandemic, the urgency of treatment is not prioritized in the patients' interest, but it is subsumed under a greater objective and for the sake of a greater good, i.e. to protect the health system from collapsing.
3 See, for instance, the reports on climate change scenarios of the Intergovernmental Panel on Climate Change (IPCC) or the European Environment Agency (EEA).

Bibliography

Agamben, G., 1998. *Homo Sacer. Sovereign Power and Bare Life*. Stanford: Stanford University Press.

Ahmed, S., 2014. *The Cultural Politics of Emotion*. Edinburgh: Edinburgh University Press.

Anderson, B., 2017. Emergency futures: exception, urgency, interval, hope. *The Sociological Review*, 65(3), pp.463–477.

Appadurai, A., 2013. *The Future as Cultural Fact. Essays on the Global Condition*. London; New York: Verso.

Arendt, F., Brosius, H-B. and Hauck, P., 2017. Die Auswirkung des Schlüsselereignisses "Silvesternacht in Köln" auf die Kriminalitätsberichterstattung. *Publizistik*, 62(2), pp.135–152.

Baldwin, A., Methmann, C. and Rothe, D., 2014. Securitizing "climate refugees": the futurology of climate-induced migration. *Critical Studies on Security*, 2(2), pp.121–130.

Bauman, Z., 2016. *Strangers at Our Door*. Malden: Polity Press.

BBC News, 2018. Climate change: report warns of growing impact on US life. *BBC News* [online] 24 November. https://www.bbc.com/news/world-us-canada-46325168 [Accessed 21 February 2021].

Bettini, G., 2013. Climate barbarians at the gate? A critique of apocalyptic narratives on "climate refugees". *Geoforum*, 45, pp.63–72.

Blom, P., 2018. The amplitude of uncertainty: some thoughts about the coming flood. *IWMpost*, (121), pp. 8– 10.

Bundesministerium für Inneres, 2016. Verordnung der Bundesregierung zur Feststellung der Gefährdung der Aufrechterhaltung der öffentlichen Ordnung und des Schutzes der inneren Sicherheit. Erläuterungen_VO BReg Feststellung der Gefährdung döO, Bundesministerium für Inneres. Wien.

Buzan, B., Wæver, O. and de Wilde, J., 1998. *Security: A New Framework for Analysis*. Boulder: Lynne Rienner Pub.

Chaturvedi, S. and Doyle, T., 2010. Geopolitics of fear and the emergence of "climate refugees": imaginative geographies of climate change and displacements in Bangladesh. *Journal of the Indian Ocean Region*, 6(2), pp.206–222.

Davies, W., 2020. Green populism? Action and mortality in the Anthropocene. *Environmental Values*, 29(6), pp.647–668.

Demos, T.J., 2013. *The Migrant Image. The Art and Politics of Documentary during Global Crisis*. Durham; London: Duke University Press.
Falk, R., 2009. The second cycle of ecological urgency: an environmental justice perspective. In: J. Ebbesson and P. Okowa, eds. *Environmental Law and Justice in Context*. Cambridge: Cambridge University Press, pp.39–54.
Foucault, M., 1980. Truth and power. In: C. Gordon, ed. *Power/Knowledge. Selected Interviews and other Writings 1972–1977*. New York: Pantheon, pp.109–133.
Gingrich, A., 2006. Neo-nationalism and the reconfiguration of Europe. *Social Anthropology*, 14(2), pp.195–217.
Greven, L., 2016. Die AfD ist nicht allein. *ZEIT Online* [online] 1 February. https://www.zeit.de/politik/deutschland/2016-02/afd-populismus-andere-parteien/komplettansicht [Accessed 21 February 2021].
Hall, S. and Massey, D., 2010. Interpreting the crisis. *Soundings*, 44(44), pp.57–71.
Hochschild, A.R., 2016. *Strangers in Their Own Land: Anger and Mourning on the American Right*. New York: The New Press.
Jones, R.P., Cox, D. and Navarro-Rivera, J., 2014. *Believers, Sympathizers, & Skeptics. Why Americans Are Conflicted about Climate Change, Environmental Policy, and Science. Findings from the PRRI/AAR Religion, Values, and Climate Change Survey*. Washington, DC: Public Religion Research Institute (PRRI).
Kappeler, F., 2008. Die Ordnung des Wissens. Was leistet Michel Foucaults Diskursanalyse für eine kritische Gesellschaftstheorie? *PROKLA*, 151(2), pp.255–270.
Klepp, S., 2018. Framing climate change adaptation from a Pacific Island perspective – the anthropology of emerging legal orders. *Sociologus*, 68(2), pp.149–170.
Kotter, J.P., 2008. *A Sense of Urgency*. New York: Ingram Publisher Services.
Ludwig Boltzmann Institut für Menschenrechte, 2016. *Stellungnahme zur Verordnung der Bundesregierung zur Feststellung der Gefährdung der Aufrechterhaltung der öffentlichen Ordnung und des Schutzes der inneren Sicherheit*. Wien. https://docplayer.org/45638038-1-art-72-aeuv-als-eu-primaerrechtliche-grundlage-zweifelhaft-ludwig-boltzmann-institut-fuer-menschenrechte-freyung-6-hof-1-stiege-wien.html
Mack, S. and Serif, W., 2016. Sie können es nicht lassen! *Mannheimer Morgen* [online] 25 August. https://www.morgenweb.de/mannheimer-morgen_artikel,-politik-sie-koennen-es-nicht-lassen-_arid,751556.html [Accessed 21 February 2021].
Marcus, G.E., 1995. Ethnography in/of the world system: the emergence of multi-sited ethnography. *Annual Review of Anthropology*, 24, pp.95–117.
Müller, J-W., 2016. *What Is Populism?* Philadelphia: University of Pennsylvania Press.
Oswald, G., 2016. Asyl-Notverordnung: "Totaler Zusammenbruch der Einrichtungen droht". *Der Standard* [online] 1 September. https://www.derstandard.at/story/2000043618337/asyl-notverordnung-totaler-zusammenbruch-der-einrichtungen-droht [Accessed 21 February 2021].
Papacharissi, Z., 2015. *Affective Publics: Sentiment, Technology, and Politics*. Oxford; New York: Oxford University Press.
Ray, C., 2019. Rejecting reality: Kiribati's shifting climate change policies. *Climate Security in Oceania*. The University of Texas at Austin. [online] 31 December. https://sites.utexas.edu/climatesecurity/2019/12/31/kiribati-policy-shift/ [Accessed 21 February 2021].
Rheindorf, M. and Wodak, R., 2018. Borders, fences, and limits – protecting Austria from refugees: metadiscursive negotiation of meaning in the current refugee crisis. *Journal of Immigrant & Refugee Studies*, 16(1–2), pp.15–38.
Richtervereinigung, 2016. Stellungnahme zur Verordnung der Bundesregierung zur Feststellung der Gefährdung der Aufrechterhaltung der öffentlichen Ordnung und

des Schutzes der inneren Sicherheit. *Vereinigung der österreichischen Richterinnen und Richter, Wien* [online] 5 October. https://richtervereinigung.at/verordnung-der-bundesregierung-zur-feststellung-der-gefaehrdung-der-aufrechterhaltung-der-oeffentlichen-ordnung-und-des-schutzes-der-inneren-sicherheit/ [Accessed 21 February 2021].

Schaller, S. and Carius, A., 2019. *Convenient Truths. Mapping Climate Agendas of Right-wing Populist Parties in Europe*. Berlin: adelphi.

Scheer, M., 2012. Are emotions a kind of practice (and is that what makes them have a history)? A Bourdieuan approach to understanding emotion. *History and Theory*, 51(2), pp.193–220.

Schirra, J.R.J. and Sachs-Hombach, K., 2013. The anthropological function of pictures. In: J.R.J. Schirra and K. Sachs-Hombach, eds. *Origins of Pictures. Anthropological Discourses in Image Science*. Köln: Herbert von Halem Verlag, pp.132–159.

Schmitz-Luhn, B., 2013. *Priorisierung in der Medizin: Kriterien im Dialog*. Berlin: Springer.

Schwell, A., 2015. The security/fear nexus. Some theoretical and methodological explorations into a missing link. *Etnofoor*, 27(2), pp.95–112.

Schwell, A., 2021. Imaginaries of sovereignty. Visualizing the loss of control. In: *Jahrbuch Migration und Gesellschaft/Yearbook Migration and Society*, 2, pp.123–138. https://www.degruyter.com/document/doi/10.1515/9783839455913/html

Sidhu, R., 2012. If it ain't broke… getting a sense of urgency for change. *ChangeQuest* [online] 7 May. https://www.changequest.co.uk/blog/if-it-aint-broke-getting-a-sense-of-urgency-for-change/ [Accessed 21 February 2021].

Skocpol, T. and Williamson, V., 2012. *The Tea Party and the Remaking of Republican Conservatism*. Oxford: Oxford University Press.

Sonntag, A., 2007. Sommernachtsträume. Eine skeptische Bilanz der Fußballweltmeisterschaften in Frankreich 1998 und Deutschland 2006. In: Deutsch-Französisches Institut, ed. *Frankreich Jahrbuch 2006. Politik und Kommunikation*. Wiesbaden: VS Verlag für Sozialwissenschaften, pp.257–278.

Trescher, T., 2018. Flüchtlinge: Notverordnung kommt, Not bleibt aus. *Kurier* [online] 6 October. https://kurier.at/politik/inland/notverordnung-die-geschichte-einer-verordnung-ohne-not/224.553.159 [Accessed 21 February 2021].

Turner, V.W., 2001. Betwixt and between. The liminal period in *rites de passage*. In: A.C. Lehman and J.E. Myers, eds. *Magic, Witchcraft, and Religion. An Anthropological Study of the Supernatural*. Mountain View: Mayfield, pp.46–55.

Voigt, A., 1891. Der Begriff der Dringlichkeit. Zeitschrift für die gesamte Staatswissenschaft. *Journal of Institutional and Theoretical Economics*, 47(2), pp.372–377.

Wahl-Jorgensen, K., 2019. *Emotions, Media and Politics. Contemporary Political Communication*. Cambridge, Medford: Polity Press.

Wodak, R., 2015. *The Politics of Fear: What Right-Wing Populist Discourses Mean*. Los Angeles/London: Sage.

Wohlgemuth, W.A., 2009. *Priorisierung in der Medizin: interdisziplinäre Forschungsansätze*. Berlin: MWV Med. Wiss. Verl.-Ges.

Wright, T., 2002. Moving images: the media representation of refugees. *Visual Studies*, 17(1), pp.53–66.

Zulianello, M. and Ceccobelli, D., 2020. Don't call it climate populism: on Greta Thunberg's technocratic ecocentrism. *The Political Quarterly*, 91(3), pp.623–631.

10 The elite as the political adversary

Neo-liberalism and the cultural politics of Hindutva

Sanam Roohi

Introduction

The question "who constitutes the elite" elicits three general if intersecting responses – first, that the elites are the governing or political class, second, that the elites control the economy or the means of production, and, third, that they constitute a minority whose members may be distinguished by their carefully cultivated taste and conspicuous patterns of consumption.[1] While interrelated, the answers reflect the disciplinary orientations of each response.[2] Taken together, it can be summarised that the elites exercise disproportionate influence or even control on the socio-political, economic and cultural landscape within any society. Yet, this standard descriptive categorising of the elite usually obfuscates more than it reveals. It does not convey the highly politicised discourses around the figure of the elite which are produced within particular socio-cultural and temporal contexts. Globally, the elite has increasingly become a polarising figure in the political discourses of the right, yet, who constitutes this elite differs across geographies.

As a relational term, studies have shown how the elite functions within and perpetuates "circles of power" (Mills, 1956). The networks they draw upon have been empirically studied in different locations and while it is hardly disputed that the presence of elites implies inequality (Khan, 2012; Jodhka and Naudet, 2019, pp.1–3), designating who the elites of any given society are – and how they are represented and constructed through processes of signification in different discursive contexts – is not only difficult but also heavily contested. Historically, both right- and left-oriented parties have strategically deployed anti-elite rhetoric at different points in time for various political goals (Curini, 2020). Yet, this rhetoric has steadfastly become an integral part of the political vocabulary of the right (Huber and Schimpf, 2017) The linkages between anti-elite populism and the salience of the nationalist project has received considerable scholarly attention in Europe and USA (Mudde, 2007; Stavrakakis et al., 2017), but in India, studies on elites, elite framing and right-wing nationalism have progressed in parallel without any meaningful conversation between them (for exceptions, see Blom Hansen, 1999; Berenschot, 2011).

DOI: 10.4324/9781003141150-13

Populist leaders claim to represent the unified general will or "will of the people", in opposition to an enemy, often embodied by the existing or preceding liberal system. To make sense of who *the people* are in this discourse, we need to pay close attention to the politics and processes behind people-making and the fashioning of the figure of the elite that is antithetical to the people. In this chapter, I contend that the discursive framings around the figure of the elite in India point to a complex interplay between a culturally embedded idea of the elite and the ideologies circumscribing these discourses. In what is to follow, the chapter foregrounds the rise of the Bharatiya Janata Party (BJP) and contextualises its attempt to hegemonise Hindutva with its recent makeover that has pegged it as a party of everyman. With Prime Minister Narendra Modi embodying the antithesis to the figure of the elite, it traces how anti-elitism as a political narrative first used – albeit in a limited way – by the left in India, travelled to become a part of the right-wing political lexicon in post-liberalised India. The chapter pairs media content analysis with digital ethnography, particularly social listening techniques (Stewart and Arnold, 2018), mostly between March 2020 and January 2021, and pays special attention to the articulation of anti-elite rhetoric during the "Harvard–hard work" debate. In doing so, it argues that the right's anti-elite discursive framings mask the contradictory nature of Hindutva politics that is helmed by "vernacular" elites – often "self-made", affluent or upwardly mobile and culturally conservative – who espouse anti-elitist ideals. This inherent contradiction of the Hindutva ideology (using anti-elite rhetoric while being led by this specific elite group) is partly mitigated by the deployment of neo-liberalism, suitably adapted to subcontinental specificities, that promises to bring "development" to Hindutva's expanding middle-class constituency.

The rise of the BJP and shifting anti-elite rhetoric

Anti-elite rhetoric in post-colonial India was hardly a major rallying point for mass mobilisation until the last decade. The Congress party – a centrist party that had been in power for over fifty years after India's independence in 1947 – had sustained its political rule by either catering to or co-opting different interest groups. The Indian left, deeply splintered within themselves (Vanaik, 1986) and with its limited federal electoral appeal,[3] did not build an overtly anti-elite populist political platform. Drawing many of its leaders from the middle classes (Dasgupta, 2005), its politics were anti-elite in that they were positioned against the big Indian industrialists and a vaguely defined transnational class of capitalists.[4]

Studies on the actual Indian elites have defined them as groups with overlapping political, economic or symbolic power (Jodhka and Naudet, 2019; Tripathy and Skoda, 2017), who were never sufficiently challenged due to India's passive revolution (Kaviraj, 1988). The entrenchment of India's elite within Indian electoral politics has also been defined as an elite capture of public welfare provisions (Panda, 2015). The use of anti-elite discourse as political platform-building by the right in India is, therefore, not only unique but worthy of some careful consideration. Interpreting the BJP's recent electoral successes through a larger framework, we need to pay attention to a discursive shift that has occurred in the country and has hegemonised the hitherto fringe ideology of Hindutva. This shift needs to be analysed in substantive ways beyond electoral mathematics and voter behaviour.

Advocating for a religiously inflected (Hindu) nationalism (Blom Hansen, 1999; Jaffrelot, 1996), the political ideology of Hindutva was first articulated in V. D. Savarkar's slim yet seminal work "Hindutva: who is a Hindu" written in 1923. Marginalised for many decades thereafter, Hindutva slowly gained currency in India's political discourse after the inception of the BJP in 1980 and its political mobilisation for the cause of building a Ram temple in Ayodhya.[5] However, it took the BJP another two decades to form a federal government, with its slow and steady rise paralleling the introduction of liberal reforms undertaken in India in the 1990s. Forming a government for the first time in 1998 (which collapsed soon after), the BJP ruled for a full term between 1999 and 2004. Despite decades of religious engineering to unite all Hindus, the BJP could not sustain its early electoral victory of 1999 and lost power after a term in 2004 – a failure attributed to its "India Shining" campaign that could not breach its upper-caste, middle-class voter base (Kaur, 2016). The loss forced it to shed its narrow constituency building approach and adopt a broad-based strategy beyond the upper-caste middle-class voter base. After a ten-year hiatus, the BJP underwent a makeover to become the largest party in the Indian parliament in the last two general elections of 2014 and 2019, entrenching the party among the voters across caste, class, gender and region (Kumar and Gupta, 2019). It is noteworthy that the BJP was initially led by upper-caste "vernacular elites" (Bhatt, 2004, p.201) who were Hindi-speaking, religiously conservative upper-caste North Indians. Its appeal electorally was limited to parts of North India (Graham, 1990). By 2014, it became a pan Indian party, garnering a vote share of 31.34% in 2014 and further consolidated its power by winning a vote share of 37.4% in 2019,[6] in a year that saw the highest voter turnout in Indian history at 67.36%. The BJP's electoral victories signal a decisive shift in the reordering of party preferences among Indian voters.

Building hegemony

The party's exceptional media campaigns (Jaffrelot, 2015a) and a strategic move to cater to constituencies beyond its *Brahmin-Baniya* (upper priestly and trading castes) base (Graham, 1990) contributed to its steep ascendance thereafter. The party expanded its reach with strong internal organisation and strategic planning, breaching earlier barriers of caste and class (Auerbach, 2015; Jha, 2017). It refurbished its image as a populist party addressing the needs of the poor, while sustaining and expanding its middle-class and economically affluent constituency.

The BJP's significant rise in Indian electoral politics since the late 1990s marks the mainstreaming of an ideology once considered "extremist"

(Graham, 1990, p.41; Jaffrelot, 2009). As a centrist party, the Congress had adopted an ideology of statism and recognition of different groups (Chhibber and Verma, 2018). By contrast, the Hindutva ideology espoused by the BJP is specific and narrow where militant nationalism converges with religion. However, in the last decade, the BJP has acquired a politically hegemonic space similar to the one the Congress party once enjoyed in the 1950s and 1960s, that led political scientists to coin monikers such as the "Congress System" (Kothari, 1964) or even "One Party Democracy" (Kochanek, 2015 [1968]). Scholars have shown that the Congress's rule was hegemonic because it built strategic alliances between the "dominant social interests" and the political leadership (Blom Hansen, 1999, p.152; Palshikar and Yadav, 2003). The BJP's decisive victories in 2014 and 2019 suggested that Hindutva ideology had replaced Congress's hegemonic position, reflected in the shift of public opinion towards religious majoritarianism (Palshikar, 2015). The 2019 election outcomes point to the further sedimentation of majoritarianism in India. The work of building this hegemony has taken several decades, having been painstakingly undertaken by the Rashtriya Swayam Sevak Sangh or RSS – the BJP's parent body. Just as the hegemony of one social group over the entire nation is exercised through organisations such as the church, educational institutions and cultural bodies (Bates, 1975; Gramsci, 1992), RSS has started *Seva vibhags* (service centres), RSS-affiliated schools, welfare forums for Adivasi (tribal) and other marginalised communities since the late 1980s. The success of the gradual penetration of the RSS to reach out to non-upper-caste voters (Thachil, 2014) was evident in the last two elections, even as their core voter base remained among the upper-caste-dominated (particularly Brahmin) constituencies (Suryanarayan, 2019).

Advocating a mixed economy in the early 1980s, the party made a programmatic shift towards neo-liberalism in the 1990s to enable the support of the middle classes "who were 'mobilizable' because of their growing disaffection" (Chhibber, 1997, p.631) with the Congress party's welfarist economic policies. The idiom of development has mainstreamed the muscular and violent Hindutva ideology among both its middle-class and lower middle-class constituencies, who imagine themselves to be its beneficiaries. However, making a further shift in the 2000s, even when the party supported less state intervention in the economy, the BJP increasingly utilised private welfare (Thachil, 2014) and state largesse (Ahmad, 2019) for the poor, while managing to retain its upper-caste middle-class constituency. The BJP's adoption of the neo-liberal development paradigm built atop the right-wing political agenda of a *Hindu Rashtra* (Hindu nation) and its pandering to the poorer constituencies through private welfare create contradictions. The inherent policy discrepancies that arise because of this are manoeuvred by a "forked tongue" approach (Kaul, 2017).

I turn the focus on that approach in the remainder of this chapter. It is an understudied discursive strategy deployed by the party to castigate those who oppose the party as "the liberals and the socialists" who are corrupt and/or elitist, while the BJP upholds "meritocracy and hard work". Centring this debate around the figure of Modi, this articulation finds echoes in media discussions to date, pointing to its longevity and recall value.

The making of anti-elite populism in India

When Narendra Modi was made the prime ministerial candidate for the 2014 election, it was touted by the media as a masterstroke. As a strong proponent of Hindutva and a RSS functionary, Modi had been the chief minister of Gujarat state for three successive terms overseeing large reforms (Jaffrelot, 2015b) in the state. It was under his first tenure when India's last large-scale organised violence against minorities took place in Gujarat in 2002. It had Modi's tacit support.[7] By deploying a politics of symbolism, Modi's popularity grew and he was perceived as a "development man" who could also put the problematic Muslims in their rightful place as secondary citizens (Sen, 2016). By 2013, the party – with RSS's support – projected Modi as a hyper-masculine strongman who could take control, bring development and was the only leader who could steer India to be a global economic and political powerhouse.

Even as the party's electoral victories have been analysed through multiple lenses,[8] an attribute that has been overemphasised in online and offline media spaces is the "Modi wave" – a phrase that conveys the impression that Indian voters made their choices based on Modi's mass cult following. But Modi is only the external façade of the "banalisation" of Hindu nationalism (Jaffrelot, 2015a). The preoccupation of the media with the "Modi wave" conceals many interlinked processes that have culminated in his reinstating as prime minister for a consecutive term.

The party's rise was precipitated by their use of violence to rectify supposed historical wrongs against the Hindu majority, ostensibly by the Muslims in medieval times (Dhattiwala and Biggs, 2012). While the othering of Muslims in India as culturally distinct has its roots in the political churnings of the late colonial period (van der Veer, 2002), it is the Hindu right from Savarkar's time that has politicised this difference as inimical to the Hindu nation. Hindutva's imagination of a homogenised religious Hindu body politic has successfully disenfranchised Muslims, thoroughly marginalising them in the political space (Shaban, 2018), while simultaneously shifting its focus to the figure of the elite, who it construes as its actual opponent.

In an attempt to redefine Hindutva as a consensual and moral force, the BJP's recent makeover has steadily worked towards a politics of encompassment, where caste, regional and linguistic schisms are subsumed to stitch a forced Hindu unity defined by Brahminical values, particularly the upholding of caste hierarchy and a prioritisation of Brahminical religious beliefs. In doing so, it has somewhat succeeded in achieving two objectives: firstly, designating the category of the Indian Muslim as the enemy and an outsider and, secondly, moulding the figure of the elite as its political adversary or foil to itself. Both of these objectives share a common ground with the global right-wing populism whose critical features are the "denial of diversity [that] effectively amounts to denying the status of certain citizens as free and equal" (Müller, 2017, p.82; Brubaker, 2017). Distilling a potent populism undercut by religious emotions, carefully overlaid on top of clientelist network politics (Berenschot, 2011), a political party such as the BJP had to shed its pro-upper-caste image while retaining them as their core base. The party particularly succeeded in mobilising many new, under-mobilised and even first-time voters (Heath, 2015). The BJP created an intricate network of supporters in urban areas, their traditional stronghold (Auerbach, 2015), while also making deep inroads into rural areas.

During its time in opposition between 2004 and 2014, the BJP relentlessly attacked Congress's secularism as elitist, on the one hand, and anti-Hindu, pro-minority, on the other (Siddiqui, 2017). It also started a persistent campaign portraying Congress's reign as corrupt misrule. Ethnographically informed studies have pushed our understanding of corruption in the subcontinent, where the acts, even when widely perceived as "bad", are understood as performing a mediatory role where informal social networks are used to improvise or build relations of clientelism (Parry, 2000; Ruud, 2000, 2001). These studies have blurred the boundaries between the private and public sphere (Gupta, 1995), unsettling the empirical distinction between civil and political society (Witsoe, 2011) and breaking down any clear distinction between virtue and vice (Jauregui, 2014). Interrogating welfare provisions for poor citizens (Gupta, 2005) or the lower caste's ascendance to power (Witsoe, 2013), scholars have described corruption as central to the functioning of the Indian state (Gupta, 2012). The expansion in retail corruption in post-reform India is seen as a direct outcome of the deeper penetration of the Indian state (Parry, 2000). Popular anti-corruption discourse in India, however, continues to be linked to cases of graft involving politicians and public officials. It was mostly Congress's mismanagement that could easily be categorised as an example of corruption in high office.

The "Congress rule as corrupt" discourse resonated most with middle-class Indians across the country during the Congress party's last reign (Sengupta, 2014). While this line of attack was not new, the BJP added a new layer of meaning to these attacks by specifying their target as the "liberal elites of Lutyens' Delhi" – constituting English-speaking, often western-educated, Congress leaders, activists or intelligentsia who hold secular and liberal values, with many of them residing in the Lutyens' area of Delhi. The BJP's vernacular elite leadership had seen the English-speaking Congress ruling elites for a long time as inheritors of secular (anti-Hindu) British values. The BJP's task was a "commitment to Hindu nationalism in the belief that this was the means of producing a new social elite which would be Hindi-speaking, austere, disciplined and traditional" (Graham, 1990, p.55). McDonnell and Cabrera's research based on interviews with key BJP figures documents the latter's critique of "a range of elites such as intellectuals, the English-language media and NGOs" but "the main bad elites were the Congress Party and the Gandhi family" (2019, p.6).

Canovan, writing on the rise of the populist right in Europe, highlights the difference between "liberalism" and a "democratic strand" in political theory and practice. The former concerns itself with "individual rights, universal principles and the rule of law, and is typically expressed in a written constitution", whereas the latter is concerned with "the sovereign will of the people, understood as unqualified majority rule and typically expressed through referendums" (Canovan, 2004, p.244). In India, these strands can be found in a similar rhetorical distinction that the political right makes between the BJP's supporters – the "true" people whose will should rule – and its detractors, with their apparent commitment to liberal principles that hinder majority rule. In this rendering, everyone from the centre to the left of the spectrum are undistinguishably clubbed together as the elite. As illustrated below, the figure of the elite converges in a gender-agnostic strawman figure who is professed to be a nominal Hindu not bound by religious norms, who is urban, foreign educated or English-speaking, influenced by western culture and an ideological adherent to liberal or socialist values. A recently coined terminology that is frequently used for this group is "Urban Naxals", used as a metonym for those nominal Hindus who engage in any kind of "antinational" activities (Singh and Dasgupta, 2019). The Naxals were originally a small group of radical left intellectuals leading uprisings against the big landlords and businesses in largely tribal rural pockets of eastern and south-eastern India from the 1970s until the 1990s. By prefixing "urban" with Naxals, the right reformulated the term from its narrow moorings to include any urban residing opponent of the "people" and the "nation" as represented by the right.

The populist and controversial leader Narendra Modi himself is antithetical to the figure of the elite. He embraces the traditional morals and values discarded by the elites. Speaking chaste Hindi, wearing Indian (if expensive) clothes, renouncing his family for public service, Modi espouses the Brahmin-led RSS values of service to the (Hindu) nation, while connecting to the masses through his non-urban, non-upper-caste background and exceptionally well-crafted "humble background" roots.

The subaltern Modi as a hegemon

The BJP's 2014 and 2019 election campaigns were the most expensive in India's history, with the 2019 election expenditure slated as the "most expensive election" in the world.[9] Its media campaign built a decisive anti-elite and pro-people rhetoric by fully exploiting the social media. Integrating pro-women, pro-poor, pro-people plank with catchy jingles, BJP centred the whole campaign around Modi, coining the following slogans: *Abki baar Modi sarkar* (this time, it's Modi's government) or its variant, *bahut hua nari pe waar, abki baar Modi sarkar* (no more attacks on women, this time it's Modi's government). It also deified Modi by raising the slogan *har har Modi ghar ghar Modi* (everywhere Modi, in every household Modi). The words *har har* meaning everyone is a war cry using the name of Lord Shiva – one of the most masculine Gods in the Hindu pantheon.

Modi, on his part, skilfully articulated his (disputable) railway station corner tea-seller roots and his lower caste and working-class background as a natural representative of the patriotic people of India. Presenting himself to the voters as a hardworking politician who, unlike many politicians, never went to an Ivy League or Russell Group university in the West, he turned his anti-elite persona into a virtue and projected himself more as a common man's representative, turning the traditional communal-secular debate between the BJP and the Congress into an elite vs. subaltern one. At one point in his campaign in 2014 when he was ridiculed by the Congress leadership for not having the necessary expertise in economics, he challenged his opponents with the remark that this country is run by hard work, not Harvard (Vijay Kumar, 2014). In subsequent years, the "hard work versus Harvard" analogy was crafted in very effective ways to reiterate Modi's hardworking strongman image that would deftly be picked up by the media. Its longevity can be gauged by its continued persistence in social media as a meme – such as in a caricature of the weak-looking, Harvard-educated then Congress Home Minister P. Chidambaram, trying to arm wrestle a grotesquely muscular Modi, with "Harvard" and "Hard Work" written on their respective biceps.

The liberal/left elite as a corrupt trope was used to make massive policy changes in India once the Modi-led BJP came to power in 2014. In this section, I recount a few remarkable instances where the "Harvard–hard work" analogy was used ostensibly as a justification for rooting out "corruption". One early instance was when Prof. Amartya Sen's tenure at the recently started Nalanda University was abruptly put to an end in early 2016 by the university's governing body. Sen, who had criticised Modi's communal and religious politics, was unceremoniously removed as his values clashed with those of the ruling regime's. Promoting its anti-intellectualism as anti-elitism, the BJP was intent on undermining Nalanda University's promising social sciences focus, which the party feared could become another breeding ground for anti-right-wing attitudes, as in other universities such as the Jawaharlal Nehru University or University of Hyderabad.

Later that year, Modi ushered in one of the most drastic financial reforms through demonetisation. In November 2016, he demonetised 85% of the currency in circulation, transacting with which was made illegal overnight. Modi argued in a televised speech that demonetisation was done in the interest of the nation to root out (deeply embedded Congress era) corruption and to force black money out of the economy, some of which was used to finance terror activities. Over the next few months, Indians with bank accounts were forced to deposit their old currency in banks standing in long queues for days. Those who did not have a bank account (daily wage earners, rural poor, homeless) were hit the hardest. Withdrawal of new notes from banks was severely limited for the next few months and a lack of cash circulation caused hardships to everyone, including the middle classes. However, the buzz in social media was that by flushing money out of people's houses, Modi injected money into the formal banking economy and, simultaneously, helped the BJP win the Uttar Pradesh (the most densely populated state of India) elections in 2017 by depriving their opponents of cash used extensively in all forms of legal and illegal election strategising. While findings in an early article following communal Hindu-Muslim riots in 1992 suggested that economic concerns take precedence during elections even when religious consciousness is high (Shah, 1994, p.1140), demonetisation with the apparent aim of weeding out corruption led to a significant contraction of the Indian economy with a detrimental impact on India's largely informal sector (Ghosh, Chandrasekhar, and Patnaik, 2017), a point re-established in later studies (Chodorow-Reich et al., 2020). However, the BJP was re-elected with a wider margin in 2019. This goes against Shah's classic analysis. What changed between 1994 and 2019 where the electorate did not punish the party that harmed them economically but rather voted it back to power?

A partial answer could lie in the powerful discursive framings of neo-Hindutva (Anderson and Longkumer, 2018), captured in phrases such as "*sabka saath, sabka vikas*" (development for all, with all) and the "Harvard–hard work" debate, which has an avowed anti-elite rhetoric and could easily be read as a pro-people gesture. After becoming prime minister, for instance, Modi renamed Race Course Road, where the residence of the prime minister is located, to Lok Kalyan Marg or the *road to public welfare*, symbolising the dismantling of not only the reminiscence of colonial past but also the Anglophile Indian elite. This name change of a road can be read as a mere symbolic gesture, but it points to the underlying politics of signification (or authority, legitimacy and meaning-making) that are fundamental to the processes of building hegemony.

Reigniting the Harvard vs. hard work debate

In June 2020, a renowned media personality announced on her Twitter account that she had been offered an associate professorship in journalism at Harvard University. She also announced her resignation from the media house where she had spent more than two decades building her career as a journalist. Seen as being critical of the BJP, both she and the media house had often attracted the ire of right-wing Twitter handles and trolls. Unsurprisingly, her announcement was met with derision from "Hindu Twitter" (a self-referential term used by Twitter users who espouse the cause of Hindutva). Six months later, in a sudden turn of events, she announced that the professorship was a prolonged hoax and an elaborate online phishing scam played on her and that she had lodged a police complaint.

Even as she later wrote a long op-ed on her foolishness for not doing enough "due diligence", she was virulently attacked on Twitter and with it the debate on "Harvard and hard work" was reignited. In a Twitter thread discussing her "fraudulent" attempt to pass herself off as a Harvard professor, a handle commented "to be fair to @xxxxx, left wing journalists like her got a job through the left-wing eco-system of contacts and phone calls in the media+academia. Easy for them to assume that even a job @harvard cud fall into their lap like that (sic)". Another Twitter handle posted the following:

"Need to do "Hard Work" to get into "Harvard". Spreader of fake news, a third grade journalist can get into Jawaharlal Nehru University or whatsapp university only (sic)". One commentator, echoing a thousand other right-wing handles, tweeted "Modiji is a genius. Long back he had advocated HARD WORK over HARVARD (sic)". The journalist's case of deception, while an outlier, aptly encapsulates the common narrative-building strategies against the "left" elites by supporters of the Modi regime on Twitter.

Since it was first formulated in 1923, Hindutva as a concept appears to be a highly robust category that has adapted to change while keeping its core ideology of Hindu nationalism intact. The entangling of development with Hindutva is one such way in which the concept has moved along with changing times. The appeal of Hindutva, for instance, is growing stronger among the Indian Hindu diaspora and Hindutva has interspersed itself with multiculturalism in its diasporic reincarnation (Anderson, 2015; Mathew and Prashad, 2000), while on Indian soil it supports the monocultural idea of Hindi-Hindu-Hindustan, superimposing Hindi as a language and Hinduism or Hindutva as a religion on India or Hindustan. It has overturned the communal-secular debate by projecting a contradictory idea of the Hindu religious fold as all-encompassing, which does not need western secularism to assimilate minorities. This anti-elite discursive framing of Hindutva takes away from the real subaltern anti-elite struggle spearheaded by marginalised groups across India.

While there are significant overlaps, where anti-elite politics in India and the West diverge is the way in which the BJP has attempted to build its political constituency, making its relationship to (anti-)elitism much more complex, thereby offering an extremely relevant yet contrasting global comparison. Relying on populist speeches and welfare measures addressed to the poor, the BJP and its parent body the RSS have primarily nurtured a Hindu, upper-caste, middle-class, educated, social media-savvy, Hindi-speaking constituency for themselves – a group which can be considered as the new vernacular elite as opposed to the old ruling elite from the Congress party. These emergent elites oppose the left or liberal ideology as a foreign import and detest the Congress for its secular, supposed pro-minorities, pro-poor measures, castigating it as a corrupt and even an antinational party. Its anti-elite rhetoric is important for contextualising why the BJP continues to work for a minority even when its discourses peg it otherwise. This seeming contradiction is important to sustain Hindutva politics in a democracy where votes matter but where the Hindutva ideology that caters largely to a upper-caste middle-class constituency needs to be perpetuated. This conundrum is partly mitigated by the deployment of neo-liberalism as a tool of general advancement for all which promises to bring "development" to Hindutva's expanding middle- and lower middle-class constituency.

Conclusion

India's political landscape in the recent past has seen a rapid ascent of the right-wing BJP to power and its subsequent consolidation. Adhering to

Hindutva ideology and affiliated to extremist right-wing groups, the BJP's constituency has visibly expanded in the last two decades. As a muscular Hindu party that stands for Hindu Rashtra (the Hindu nation), its rise can be attributed to a constellation of factors including a liberalising economy, the growth of an aspirational upper-caste middle class, and political strategies adopted by different right-wing groups affiliated to it (Corbridge, Harriss, and Jeffrey, 2013). Similar to other forms of extremist nationalism, the Hindutva ideology builds upon the politics of exclusion, often with violent consequences for Muslims, women, Dalits and non-upper castes, which together make up the bulk of the Indian population. Shedding its overt violent avatar, however, and forging an explicit development agenda mixed with national security jingoism, the BJP has relied on populist speeches and appeals addressed to the poor, creating newer vote banks among non-upper-caste, lower middle-class constituencies. Nevertheless, it has sustained and nurtured its traditional base constituted from the educated but religiously conservative upper-caste middle classes who can be considered as vernacular elites.

Attempting to understand the entanglement of right-wing populism with anti-elite discursive framing, in this chapter, I have explored the rise of right-wing politics in India using Gramsci's concepts of hegemony and signification. Much like in the West, right-wing politics and political articulation in India has also targeted entrenched power elites. Similar to its western counterparts, elite politics in India is equated with endemic corruption, which is challenged by the right-wing BJP and its allies. The electoral successes of the BJP have, in no small measure, been projected as the victory of ordinary common Indians, as illustrated by its leader Narendra Modi and his rise from a roadside tea-seller to the post of prime minister. Tracing the highly adaptive Hindutva ideology from an overtly violent to a mainstream political ideology, I argue that Hindutva's expanding constituencies work towards hegemonising its ideology, even as it keeps producing the notion of the Muslim man as a problem figure, in need of fixing and disciplining. This is not to say that the hegemonising discourse of Hindutva is not contested[10] in an increasingly shrinking democratic and dialogic space but, in the absence of a nation-wide alternate encompassing ideology, the largest ground is ceded to the Hindutva ideology. Moreover, the coalescing of interests of an aspirational middle class and an ascendent state intent on becoming a global power has made Hindutva an ideology that is amenable to large sections of the Indian electorate, aided amply through the use of social media.

Notes

1 I would like to thank the editors of the book for their insightful comments and suggestions on the draft that helped me to refine my arguments. Any shortcoming in the chapter, however, is entirely mine.

2 For an overview of the discussion around the political elite, see Zuckerman (1977); on consumption, see Trigg (2001). Giddens (1972) provides an early schema to disentangle who the elite is, particularly in relation to class. See additionally the chapters by Gerbaudo and the Introduction in this volume.

3 The left parties in India have only been in power in 3 out of 29 states of India – viz. Tripura, West Bengal and Kerala and have been a minor coalition partner at the federal level sporadically.
4 The notable exception is the now banned Communist Party of India (Maoist).
5 Ayodhya is considered as the birthplace of Lord Ram, and the Babri mosque built by Babur, the Mughal ruler in the 15th century, was believed to have been built after breaking down a temple. In order to set right this historical wrong (with little historical evidence), Hindutva supporters called "*kar sevaks*" demolished the Babri mosque on December 6, 1992.
6 The election commission data can be accessed here: http://results.eci.gov.in/pc/en/partywise/index.htm Accessed June 1, 2019].
7 The 2002 Gujarat riots in which some 1200 people were killed (a conservative estimate, of which a quarter were Muslims) was a state-sponsored pogrom. Modi, the then chief minister of Gujarat, is said to have asked the police force to let the Hindus vent their anger against Muslims who had allegedly burned a train compartment after being provoked by *kar sevaks* (Jaffrelot, 2003).
8 Gilles Vernier, Christophe Jaffrelot and others have made a first-cut attempt to understand voting patterns in a series of news articles on the 2019 election, using the data available from the Election Commission of India. These articles can be found following this link: https://tcpd.ashoka.edu.in/press-articles/ [Accessed June 15, 2019].
9 The party spent 45% of the total election budget (Scroll, 2019).
10 Hindutva ideology saw a very strong backlash from Muslims after the 2019 Citizenship Act was amended to make religion a criterion for granting asylum to illegal migrants. More recently, India is seeing prolonged agitation by farmers against the passage of Farm Bills that aim to corporatise agriculture.

Bibliography

Ahmad, I., 2019. Democracy as rumour: media, religion, and the 2014 Indian elections. In: I. Ahmad and P. Kanungo, eds. *The Algebra of Warfare-Welfare*. New Delhi: Oxford University Press, pp.55–90

Anderson, E., 2015. "Neo-Hindutva": the Asia House MF Husain campaign and the mainstreaming of Hindu nationalist rhetoric in Britain. *Contemporary South Asia*, 23(1), pp.45–66.

Anderson, E. and Longkumer, A., 2018. "Neo-Hindutva": evolving forms, spaces, and expressions of Hindu nationalism. *Contemporary South Asia*, 26(4), pp.371–377.

Auerbach, A.M., 2015. India's urban constituencies revisited. *Contemporary South Asia*, 23(2), pp.136–150.

Bates, T.R., 1975. Gramsci and the theory of hegemony. *Journal of the History of Ideas*, 36(2), pp.351–366.

Berenschot, W., 2011. *Riot Politics: Hindu-Muslim Violence and the Indian State*. New York: Columbia University Press.

Bhatt, C., 2004. "Majority ethnic" claims and authoritarian nationalism: the case of Hindutva. In: E.P. Kaufmann, ed. *Rethinking Ethnicity: Majority Groups and Dominant Minorities*. London/New York: Routledge. pp.198–220.

Blom Hansen, T., 1999. *The Saffron Wave: Democracy and Hindu Nationalism in Modern India*. Princeton, NJ: Princeton University Press.

Brubaker, R., 2017. Why populism? *Theory and Society*, 46(5), pp.357–385.

Canovan, M., 2004. Populism for political theorists? *Journal of Political Ideologies*, 9(3), pp.241–252.

Chhibber, P., 1997. Who voted for the Bharatiya Janata Party? *British Journal of Political Science*, 27(4), pp.631–639.
Chhibber, P.K. and Verma, R., 2018. *Ideology and Identity: The Changing Party Systems of India*. New Delhi: Oxford University Press.
Chodorow-Reich, G., Gopinath, G., Mishra, P. and Narayanan, A., 2020. Cash and the economy: evidence from India's demonetization. *The Quarterly Journal of Economics*, 135(1), pp.57–103.
Corbridge, S., Harriss, J. and Jeffrey, C., 2013. *India Today: Economy, Politics and Society*. Hoboken, NJ: John Wiley & Sons.
Curini, L., 2020. The spatial determinants of the prevalence of anti-elite rhetoric across parties. *West European Politics*, 43(7), pp.1415–1435.
Dasgupta, R., 2005. Rhyming revolution: Marxism and culture in colonial Bengal. *Studies in History*, 21(1), pp.79–98.
Dhattiwala, R. and Biggs, M., 2012. The political logic of ethnic violence: the anti-Muslim pogrom in Gujarat, 2002. *Politics & Society*, 40(4), pp.483–516.
Giddens, A., 1972. Elites in the British class structure. *The Sociological Review*, 20(3), pp.345–372.
Graham, B., 1990. *Hindu Nationalism and Indian Politics: The Origins and Development of the Bharatiya Jana Sangh*. Cambridge: Cambridge University Press.
Gramsci, A. 1992. *Selections from Prison Notebooks*. Edited and translated by Q. Hoare and G.N. Smith. New York: International Publishers.
Gupta, A., 1995. Blurred boundaries: the discourse of corruption, the culture of politics, and the imagined state. *American Ethnologist*, 22(2), pp.375–402.
Gupta, A., 2005. Narratives of corruption: anthropological and fictional accounts of the Indian state. *Ethnography*, 6(1), pp.5–34.
Gupta, A., 2012. *Red Tape: Bureaucracy, Structural Violence, and Poverty in India*. Durham: Duke University Press.
Ghosh, J. Chandrasekhar, C.P. and Patnaik, P., 2017. *Demonetisation Decoded: A Critique of India's Currency Experiment*. London: Routledge.
Heath, O., 2015. The BJP's return to power: mobilisation, conversion and vote swing in the 2014 Indian elections. *Contemporary South Asia*, 23(2), pp.123–135.
Huber, R.A. and Schimpf, C.H., 2017. On the distinct effects of left-wing and right-wing populism on democratic quality. *Politics and Governance*, 5(4), pp.146–165.
Jaffrelot, C., 1996. The genesis and development of Hindu nationalism in the Punjab: from the Arya Samaj to the Hindu Sabha (1875–1910). *IndoBritish Review*, 21, pp.3–40.
Jaffrelot, C., 2003. Communal riots in Gujarat: the state at risk? In: S.K. Mitra, ed. *Heidelberg Papers in South Asian and Comparative Politics*. Heidelberg: Department of Political Science, South Asia Institute, no. 17.
Jaffrelot, C., 2009. *Hindu Nationalism: A Reader*. Princeton, NJ: Princeton University Press.
Jaffrelot, C., 2015a. The Modi-centric BJP 2014 election campaign: new techniques and old tactics. *Contemporary South Asia*, 23(2). pp.151–166.
Jaffrelot, C., 2015b. What "Gujarat model"? – growth without development – and with socio-political polarisation. *South Asia: Journal of South Asian Studies*, 38(4), pp.820–838.
Jauregui, B., 2014. Provisional agency in India: Jugaad and legitimation of corruption. *American Ethnologist*, 41(1), pp.76–91.
Jha, P., 2017. *How the BJP Wins: Inside India's Greatest Election Machine*. New Delhi: Juggernaut Books.

Jodhka, S. and Naudet, J., eds., 2019. *Mapping the Elite: Power, Privilege, and Inequality*. Oxford: Oxford University Press.

Kaul, N., 2017. Rise of the political right in India: Hindutva-development mix, Modi myth, and dualities. *Journal of Labor and Society*, 20(4), pp.523–548.

Kaur, R., 2016. "I am India shining": the investor-citizen and the indelible icon of good times. *The Journal of Asian Studies*, 75(3), pp.621–648.

Kaviraj, S., 1988. A critique of the passive revolution. *Economic and Political Weekly*, 23(45–47), pp.2429–2444.

Khan, S.R., 2012. The sociology of elites. *Annual Review of Sociology*, 38, pp.361–377.

Kochanek, S.A., 2015 [1968]. *The Congress Party of India: The Dynamics of a One-Party Democracy*. Princeton, NJ: Princeton University Press.

Kothari, R., 1964. The Congress "system" in India. *Asian Survey*, 4(12), pp.1161–1173.

Kumar, S. and Gupta, P., 2019. Where did the BJP get its votes from in 2019. *Mint* [online] 3 June. https://www.livemint.com/politics/news/where-did-the-bjp-get-its-votes-from-in-2019-1559547933995.html [Accessed 6 March 2021].

Mathew, B. and Prashad, V., 2000. The protean forms of Yankee Hindutva. *Ethnic and Racial Studies*, 23(3), pp.516–534.

McDonnell, D., and Cabrera, L., 2019. The right-wing populism of India's Bharatiya Janata Party (and why comparativists should care). *Democratization*, 26(3), pp.484–501.

Mills, C.W., 1981 [1956]. *The Power Elite*. New York: Oxford University Press.

Mudde, C., 2007. *Populist Radical Right Parties in Europe*. Cambridge: Cambridge University Press.

Müller, J-W., 2017. *What Is Populism?* London: Penguin UK.

Palshikar, S., 2015. The BJP and Hindu nationalism: centrist politics and majoritarian impulses. *South Asia: Journal of South Asian Studies*, 38(4), pp.719–735.

Palshikar, S. and Yadav, Y., 2003. From hegemony to convergence: party system and electoral politics in the Indian states, 1952–2002. *Journal of Indian Institute of Political Economy*, 15(1/2), pp.5–44.

Panda, S., 2015. Political connections and elite capture in a poverty alleviation programme in India. *The Journal of Development Studies*, 51(1), pp.50–65.

Parry, J., 2000. "The crisis of corruption" and the "idea of India": a worm's eye view. In: I. Pardo, ed. *The Morals of Legitimacy: Between Agency and System*. Oxford: Berghahn Books, ch. 2.

Ruud, A.E., 2000. Corruption as everyday practice. The public-private divide in local Indian society. *Forum for Development Studies*, 27(2), pp.271–294.

Ruud, A.E., 2001. Talking dirty about politics: a view from a Bengali village. In: C.J. Fuller and V. Benei, eds. *The Everyday State and Society in Modern India*. London: C. Hurst & Co. Publishers Ltd., pp.115–136.

Savarkar, V., 1984 [1923]. *Hindu rashtra darshan*. New Delhi: Prabhat Prakashan.

Scroll, 2019. BJP spent nearly 45% – or Rs 27,000 crore – of total expenditure for 2019 Lok Sabha polls: report. *Scroll.in* [online] 4 June. https://scroll.in/latest/925882/bjp-spent-nearly-45-or-rs-27000-crore-of-total-expenditure-for-2019-lok-sabha-polls-report [Accessed 6 March 2021].

Sen, R., 2016. Narendra Modi's makeover and the politics of symbolism. *Journal of Asian Public Policy*, 9(2), pp.98–111.

Sengupta, M., 2014. Anna Hazare's anti-corruption movement and the limits of mass mobilization in India. *Social Movement Studies*, 13(3), pp.406–413.

Shaban, A., ed., 2018. *Lives of Muslims in India: Politics, Exclusion and Violence.* London/New York: Routledge.

Shah, G., 1994. Identity, communal consciousness and politics. *Economic and Political Weekly*, 29(19), pp.1133–1140.

Siddiqui, K., 2017. Hindutva, neoliberalism and the reinventing of India. *Journal of Economic and Social Thought*, 4(2), pp.142–186.

Singh, M. and Dasgupta, R., 2019. Exceptionalising democratic dissent: a study of the JNU event and its representations. *Postcolonial Studies*, 22(1), pp.59–78.

Stavrakakis, Y., Katsambekis, G., Nikisianis, N., Kioupkiolis, A. and Siomos, T., 2017. Extreme right-wing populism in Europe: revisiting a reified association. *Critical Discourse Studies*, 14(4), pp.420–439.

Stewart, M.C. and Arnold, C.L., 2018. Defining social listening: recognizing an emerging dimension of listening. *International Journal of Listening*, 32(2), pp.85–100.

Suryanarayan, P., 2019. When do the poor vote for the right wing and why: status hierarchy and vote choice in the Indian states. *Comparative Political Studies*, 52(2), pp.209–245.

Thachil, T., 2014. *Elite Parties, Poor Voters.* Cambridge: Cambridge University Press.

Trigg, A.B., 2001. Veblen, Bourdieu, and conspicuous consumption. *Journal of Economic Issues*, 35(1), pp.99–115.

Tripathy, J. and Skoda, U., 2017. Cultural elites and elite cultures in contemporary India and South Asia. Constructions and deconstructions. *International Quarterly for Asian Studies*, 48(1–2), pp.5–13.

Vanaik, A., 1986. The Indian left. *New Left Review*, 159, pp.49–70.

Veer, P.V.D., 2002. Religion in South Asia. *Annual Review of Anthropology*, 31(1), pp.173–187.

Vijay Kumar, S., 2014. Hard work matters, not Harvard: Modi. *The Hindu* [online] 9 February. https://www.thehindu.com/news/national/hard-work-matters-not-harvard-modi/article5668557.ece [Accessed 6 March 2021].

Witsoe, J., 2011. Corruption as power: caste and the political imagination of the post-colonial state. *American Ethnologist*, 38(1), pp.73–85.

Witsoe, J., 2013. *Democracy against Development: Lower-caste Politics and Political Modernity in Postcolonial India.* Chicago, IL: University of Chicago Press.

Zuckerman, A., 1977. The concept "political elite": lessons from Mosca and Pareto. *The Journal of Politics*, 39(2), pp.324–344.

"AGAINST THE ELITES!"

ANTI-ELITISM AND THE (NEW) RIGHT

PART IV

SEBASTIAN DÜMLING, RICHARD GEBHARDT, DIANA WEIS

11 The heroic deed, the wrong word and the utopia of clarity

The discourse of Germany's *New Right* on elites and its links to popular culture

Sebastian Dümling

Introduction

The current conjuncture of populism is often interpreted as an insurrection against elites (Taggart, 2004).[1] The trope "against the elites!" works as a communicative conduit connecting different fractions, interests and ideologies. However, this assumption needs an explanation, at least for the German case. Finally, German right-wing populism with its anti-elite rhetoric intertwines closely with the so-called *New Right* (Rosellini, 2019). Some researchers even recognise the *New Right* as the intellectual spearhead – not to say the *elite* – of German populism in the form of the *Alternative für Deutschland* (AfD). New-Right's publisher Götz Kubitschek is considered the *spiritus rector* for the *völkisch* (ethnic-nationalist) fraction of the AfD represented by Björn Höcke. However, this connection between the AfD and the *New Right* contradicts the trope "against the elites!" since the *New Right* uses the term "elite" in a positive sense, unlike the AfD. Moreover, Germany's *New Right* draws programmatically on a concept of a true elite opposing the actual, corrupted one. In this regard, the claim is less "against the elite!" than rather "against a wrong elite – for a legitimate one". The following article attempts to shed light on this tension and to classify it analytically.

There could be several possible approaches to investigate this programmatic tension. An approach based on the history of ideas would be particularly obvious. There already are several studies on the *New Right*'s intellectual history (Griffin, 2000). Indeed, these studies are helpful for understanding the genealogy of certain elaborated concepts and ideas. The historian Volker Weiss can trace the current *New Right*'s ideational programme back to more or less great thinkers – from Alain de Benoist via Carl Schmidt and Martin Heidegger to Friedrich Ludwig Jahn (Weiß, 2017). This approach may be helpful in understanding the development of a reflected idea but, in other respects, must disappoint (Collini et al., 1988): Intellectual history's main weak spot is that it is usually blind to unreflexive entanglements, to the overlapping of different cultural registers. Texts make use of a number of heterogeneous resources for that – of which the authors are usually not fully aware (Eco, 1979; Link, 1999).

DOI: 10.4324/9781003141150-15

My article tries to understand the *New-Right*'s contradictory relationship to elites and elitism by keeping in mind that most texts are not bound to one single discursive, generic and semantic register. Instead, they are meandering and proliferating. That means empirically that a free play of floating signifiers connect the – as right-wing intellectual circles are wont to say – *respublica literari*, i.e. high literature, and the realm of trash and pulp, i.e. popular culture, as well as narrative and descriptive accounts, whether factual or fictional (Derrida, 1978). After all, meanings do not care where they belong. Nevertheless, where they belong does mean *something*. Bearing this in mind, I want to zone in on the elitism discourse within Germany's *New Right*. I argue that the elitism discourse has a share in the storage space where society keeps understandable accounts about the social sphere. This storage space may be termed as "archive" (Foucault, 1982) or "cultural memory" within very different theoretical perspectives (Assmann and Assmann, 1994). The analysis developed in the article connects the right-wing accounts to this general storage space by exploring its interdiscursive links to very different spaces, such as global popular culture and European intellectual history. After all, it is necessary to refer symbolically to the storage space mentioned in order to make accounts comprehensible. To put it a bit mechanically: The more an account resembles the contents of culture's storage space, the more its validity can be extended (Schneider, 2010).

Regarding the current wave of right-wing populism, this approach prevents one from seeing right-wing discourses as the dark *other* of a bright and progressive liberal mainstream culture. Instead, this approach highlights that right-wing "special discourses" (Link, 1999) intrinsically belong to and intertwine with broader culture, including mass-market TV series, films and popular literature.[2] In order to analyse these connections and exchanges, I will speak of "observations" instead of using similar terms like "interpretation" or "statements". This usage refers to the Luhmannian system theory. Here, "to observe" generally means that somebody/something (an actor/a discourse) makes distinctions towards someone/something else from an alien perspective (Luhmann, 1995). An observation is any communicative act that makes meaningful distinctions within the world.

Regarding the *New Right*'s discourse on elitism, this perspective fits very well, because speaking of an "elite" semantically is a very strong distinction *per se*. The statement implies that there is an opposition – the common people, dishonourable figures or, even worse, a false elite. In this way, speaking of an elite should be understood as the basic operation for imaging a differentiated society. It is almost impossible to talk about modern society without referring to elites (Staehli, 2011). As I will further show, critical observations of society and its elite draw on an opposition: On the one hand, they discredit current society's logic of the distribution of power, wealth and merit as inscrutable and, on the other hand, they proclaim an alternative code, regulating society completely comprehensibly. This tension within right-wing discourses operates through the semantics of an illegitimate elite consisting merely of loudmouths (*Maulhelden* in German, lit. mouth-heroes) and a legitimate elite consisting of men[3] of action (*Tatelite* in German).

My article's outline is in two stages. The following, second section establishes context by looking at broad discursive patterns used for observations of elites by reading a text important for global popular culture. In the third section, I take a closer look at recent and early 20th century radical-right discourses on elites. Accordingly, I will extract the basic semantics that appear in both fields that I describe as an expression of a political desire connecting right-wing special discourse with some aspects of culture's general discourse. The methodological starting point in all this is that concrete actors are deliberately ignored – even where I refer to "authors". Instead, the focus in a controlled reductionism is mainly on the linguistic-symbolic economy regulating this field. Finally, the ultimately structuralist basic intention of the following argumentation is to show the agency or, more precisely, the irreducibility (*Eigenmächtigkeit*) of language in the processes of world-making.

Popular culture observes elites

Thanks to the interdiscursive mechanisms of culture, we do not have to read annoying or, at least, boring manifestos of populist parties to find evidence that the current wave builds on scepticism towards the elite. It is sufficient to watch the TV: The omnipresent superhero films show a world in which the political and administrative institutions and the pending functionary elite have failed – the superheroes feed on the audience's desire for a true, functioning elite. This is a similar story that has been told in the action film genre for some time and currently forms the core of the successful *The Fast & the Furious* franchise. In the face of global elite failure, the sole revenger, acting beyond state institutions, the hero-figure in the tradition of Bronson's *Death Wish* (1974), is no longer enough. Instead, a whole team of *avengers* has to form an alternative elite to remedy the grievances caused by the incompetent functional elite (Seeßlen, 2017; Purse, 2019).

However, based on all the possible pop-cultural accounts one could consult, I want to concentrate on the very TV show paradigmatically rendering popular discourses on elitism: *Game of Thrones* (GoT) (Dümling, 2017; Koch, 2017). Significantly, not only does GoT observe a decadent, aloof, corrupt elite betraying the people, it also shows an alternative model of a legitimate elite that could lead society rightfully. The decadent elite is formed by the sadomasochists, incestuous siblings and whorehouse owners gathered in King's Landing, the realm's capital, which can be seen as equivalent to corrupt Washington or Berlin. On the other hand, the series shows the members of the Stark family as the legitimate elite, whose expulsion from Castle Winterfell – their paradise – opens the narrative. In turn, this differentiation builds on a long motif history: Medieval literature already knew it by the antinomy of an *aristocracy of injustice* and an *aristocracy of virtue* (Habinek, 1992).

However, the main reason why the King's Landing elite is diegetically observed as illegitimate is not their moral wrongness in the first place but the ambiguity of their real motives and goals. Whereas the Stark family is undoubtfully righteous, the diegetic actors and the *viewers* must always view the King's Landing personnel with doubt: The whorehouse owner may save raped women, the sly dwarf may be a great wise man and the evil queen might not be evil but just suffering from severe family issues. On the contrary, the Starks are simply good! In the terminology of ethnomethodology, the illegitimate elite stands for a crisis of indexicality; in this world it is impossible to attach a particular phenomenon to a distinct cultural system. Nothing is necessary and everything is possible. In other words, the illegitimate elite marks contingency (Garfinkel and Sacks, 1970).

That is the core of every observation of elites: The figure of the elite helps social actors to cope with contingency. By this, I refer to large-scale sociological interpretations: As Luhmann and other sociological theorists argue, modernity generally conveys the experience that we cannot fully figure out an institution's principles, the experience that human behaviour follows different social roles and, finally, the understanding that there is no single observation deck making holistic sense out of all social fields (Luhmann, 1994; Eisenstadt, 1998). In the words of Luc Boltanski (2014), modernity means to suspect that everybody and everything is an iceberg of which one can only see its peak. Observations of elites are a particular and crucial mode of dealing with this suspicion. Accordingly, illegitimate elites are often highly contingent, even liquid figures, in popular narratives such as GoT, whereas legitimate elites are predictable, even solid. I would like to take this as my first hypothesis: We can understand the normative judgement both in narratives and political debates – the elite is bad, evil and unrighteous – as the linguistic coding of an experience of contingency because it enriches explanations of social dynamics with necessity. Through the figure of the elite, observers can explain social dynamics well and, above all, convincingly, by being able to name clear motives, purposes and interests that drive these dynamics. Finally, the history of these semantics supports this interpretation. Dante, in his *Convivio*, separated righteous nobility from unrighteous by focusing on the point of predictability, respectively, contingency: Those noblemen who are driven by passion are unrighteous, so that no one can predict how they will reign – least of all, the noblemen themselves (Jorde, 1995).

Let me illustrate this with the differentiation regarding the elites in *GoT*. The show contrasts the righteous Stark family with the unrighteous King's Landing personnel in an almost structuralist way, at least in the early seasons. Lord Stark rules in his small lordship in keeping with the old law of his forefathers – he summarises his house's ideology as "Our way is the old way" – whereas the ruling parvenus in the capital only follow their passions (Dümling, 2019). He justifies his decisions transparently to his sons and followers, whereas the rulers in the capital – Joffrey, Tywin, Cersei – deny every sociality of decision-making by virtually incubating decisions on their own. The first scene with Lord Stark shows him judging a delinquent by the sword; the first scene of a King's Landing ruler shows him talking drunken nonsense. Finally, Lord Stark acts in an authentic and even holistic manner: He always acts in accordance with his morality and sense of justice, whether it is as father, husband, lord or councillor. On the contrary, the King's Landing characters behave like dice with covered sides – to quote Dante's account on unrighteous nobility (Jorde, 1995). The Starks come from the countryside. Remarkably, when we see Winterfell Castle for the first time, it is through an extremely long camera shot. Hence, we, the viewers, have an overview of literally everything. By contrast, the arrival of the Stark family at a city gate delivers the first picture of King's Landing, the capital. What lies behind the gate is the confusing, chaotic, cacophonic muddle of the city only a sympathetic social researcher would describe as a well-ordered network.

In sum, these representations draw on a set of binary semantics: visibility vs. inscrutability, social responsibility vs. social ruthlessness, authenticity vs. artificiality and ordered history vs. contingent future. I suggest this opposition renders almost every observation of the elite in both popular culture and politics. In my next point, I explore the latter based on right-wing discourses on the elite.

The right observes elites

In 2015, Swiss media reported on the notorious Pegida demonstrations in Dresden, Germany, with outrage. The outrage was not caused primarily by the demonstration's open racism but by a kind of cultural appropriation: Not only were some demonstrators waving Swiss national flags but some of them also wore T-shirts referring to William Tell, the Swiss national hero (Blick, 2018). On the front, one saw the so-called *Gesslerhut* (Gessler hat) with the motto *civil disobedience* and on the back, a writing quoting Friedrich Schiller's 1804 play *William Tell.* Thor Steinar, an important clothing brand within Germany's radical right-wing culture, makes these T-shirts. The choice of Tell as a T-shirt hero was very logical in terms of narrative history: He delivers the plot that occurs in nearly every right-wing observation of society and its elite. Following a very postmodern hypothesis, I suggest that the *New Right* would have to act completely differently today if such stories had not been related for hundreds of years – it is very easy to replace William with other folk-tale heroes, such as Robin Hood, Toni Bajada or William Wallace. This heroism, therefore, is not only open to radical right-wing appropriations in the present but also indicates a political desire that has been discursively virulent for a very long time – already at a time when *right* and *left* did not even exist.

In Schiller's play, citizens from Swiss cities form an alliance against the Hapsburg occupation, represented by Gessler, the reeve and tyrant (Dümling, 2020). While the allied citizens discuss the future laws, state institutions and even economy, the audience learns that these honourable men are indeed proper talkers but actually nothing more than paper tigers. William Tell then appears. He is a simple farmer from the countryside without any civic rights and lacking any interest in politics. Nevertheless, one day, Gessler's henchmen try to force him to bow to Gessler's hat, which is perched on a pole. Tell declines, is forced to shoot an apple from his son's head and, finally, kills Gessler, the representative of the alien elite. The point of the matter is that the oath-taking, well-read and distinguished citizens cannot found the Swiss Confederation by themselves; they need the archaic deed of a simple and common man, Tell, to release the power to build the nation state. Furthermore, the narrative needs Tell in order to avoid the impression that the new nation is just a product of the Swiss urban elite. Tell's deed is the symbolic ingredient eliminating any elitism within the new nation. Thus, it is important that one of the oath-taking citizens declares that only Tell's true and authentic deed, as Schiller puts it, enabled the founding act (Schiller, 2013, p.l.1122). What is important here is that Tell's deed is accompanied by the social idea of complete clarity: After his deed, everything will be in the right place (Koschorke, 2004). This holistic utopian space comes very close to the dreams of simplicity envisaged by Dresden's Pegida marchers, who are convinced that all problems will be solved when the traitors to the people (i.e. Angela Merkel and her cabinet) are in prison and the pure and direct will of the people shapes policy.

These dreams suspend modern society's ambiguity. Instead, they simulate a homogenous and holistic alternative working by the binary code of cultural, ethnic and national affiliation. While the saturated citizens just chat, Tell, who can no longer wait, takes the floor: Only those who follow him in his deed would be the true Swiss (Schiller, 2013, pp.1303–1306)! All the complicated institutions that the urban Swiss elite has concocted will be forgotten: It is only about following Tell. The deed – *die Tat* – is a very simple and clear decision that can be seen. The deed is *transparent*.

By this, Schiller enters a semantic path which was very crowded at his time and that revaluated the *deed* immensely. Immanuel Kant, for example, understands "the deed" as an action "as long as it is subject to the laws of liability", and Johann Gottfried Herder defines the Germans as *Tatvolk*, as a people of the deed, distinguished by their authentic (*echten*) deeds, referring directly to the time of their origin in mythical prehistoric times when the *Volksgeist* (spirit of the people or national character) was still completely visible.[4] Later, in 1915, the sociologist Werner Sombart enters – and of course, varies – this path with his book *Traders and Heroes*, in which the term of the German "heroes of the deed" (*Tathelden*) occurred very prominently for the first time.[5] Here, it distinguishes them from Anglo-Saxon traders. From a semantic perspective, Sombart traces the German deed back to the forests of primordial Germania: German deeds, like the leaves, branches and trunks of a German oak, lie open, have never been hidden and were organically planted in German soil (Sombart, 1915).[6] By contrast, he locates the birthplace of the Anglo-Saxon trading character in the desert. Sombart blames the Jews, the desert's people, for having invaded England and introduced the almightiness of trade. He thinks of Jews and Anglo-Saxons as good traders because they are good talkers and, like the grains of desert sand drifting in the wind, this talking drifts unbound and without any fixed basis. Hence, all this talking and trading cannot be anything else than lies. This semantic path is not forgotten within German right-wing popular culture, as one can see by the German neo-Nazi band *Nordglanz*, who called an album *Von Helden und Händlern* (*Of Heroes and Traders*).

However, the Pegida demonstration was not the only context where actors from the right remembered William Tell. The latter is also mentioned in Mario Müller's book *Kontrakultur* (Müller, 2017), a kind of encyclopaedia for the German-speaking Identitarian Movement.[7] Müller indexes Tell here next to other *New Right* historical heroes such as Joan of Arc or Leonidas, the Spartan King. The communicative goal of the book is to celebrate the elite of the deed – the *Tatelite* – and to condemn the contemporary, merely "talking" elite – the *Maulhelden*. For this purpose, Müller mainly exploits the writings of Weimar Republican intellectuals, which should be understood as a particularly meaningful reference: He tries to imitate the sound of Weimar Republic literature. Ernst Jünger – a *New Right*'s T-shirt hero in his own right – expresses this sound in his *Storm of Steel*, in a famous phrase expressing a physical tension that propels to immediate action: "I had the sensation of a sort of supreme awakeness – as if I had a little electric bell going off somewhere in my body" (Jünger, 2004, p.88). Jünger's lyrical "I" seems paradigmatic for the New Right's image of itself today as well: He is always ready to intervene in a declining world with a deed. By that, Jünger's narrator seems to be a precursor of Tyler Durden, the lead role of David Fincher's film *Fight Club*, another *New Right* favourite. Müller quotes Tyler Durden at length and programmatically in his book: "I see in Fight Club the strongest and smartest men who've ever lived. I see all this potential, and I see squandering. [...] They sleep and must wake up. [...]" (2017, p.89). Durden already has Jünger's electric bell in his body and wants to pass it on to the men still sleeping in the *Fight Club* – the new elite's meeting place. In addition, this narrative programme leads nearly every pop-cultural text within the community of the *New Right*: In the movie *300*, Leonidas awakens his people to fight against the treasonous domestic elite and the invading Persians. The French Identitarian band *Hotel Stella* sings in their most famous song – *Love, absinth, revolution!* – about a youth betrayed by liberal politicians, brainwashed by the mass media, nearly alienated by capitalism, but then this youth wakes up due to the Identarian Movement, which results in the chorus:

> Let us create a world, we want.
> Love, absinth, revolution!
> We will never subordinate
> Love, absinth, revolution![8]

However, let me take one further look at Weimar Germany. One of the most important gazettes of the so-called conservative revolutionaries at this time was the journal *Die Tat*, which frequently attacked the political elite of the Weimar Republic (Hanke and Hübinger, 1996). It was also in *Die Tat* that a term occurred very often that is currently celebrating a renaissance in the German right-wing populist context: The parliament was a *Quasselbude* – a mere talking-shop. In the observations of the parliament, we can reencounter the semantics we know from Sombart – the parliament as the embodiment of a western, even Jewish, elite conspiracy against the German people. I think that tropes of anti-elitism have generally a lot in common with those of anti-Semitism, even if this is not done intentionally and does not mean that this criticism would be *per se* anti-Semitic. It is more about the fact that these discourses use a similar register of language, images and arguments. Both discourses have in common that they semantically emphasise the aloof, the non-fixed, the intransparent. What anti-elitists – if one can call them so – and anti-Semites have in common is that they imagine a conspiracy of an exclusive class or even race directed against the holistically conceived people, the *Volk*.[9]

Remarkably, many of the attacks against the political elite refer to the Reichstag building. It is repeatedly lamented that no one from the outside can know what happens inside the massive building. The glass dome in the current newly constructed building responds to these allegations. This is a master trope for observations of elites in all sorts of contexts. Just note the book cover's iconography of the latest critical book on economic elites by German social researcher Michael Hartmann (not a *völkisch* or anti-Semitic author by any stretch of the imagination!), where the elite in its bubble is so aloof that we cannot even identify their faces (Hartmann, 2018).[10] The point here is not to claim a political tendency in the choice of such an image. Instead, I want to stress that such an image contains and stabilises a very robust collective knowledge in the symbolic economy of our culture.

When Björn Höcke, the leader of the far-right fraction within the AfD, demanded in an interview that a new *Tatelite* should replace Germany's contemporary "manic pseudo elite" (Höcke, 2015) – personified by Angela Merkel – he refers to this Weimarian discourse and especially to its *aesthetic of suddenness*.[11] Karl Heinz Bohrer, the literary theorist, describes Weimar's main cultural form by this term: Observing their political and cultural environment dissolving, betrayed by an unbound elite, without any solid institutions, the authors from the 1920s have celebrated the sudden *deed* as the only possibility to break the disintegration (Bohrer, 1981). That is the very template for the right wing observing the current elite and society. Höcke takes up this thought again in his latest book and differentiates it: "There is no life of the deed (*Leben der Tat*) without injustice, error and guilt. Those who shy away from it should remain in the realm of passivity" (Höcke and Hennig, 2018, p.156).[12]

Thus, Götz Kubitschek, Germany's most influential *New Right* intellectual today, highlights this by emphasising two models of the elite. He defines – and condemns – the current left-liberal elite according to their alleged (constructivist) idea that human beings have no solid, natural bases. This is why left-liberal politics have believed that human beings could be regulated arbitrarily. On the contrary, a future right-wing elite would act on the assumption of an "anthropological constancy", binding human beings in time and space. Kubitschek sees this constancy, for example, in the traditional family as the nucleus of a harmonious social order and in a clearly fixed gender identity (Kubitschek cited in Nassehi, 2015, pp.317–330). Again, fluidity vs. solidity, contingency vs. predictability, the utopia of controlling temporality and spatiality. In brief, the utopia of clarity.

This binary code is so powerful that it even regulates the media reports on Kubitschek himself. The *Neue Zürcher Sonntagszeitung* described him and his living on a farm in a detailed feature: "Kubitschek is a tall, strong man. [...] There might be few other people who can sit in a chair as straight as he, the former army officer, can." He seems to sit in the same way he wants society to have been created: clearly ordered, strictly contoured, always under tension. The report ends by showing Kubitschek as a man of action, as a *Tatheld* in a way of his own: "His daughter comes back with geese from the village's meadow. Ear-piercing noise. That's too much for Kubitschek, the goose-whisperer: 'Okay guys, enough is enough!' he says and locks up the poultry" (Mertens, 2018).

The evidence that the elite paradigm of the *New Right* plays out on a decidedly physical level is also made clear by Höcke in his memories of his early Nietzsche studies, where he describes his reading experience as a higher body experience:

> I was still more interested in activities, was life-affirming and full of zest for action [*Tatendrang*]. The correspondence with the world was mainly through my body. It was the youthful-pagan 'throwing-spears and honouring-the gods' [...]. The reading of a yellow [...] booklet [...] fit into this mood: It was Friedrich Nietzsche's *On the Genealogy of Morality* [...] [H]e [Nietzsche] gave me important spiritual impulses and expressed with the power of speech what I felt as a young man. My enthusiasm for sport, my youthful body experience. This was my expression of the positive will to live and the masculinity Nietzsche demanded.
>
> (Höcke and Hennig, 2018, p.74; author's translation)[13]

Whoever opens up the world through such plots must always be prepared to act. I assume that this provides the basic programme for the *New Right*'s subject cultures: They configure themself to the world in the *mode of necessity* (Dümling, 2018). Processes and actions in this world can only be observed as necessities but not as possibilities. For this reason, *New-Right* actors also currently believe themselves to be completely surrounded by the *thymos*. The Platonic concept of *thymos* in the *New Right's* reading means: When the wrath of the individual turns into anger in the face of the depravity of the world, then superhuman deeds can be done to heal that depravity (Hindrichs, 2019).[14] In the face of rage, all contingency will dwindle. Ultimately, this is what the new right-wing elite concept is all about: It provides for an elite that is designed to close spaces of possibility and translate all contingent processes into necessity, for instance, by operating wrath as programme code; because everything is clear for someone who is wrathful, and wrath is a powerful motive for a necessary act. This is diametrically opposed to an elite concept that current organisational sociology favours in relation to the higher political and administrative functional elite of experts: It is their social function to remind us that there are social questions that can be formulated but not answered (Maranta et al., 2003).

Conclusion

In my conclusion, I want to refine a hypothesis that underlined my considerations. I was trying to show that right-wing observations of the elite are actually a specific type of observations of sociality in general. This observation type declares that society can be clearly differentiated, metaphorically into a top and a bottom, and that this differentiation creates considerable tension, which is why it must be overcome.

The discomfort towards elites seems to be one regarding this basic condition of modern society. If this were true, the figure of the elite would be a placeholder for broad observations of society. Expressed in system theoretical terms, the observation of the elite is a mode enabling general observations of society at a minimum cost. The latter means that no major expositions or explanations are needed to make a diagnosis of society; those who use the term "elite" have usually already said enough to make their view of society clear (Lehmann, 2017).

Hence, I think that the main feature of the current conjuncture is not the scepticism towards elites but the omnipresence of general observations on society – for which elites are necessary.[15] If one looks at debates that are currently being held particularly hotly, it seems to me that they have one thing in common: The question is what kind of society do we actually live in. Whether it is debates about climate policy, digitalisation, identity politics or populism.[16] All these cases, and this seems important to me, are framed by questions of temporality: How does society orient itself in the present in order to work on certain futures? More precisely, they are reflections on how to control the temporality of society. However, contrary to a liberal idea that tends to interpret the openness of social processes positively and even promotes it, right-wing social imaginaries want to control the future and make it possible to plan it exactly in the present.

In order to be able to take on such macro perspectives linguistically and cognitively, it is necessary to make reductions without neglecting to grasp the whole. In the context I have looked at, this means that rhetorically, it is laborious to plausibilise that the abstractum called *society* is unrighteous, but it is very easy to denote the elite's unrighteousness – as the *New Right* does. Quoting the historian Hayden White, who is one of the linguistic turn's most influential figures, one can call this rhetorical operation the "politics of the synecdoche" – a part of a thing stands for the whole (White, 1973, p.32). Part of this economy of signs is also that the *New Right* does not stand still in idiosyncratic ideological homogeneous self-talks. Instead, it is in active narrative and semantic exchange with other discourses, such as pop culture. Therefore, I repeat, one should not make the mistake of separating a supposedly gloomy, normatively questionable right-wing discourse from a liberal, bright mainstream culture. Instead, one should understand the interplay that ultimately consists here and there of making sense of the world.

Notes

1 That this conjuncture has already existed for some time shows that a very early analysis of it was presented by Taggart in 2004.
2 The German discourse theorist Jürgen Link (1999) understands by the term "special discourse" an institutionally, disciplinarily and linguistically self-contained registers of language.
3 Regarding the gender aspect, see Dümling (in press).
4 Quotes come from the entry: *that, tat*, in Grimm and Grimm (1971).
5 On Sombart, as the most important sociological representative of an anti-Semitic, anti-liberal critique of capitalism in the age of Wilhelminism, see Schmoll (2003).
6 More exactly regarding the semantics of the forest, desert and grain of sand, see Sombart (1934, pp.17–20).
7 Müller's book is arranged alphabetically by keywords that introduce the Identitarian Movement's canonical texts and figures – historical heroes, authors, film characters and musicians. See Dümling (in press).
8 My translation of the French original: "*Créons le monde que nous voulons / Amour, Absinthe, Révolution! / Jamais nous ne nous soumettrons / Amour, Absinthe, Révolution!*", see Hotel Stella (2008). The manifest-concrete song text merely tells of a nocturnal carousal with absinthe. Müller again presents the decidedly political interpretation (2017, pp.16, 124).
9 Regarding this interpretation, see classically Horkheimer and Adorno (2009), Bauman (1989). For a historical analysis, see Wein (2014). How this discourse shapes the ideology of the AfD or the New Right, see Wildt (2017).
10 Loosely translated, the German title is: Those who are aloof. How the elites endanger democracy (my translation).
11 Remarkably, the video was uploaded by a user who calls himself "Wilhelm Tell" on YouTube.
12 In German: "*[Es gibt] kein Leben der Tat ohne Unrecht, Irrtum und Schuld. Wer davor zurückschreckt, sollte im Reich der Passivität verharren*" (Höcke and Hennig, 2018, pp.156).
13 "*Aber er gab mir wichtige geistige Impulse und brachte das sprachgewaltig zum Ausdruck, was ich als junger Mann empfand: Meine Begeisterung für den Sport, die jugendliche Körpererfahrung – das galt mir als Ausdruck des positiven Lebenswillens und der Männlichkeit, die Nietzsche forderte*" (Höcke and Hennig, 2018, pp.74; author's translation).
14 See, for example, Marc Jongen (2017), AfD intellectual and Member of the Bundestag.
15 Regarding the assumption that contemporary society is characterised by the fact that it permanently observes itself, see Nassehi (2015).
16 See the summary in Nassehi (2019).

Bibliography

Assmann, A. and Assmann, J., 1994. Das Gestern im Heute: Medien und soziales Gedächtnis. In: K. Merten, S.J. Schmidt and S. Weischenberg, eds. *Die Wirklichkeit der Medien: Eine Einführung in die Kommunikationswissenschaft*. Wiesbaden: VS Verlag für Sozialwissenschaften, pp.114–140.

Bauman, Z., 1989. *Modernity and the Holocaust*. Cambridge: Polity Press.

Blick, 2018. So sieht Pegida die Schweiz: Älpler und Tell vs. Schleier und Schächten. *Blick* [online] 9 September. https://www.blick.ch/news/schweiz/so-sieht-pegida-die-schweiz-aelpler-und-tell-vs-schleier-und-schaechten-id3417368.html [Accessed 20 February 2021].

Bohrer, K-H., 1981. *Plötzlichkeit: Zum Augenblick des ästhetischen Scheins*. Frankfurt am Main: Suhrkamp.

Boltanski, L., 2014. *Mysteries and Conspiracies: Detective Stories, Spy Novels and the Making of Modern Societies*. New York: Wiley.

Collini, S., Skinner, Q., Hollinger, D.A., Pocock, J.G.A. and Junter, M., 1988. What is intellectual history…?. In: J. Gardiner, ed. *What Is History Today?* London: Macmillan Education, pp.105–119.

Derrida, J., 1978. *Writing and Difference*. London: Routledge & Paul.

Dümling, S., 2017. "Game of Thrones" als politische Krisenerzählung: Wo sich Frauke Petry und Daenerys Targaryen treffen. *Spiegel Online* [online] 15 July. https://www.spiegel.de/kultur/gesellschaft/game-of-thrones-als-politische-krisenerzaehlung-essay-a-1157181.html [Accessed 27 February 2021].

Dümling, S., 2018. AfD (um 1500). *Merkur. Zeitschrift für Europäisches Denken*, 11(834), pp.72–77.

Dümling, S., 2019. Ordnung, Kontingenz, Krise – Zur Narratologie des Schlosses. *Fabula*, 60, pp.38–49.

Dümling, S., 2020. Der *Tell*, das Dieselverbot und das *Blackwashing* Europas. Überlegungen zu einer Grammatik der Verschwörungsbeobachtung. In: B. Frizzoni, ed. *Verschwörungserzählungen*. Würzburg: Königshausen und Neumann, pp.89–103.

Dümling, S., 2021. Das Geschlecht der Geschichte – Historie als antifeministische Ressource der Neuen Rechten. *GENDER. Zeitschrift für Geschlecht, Kultur und Gesellschaft* (Special Issue 6), pp. 59–75.

Eco, U., 1979. *The Role of the Reader: Explorations in the Semiotics of Texts*. Bloomington: Indiana University Press.

Eisenstadt, S.N., 1998. *Fundamentalism, Sectarianism and Revolutions: The Jacobin Dimension of Modernity*. Cambridge: Cambridge University Press.

Foucault, M., 1982. *The Archaeology of Knowledge*. New York: Pantheon Books.

Garfinkel, H. and Sacks, H., 1970. On formal structures of practical action. In: J.C. McKinney and E.A. Tiryakian, eds. *Theoretical Sociology: Perspectives and Developments*. New York: Appleton-Century-Crofts, pp.338–366.

Griffin, R., 2000. Between metapolitics and apoliteia: the Nouvelle Droite's strategy for conserving the fascist vision in the "interregnum". *Modern & Contemporary France*, 8(1), pp.35–53.

Grimm, J. and Grimm, W., 1971. Deutsches Wörterbuch von Jacob und Wilhelm Grimm. 16 Vol. Leipzig 1854–1961. Leipzig 1971. *Wörterbuchnetz* [online]. https://www.woerterbuchnetz.de/DWB [Accessed 20 February 2021].

Habinek, T.N., 1992. An aristocracy of virtue: Seneca on the beginnings of wisdom. In: F.M. Dunn and T. Cole, eds. *Beginnings in Classical Literature*. Cambridge: Cambridge University Press, pp.187–204.

Hanke, E. and Hübinger, G., 1996. Von der "Tat"-Gemeinde zum "Tat"-Kreis. Die Entwicklung einer Kulturzeitschrift. In: G. Hübinger, ed. *Versammlungsort moderner Geister: Aufbruch ins Jahrhundert der Extreme*. München: Diederichs, pp.299–334.

Hartmann, M., 2018. *Die Abgehobenen: Wie die Eliten die Demokratie gefährden*. Frankfurt am Main: Campus Verlag.

Hindrichs, G., 2019. Thymos. *Merkur. Zeitschrift für Europäisches Denken*, 6(841), pp.16–31.

Höcke, B., 2015. Asyl. Eine politische Bestandsaufnahme – Höcke beim Institut für Staatspolitik. *YouTube* [online] 12 December. https://www.youtube.com/watch?v=ezTw3ORSqlQ [Accessed 20 February 2021].

Höcke, B. and Hennig, S., 2018. *Nie zweimal in denselben Fluß: Björn Höcke im Gespräch mit Sebastian Hennig*. Lüdinghausen: Manuscriptum.
Horkheimer, M. and Adorno, T.W., 2009. *Dialectic of Enlightenment: Philosophical Fragments*. Stanford, CA: Stanford University Press.
Hotel Stella, 2008. Lyrics: Absinthe. *Skyrock* [online] 14 April. https://feverte.skyrock.com/1612604802-Hotel-Stella-Absinthe.html [Accessed 21 February 2021]
Jongen, M., 2017. Migration und Thymostraining. *Marc Jongen* [online] 24 February. https://marcjongen.de/migration-und-thymostraining-dr-marc-jongen-beim-ifs-2/ [Accessed 21 February 2021].
Jorde, T., 1995. *Cristoforo Landinos "De vera nobilitated": Ein Beitrag zur Nobilitas-Debatte im Quattrocento*. Beiträge zur Altertumskunde. Stuttgart, Leipzig: Teubner.
Jünger, E., 2004. *Storm of Steel*. New York: Penguin Books.
Koch, L., 2017. 'Power resides where men believe it resides.' Die brüchige Welt von 'Game of Thrones'. In: A-K. Federow, K. Malcher and M. Münkler, eds. *Brüchige Helden – brüchiges Erzählen: Mittelhochdeutsche Heldenepik aus narratologischer Sicht*. Berlin, Boston: De Gruyter, pp.199–216
Koschorke, A., 2004. Brüderbund und Bann. Das Drama der politischen Inklusion in Schillers Tell. *Goethezeitportal* [online]. http://www.goethezeitportal.de/db/wiss/schiller/tell_koschorke.pdf [Accessed 20 February 2021].
Lehmann, M., 2017. Wo ist 'unten'? In: M. Lehmann and M. Tyrell, eds. *Komplexe Freiheit. Komplexität und Kontingenz*. Wiesbaden: Springer VS, pp. 167–184.
Link, J., 1999. As-Sociation und Interdiskurs. *Kulturrevolution*, 38(39), pp.13–22.
Luhmann, N., 1994. The modernity of sciences. *New German Critique*, 61, pp.9–23.
Luhmann, N., 1995. The paradoxy of observing systems. *Cultural Critique*, 31, pp.37–55.
Maranta, A., Guggenheim, M., Gisler, P. and Pohl, C., 2003. The reality of experts and the imagined lay person. *Acta Sociologica*, 2(46), pp.150–165.
Mertens, S., 2018. Zu Besuch beim Vordenker der AfD: Götz Kubitschek ist der einflussreichste Intellektuelle der rechtspopulistischen Bewegung in Deutschland. Ein Besuch in der ostdeutschen Provinz. *Neue Zürcher Sonntagszeitung* [online] 13 October. https://nzzas.nzz.ch/international/afd-goetz-kubitschek-zu-besuch-beim-vordenker-ld.1428062?reduced=true [Accessed 21 February 2021].
Müller, M., 2017. *Kontrakultur*. Schnellroda: Verlag Antaios.
Nassehi, A., 2015. *Die letzte Stunde der Wahrheit: Warum rechts und links keine Alternativen mehr sind und Gesellschaft ganz anders beschrieben werden muss*. Hamburg: Murmann Verlag.
Nassehi, A., 2019. *Muster: Theorie der digitalen Gesellschaft*. München: C. H. Beck.
Purse, L., 2019. The new dominance: action-fantasy hybrids and the new superhero in 2000s action cinema. In: J. Kendrick, ed. *A Companion to the Action Film*. Hoboken, NJ: John Wiley & Sons Ltd., pp.55–73.
Rosellini, J.J., 2019. *German New Right: AfD, Pegida and the Re-imagining of National Identity*. London: Hurst & Company.
Schiller, F., 2013. *Wilhelm Tell: Schauspiel; Reclam XL – Text und Kontext* Ditzingen: Reclam.
Schmoll, F., 2003. Die Verteidigung organischer Ordnungen: Naturschutz und Antisemitismus zwischen Kaiserreich und Nationalsozialismus. In: J. Radkau and F. Uekötter, eds. *Naturschutz und Nationalsozialismus*. Frankfurt am Main: Campus Verlag, pp.169–182.
Schneider, I., 2010. Strategien zur Erlangung von Glaubwürdigkeit und Plausibilität in gegenwärtigen Sagen und Gerüchten. In: U. Marzolph, ed. *Strategien des*

populären Erzählens: Kongressakten der Bursfelder Tagung der Kommission Erzählforschung in der Deutschen Gesellschaft für Volkskunde. Studien zur Kulturanthropologie, Europäischen Ethnologie. Berlin: Lit-Verl, pp.127–149.

Seeßlen, G., 2017. *Trump! Populismus als Politik*. 2nd ed. Berlin: Bertz und Fischer.

Sombart, W., 1915. *Händler und Helden: Patriotische Besinnungen*. München: Duncker & Humblot.

Sombart, W., 1934. *Deutscher Sozialismus*. Berlin: Buchholz & Weisswange Verlagsbuchhandlung.

Staehli, U., 2011. Seducing the crowd: the leader in crowd psychology. *New German Critique*, 38(3), pp.63–77.

Taggart, P., 2004. Populism and representative politics in contemporary Europe. *Journal of Political Ideologies*, 9(3), pp.269–288.

Wein, S., 2014. *Antisemitismus im Reichstag. Judenfeindliche Sprache in Politik und Gesellschaft der Weimarer Republik*. Frankfurt am Main: Lang Verlag.

Weiß, V., 2017. *Die autoritäre Revolte: Die Neue Rechte und der Untergang des Abendlandes*. Stuttgart: Klett-Cotta.

White, H.V., 1973. *Metahistory: The Historical Imagination in Nineteenth-Century Europe*. Baltimore: John Hopkins University Press.

Wildt, M., 2017. *Volk, Volksgemeinschaft, AfD*. Hamburg: Hamburger Edition.

12 "Unpolitical in this time/truly one can no longer be so"

The raw anti-elitism of hooligans in Germany

Richard Gebhardt

Introduction: hooliganism – an "apolitical" subculture?

Hooligans have formed the militant arm of the (extreme) right-wing protest movement in the Federal Republic of Germany since October 2014. This finding may be irritating at first glance because, after all, an "apolitical" self-image has been propagated in the subculture of hooliganism for decades, not only in Germany – "*Fußball ist Fußball und Politik bleibt Politik* (Football is football and politics remains politics)" is an exemplary line from the song "Ha Ho He"[1] by the cult combo *Kategorie C*, which affirmatively adopts the police classification of the "Zentrale Informationsstelle Sporteinsätze (ZIS; Central Information Centre for Sport Operations)" responsible for dealing with violence in German professional football in its band name (ZIS, 2019). *Kategorie C* includes the "violence-seeking 'fans'" called "hooligans", who seek physical confrontation with opposing groups (so-called *firms*) in football stadiums as well as in the context of the so-called "third half" or "third place encounter". Violence, as well as martial masculinism, is a central feature of hooliganism that has emerged in the football environment. The first verse and the refrain of the *Kategorie C* song "*So sind wir* (That's how we are)" make this clear:

> Warst du schon einmal selber dabei?/In der dritten Halbzeit, der Fußballkeilerei?/Die Hände schwitzen/die Augen werden groß/wir haken uns ein und dann geht es los: So sind wir/und das ist unser Leben/so wird es immer weiter gehen/für immer Kategorie C./Wir sind Hools/und werden uns ewig jagen/gegenseitig auf die Schnauze schlagen/für immer Kategorie C/So sind wir [...].

> Have you ever been there yourself?/In the third half of the football fight?/ The hands sweat/the eyes get big/we hook up and then it starts: That's how we are/and that's our life/that's how it will always go on/forever *Kategorie C*./We are hooligans/and we'll chase each other forever/smash each other on the nose/forever *Kategorie C*/That's how we are [...].[2]

At the partly clandestinely organised concerts of *Kategorie C*, which play a significant role in the *Hool* scene, you see an uninhibited pack of men

DOI: 10.4324/9781003141150-16

dancing pogo, throwing punches, shoving. The everyday patina is washed down by showers of beer, the band's lyrics are bawled along. *Kategorie C* frontman Hannes Ostendorf does not sing his songs, he bawls, rattles and grunts. The guttural rolling "R" sounds from his mouth almost like a satire of a Hollywood Nazi. This subculture defines itself in self-perception and the perception of others through the search for the adrenaline rush, for the storms of steel that liberate the protagonists from the boredom of the administered world. Football has always been seen by the hooligans as a vehicle for the *kick*, which, however, is not generated by the game but by their own fighting activities. Bill Buford succinctly formulates the barely concealed leitmotif of the scene in his brilliant literary field study on hooligans, published under the title *Among the Thugs. Face to Face with English Football Violence*: "Violence is one of the most intensely lived experiences and, for those capable of giving themselves over to it, is one of the most intense pleasures" (Buford, 1991, p.207). The fact that hooligans, as part of a thoroughly heterogeneous subculture, also appear at demonstrations as political actors with a crude critique of the elite borne by extreme right-wing ideology – which is in fact an *enmity* against the elite – is, with a few exceptions, a new phenomenon, as will be shown in this article.

According to a handy definition, a hooligan is "a usually young man who engages in rowdy or violent behavior especially as part of a group or gang".[3] The term in its current usage originated in the late 19th century England as a "pejorative term for a mixture of male 'thugs' and juvenile 'petty criminals'" and was already "a media buzzword at the time which drew state intervention" (Wellgraf, 2017, p.221). Etymologically, the term "hooligan" is presumably derived from an "Irish family of the same name", "whose members are said to have been notorious bullies", according to the internet site of the German language dictionary Duden.[4] Other sources, however, point to the likely fictionality of this "Irish family" sung about in the music hall songs of the 1890s, featured in comic books[5] and not infrequently paraded with racist undertones as the stereotypical caricature of the crude, uncivilised Irish (Sonnad, 2018).

However, irrespective of the actual etymological origin, which is not to be investigated here, the mere reduction to "rowdyism" would by no means cover every dimension of the current hooliganism in the Federal Republic of Germany (and in Europe). From the very beginning, their habitus and appearance were aimed at distinguishing themselves from the *Kuttenfans* (fans dressed, *inter alia*, in cut-off denim jackets or waistcoats with the club logo sewn on them) who were frequently encountered in the 1970s and 1980s, especially in Germany, and who, according to a popular portrayal of fan cultures, "often looked like yelling fir trees with their scarf skirts" (Grüne, 2013, p.80). The often fashion-oriented hooligans (who sometimes also appear under the term *casuals*), on the other hand, rarely wear or carry club paraphernalia, and flags and banners are also much less common in this scene. Like any subculture, hooligans communicate through a particular style that aims for distinction and originality, which in this case is often enough itself. Hooliganism, which was initially mostly related to the football environment, has expanded its zones of influence since 2014 and also finds an important forum at Mixed Martial Arts festivals or in the context of other martial arts celebrated in front of an audience, the significance of which has been vividly illustrated by fan researcher Robert Claus in recent years (Claus, 2017, 2020). A professionalisation or commercial exploitation of the now public staging of violence of this subculture can generally be noted here, which has financial advantages for the organisers of the events. Exemplary here are sports and right-wing rock events, such as the "*Kampf der Nibelungen* (Battle of the Nibelungs)" in Ostritz in Saxony in 2018, in the context of which the current alliances between neo-Nazis and hooligans became abundantly clear. The event in Ostritz, which mobilised around 1,000 participants in 2018, was largely organised by members of the leadership of the extreme right-wing National Democratic Party of Germany (NPD), and *Kategorie C* was one of the headliners of the music programme (Alshater, 2018).

Hooligans as political actors – questions and methods

This chapter is devoted to contemporary hooliganism in the Federal Republic of Germany in the form of a journalistic-scientific essay. The aim is to show that German hooligans are by no means only active in public as a subculture in the context of football but have, for example, been *political subjects* in their own right for a long time – and not, as was diagnosed in the early international research on the subject conducted by leading representatives of cultural studies, essentially *objects* of measures of security policy (Hall et al., 1978). Hooligans act in the political field outside the official protocol, so to speak, and still largely outside the perception thresholds of academic studies of social and political movements. But it is precisely the "politically incorrect" self-dramatisation of hooligans that has an immensely political dimension, even if the subjects themselves may not always be aware of it. A research group led by the sociologist John Williams, which had been conducting research on the topic of hooliganism since the late 1970s, emphasised in their 1984 study "Hooligans Abroad" that the specific character of this subculture is misjudged by moralising rejection. Williams et al. therefore wrote clear-sightedly as early as the 1980s:

> The tendency to dismiss football hooligans as "morons" and "thugs", and to describe their behaviour as "meaningless" involves the labelling by outsiders of others whose behaviour they fail to understand because it is based on values that are different from their own.
>
> (Williams, Dunning, and Murphy, 1984, p.10)

According to this perspective, hooligans are not only passive delinquents of state regulations or mere spectators in the stadium ring but, in many respects, conscious actors who, contrary to their long-cultivated "apolitical" self-image, take offensive political positions both in the stadium and on the street. The demonstrations of the "Hooligans gegen Salafisten (Hooligans against Salafists)" (HogeSa) or the presence of the hooligans at the rallies of Pegida (Patriotic Europeans against the Islamisation of the Occident) and their offshoots are examples of this. As will be analysed here, they appear as actors at demonstrations against government policy or the "establishment" as well as in the fight on the terraces (Gebhardt, 2017). Our topic, then, is the idiosyncratic political practice of hooligans – a topic that the author of this text has worked on and analysed in recent years not only based on a review of the current literature on the subject but also within the framework of participant observation in stadiums and at demonstrations. For the author – a regular visitor to football stadiums since 1980, especially in the Federal Republic and later also in England, and a long-time observer of the public interventions of the extreme right – the sub-scene of hooliganism has been an integral part of more than just football culture for decades. What is new, however, is the phenomenon of a hooliganism that was previously only partially visible outside the stadium. The author has observed and evaluated explicitly political demonstrations and events of the scene in German cities, such as Cologne, Düsseldorf, Essen and Dortmund, or rural towns, such as Ostritz, the significance and impact of which independent filmmaker Fred Kowasch has captured in his long-term documentary "Inside Hogesa" (interpool.tv, 2018) as a bizarre phenomenon in the recent contemporary history of the Federal Republic of Germany. The author of this article was involved here as an accompanying observer and provider of sound bites and has, thus, devoted himself to the topic as part of intensive journalistic research on site. What is new, then, is the presence of the hooligans outside the peripheral battlefields and the public spaces temporarily charged with meaning, which have been decisively investigated in fan research by Richard Giulianotti and Gary Armstrong.

> Hooligans themselves have many other social identities, so they will try to define or redefine the meaning of specific places according to the presentations of self that they are seeking to deploy. Outside match-days, and in the company of non-hooligans, a hooligan "place" might have no significant meaning.
>
> (Giulianotti and Armstrong, 2002, pp. 233f)

This is how the researchers summed up the socio-spatial dimension of hooliganism at the time. However, it is not only match day that counts in the current practice of politicised hooliganism; here, the entire *polis* now has a special significance as a place of "resistance". The question to be analysed is in which form hooligans appear as political actors and how their public interventions oscillate between self-dramatisation as protest avant-garde and an anti-elitism that is as ambivalent as it is crude but decisive for the recent formation of ideology and the habitus of hooligans in Germany. The specifics of this anti-elitism are the focus of these deliberations.

Extreme right-wing criticism of the elite by the hooligans

Since 2014, the phenomenon of hooliganism in Germany has been negotiated outside the football stadiums, where groups have emerged that are waging a decidedly political battle for the streets. In this respect, the increasing interest in martial arts can also be interpreted as individual armament for political confrontations. Actors such as the HogeSa or the Football Lads Alliance (FLA) in England (Allen, 2019) represent decidedly Islamophobic positions and act as part of a right-wing political movement, whose effect reaches far beyond the classic but now diversely fragmented football fan scenes (Gebhardt, 2017). In many respects, hooligans today see themselves as an anti-elite agitating with radical populist topoi, which presents itself in football and in the "mixed scenes" of the political protest movement from the right characterised by rockers, bouncers, neo-Nazis and so-called "angry citizens". At the same time, hooligans articulate and radicalise large parts of the right-wing populist critique of the elite, which admittedly does not come in the form of factually presented arguments or elaborate postulates covering many political fields. Apparently, hooligans almost instinctively understand the anti-elitist character of (right-wing) populism, which promotes a reductionist world-view. Theorists of populism such as Michael Freeden (1998, p.750) have pointed out its ideologically "thin" character, i.e. only encompassing important individual points. Unlike the traditional ideological families of socialism or liberalism, populism does not have multiple interpretative patterns and intellectual facets. Populism has a central ideological core: The distinction between a "people", conceived as anti-pluralistic and homogeneous, and a frequently conspiracy-ideologically charged concept of "elite" is crucial. Cas Mudde also adopts the term "thin ideology" and gives the following precise definition in his discussion of the term:

> I define populism as *an ideology that considers society to be ultimately separated into two homogeneous and antagonistic groups, "the pure people" versus "the corrupt elite", and which argues that politics should be an expression of the* volonté générale *(general will) of the people.*
>
> (Mudde, 2004, p.543, emphasis in the original)

> Populism, so defined,", Mudde continues "has two opposites: elitism and pluralism. Elitism is populism's mirror-image: it shares its Manichean worldview, but wants politics to be an expression of the views of the moral elite [...].
>
> (Mudde, 2004, p.543)

Hooligans, whose Manichean world-view is being analysed here, often act like the collective performers of populism – but it is not individual, charismatic personalities that are central here, it is the effective power of the masses, whose anti-elitist agitation is supposed to achieve the media effect. The hooligans' accusation of competing fan groups or opposing political actors of forming an "elite" is, of course, not a term defined by social science, but a pejorative and projectively charged term. The elite criticism outlined here is articulated in the stadium in the form of mocking cries – long since not limited to hooligans but also articulated by Ultras – against the "fucking millionaires" or the "football mafia DFB". This refers to the *Deutscher Fußballbund*, the central institution of organised club and professional football in Germany, which is also responsible for the national team. It is precisely the regulation of their actions in the stadium or its surroundings that has led in rare cases to public manifestations. In 2000, for example, Berlin hooligans organised a demonstration under the slogan "*Reisefreiheit für alle* (Freedom to travel for all)", which was directed against the security regulations of the time and in which 350 people took part, according to the press (Spiegel Sport, 2000).

Polemic slogans such as "traitors to the people", "lying press" or "those who do not love Germany should leave Germany" have been heard on German streets in recent years during demonstrations or rallies with significant hooligan participation. As has already been mentioned, the scene, which is hardly known for theoretical elaboration, does not have a sophisticated concept of the elite that goes beyond resentment against "those up there". Its criticism of elites is often an expression of a "gut feeling" that is invoked and duplicated by (extreme) right-wing discourses. The question of elites or criticism of elites also touches on the question of obtaining, distributing and maintaining power. Of course, hooliganism is not only about power *criticism* but also about the formulation of one's own claim to power, for example, in the stadium, as well as the self-dramatisation as an anti-elite. Similar to the ultra-groups that are very present in German and European football, hooligans claim, in a certain way, to form a specific elite, but this has been denied them – according to their own self-perception – by the *establishment* represented by club management and politics, which is marked as the enemy, as well as by competing fan groups. However, this anti-elitism itself remains ambivalent. The concept of "anti-elitism" introduced in the context of these reflections can be specified as being less about forming a competing counter-elite on the horizontal level and more about the claim to be the true "voice of the people" in the stadium and on the street as the loudspeaker of the general will and, in accordance with this ideologically connoted imagination, to speak authentically and originally not only for the sport of football, its values and the fan community but also for the "people". In this way, hooligans are supposed to assume a spokesperson or subject position that formulates the correct – and not the establishment's distorted – view of society or organised football. Hooligans in Germany also count parts of the Ultra movement as belonging to the establishment because they (the Ultra members) actually claim to form a fan elite within fan culture and stage their support with advanced and internationally inspired ceremonies (see below).

Fight against the "establishment"

The opponents marked by the hooligans are first and foremost the associations of the DFB and the *Deutsche Fußball Liga*, competing fan groups and the commercially oriented "fashion fans" equipped with fan merchandising, including club officials. The stadium regulations are not the only manifestation of highly relevant restrictions for hooligans in the *Bundesliga* which make physical confrontations – the vitamin of the scene – difficult or even impossible. In addition, especially in the *Bundesliga*, there is close co-operation with the stadium's own security service and the police, who also use so-called officers with a knowledge of the scene for surveillance. The police, thus, play a prominent role as the enemy.

Since the 2006 World Cup, football associations in Germany have also been conducting campaigns against racism – from the hooligans' point of view: "politically correct". The de facto status of the national team as a "transnational team" was later accommodated by the DFB eleven officially trading as "Die Mannschaft (The Team)". It is debatable whether this renaming was the result of a marketing concept, i.e. an internationally comprehensible branding comparable to the "Les Bleus" reserved for the French national team, or a politically conscious concession to the intercultural character of the team The official renunciation of the emphasis on the "national" was an affront especially for the hooligan subculture, which often accompanies the national team to away games under the slogan "Divided in colours, united in cause". *Kategorie C* satirise the renaming in their own way in their song "Die Mannschaft (The Team)":

> Ein Hoch auf die Mannschaft/Ein Hoch auf dieses Land/Mir kommt gleich das Kotzen/Von diesem Affentanz/Sollen denn alle Traditionen am Boden liegen/Nationalmannschaft, Ruhe in Frieden (Hurray for the team/Hurray for this country/I'm about to puke/From this monkey dance/Should all traditions lie on the ground/National team, rest in peace).[6]

The new image of the DFB selection is not only considered "politically correct" here, but also oblivious to tradition and history.

In the context of their political actions outside the stadium, hooligans stage themselves as the militant wing of the protests, for example, against the pandemic requirements of the federal government. Right-wing-groups dominated by bike gang members, neo-Nazis and hools, such as the *Bruderschaft Deutschland*, have been a central part of the protests and not only in cities like Düsseldorf (Virchow and Häusler, 2020). At the demonstrations against migration policy – such as in Chemnitz in 2018 – hooligans presented themselves as the mouthpiece of popular anger, which was also supported by parts of the bourgeois public. The online edition of the liberal German weekly newspaper *Die Zeit* chose the apt headline "Neo-Nazis, hooligans and their bourgeois friends" for an article about the demonstrations in Chemnitz (Weiß, 2018). In 2018, an alliance became apparent that was already visible elsewhere. HogeSa activists and hooligans were present in isolated cases during the so-called storming of the Reichstag in Berlin in late summer 2020, as well as during the large demonstration in Leipzig in mid-November 2020, which was accompanied by clashes with the police (Virchow and Häusler, 2020). What is new against this background is that hooligans, for example, in cities of North

Rhine-Westphalia, the most populous German state, present themselves as a hierarchically led group in the style of "citizens" militias' (such as in Essen within the framework of the *Steeler Jungs*). Hooligans here propagate a self-image as a law-and-order elite, as a kind of "alternative security service" of the Federal Republic – although in reality it should be noted that the hooligans consciously turn against the established state order in their collaborations with neo-Nazis. This new role of hooligans – sometimes comparable to vigilantism – in the context of the political protest movement from the right has hardly been analysed academically so far and represents a research desideratum (Erhard, Leistner, and Mennicke, 2019, is one of the exceptions). However, there is no mistaking the relevance of this subculture in the recent protests. In the sense of a right-wing critique of the elite, the demonstrations against the pandemic policy are directed against a large number of those who are counted as part of the establishment: Against virologists as the stooges of the pandemic requirements, against politicians as their enforcers, against the media as their intermediaries and the security and public order services (public order office, police) as their vicarious agents. However, the protest has hardly found any significant articulated expression. The catchy slogan used to demonstrate in Düsseldorf at the beginning of November 2020, for example, was simply "*Pandemie/gab es nie!* (Pandemic/never existed!)" On the spot, as in a panopticon of political culture, alternative doctors held esoteric speeches in front of an audience consisting mainly of hooligans.

The politics of hooliganism: historical and current developments

In contrast to the long-held – and now partly revised – self-image, the links between hooliganism and the camp of the extreme right have been relatively close from the beginning. In England, the country of origin of hooliganism, where the different fan groups have sometimes fought spectacular battles inside and outside football stadiums since the end of the 1960s, members of the neo-fascist British National Party or the National Front have sometimes had an important influence on the supposedly "apolitical" milieus of militant fans (Buford, 1991, pp. 129ff.). Hooligans have represented the violent face of football, not only on the island, for decades. The catastrophe at the Heysel Stadium in Brussels in 1985 and the attack on the police officer Daniel Nivel by German hooligans during the World Cup in France in 1998 were the turning points. Thirty-nine people were killed in Brussels as a result of a mass panic that broke out after riots between supporters of Liverpool FC and Juventus Turin in the final of the European Champion Clubs' Cup in 1984/85. After Heysel, English football clubs were excluded from all European competitions for five years, and hooliganism became internationally known as the English disease.

Hooligans as a "dangerous crowd"

The theoretical tradition of cultural studies also devoted itself to this subculture early on, not in detail but, nevertheless, in important works written by

John Clarke or Stuart Hall *inter alia* (exemplified in Ingham, 1978). Stuart Hall and his co-authors in their analysis of law-and-order policy in England since the 1970s entitled *Policing the Crisis* often refer to the importance of police-media discourse about "lunatic hooliganism" (Hall et al., 1978, p.300) – hooligans have been seen as a symbol of the threat to public order that could only be maintained with repressive measures. The negative public image was used for self-dramatisation: "No One Likes Us, We Don't Care" was the slogan in the 1970s at the Millwall FC stadium in south-east London, which was (and still is) notorious for its hooligans (Gebhardt, 2015, p.19). Fan scenes in Germany have also increasingly become objects of repressive police measures since the beginning of the 1980s, and the security architecture in stadiums has been tightened accordingly. The hooligans disappeared from public perception in Germany after the 1998 World Cup but were still present in the stadium environment, at least in part, as the *elder statesmen* of the fan scene, drawing on the glory of old battles. In this context, football stadiums should be understood as political places. They are spaces of socialisation for young fans, similar to youth centres elsewhere; they serve as forums and places of contact. In the event of deviant behaviour (rioting, etc.) in the stadium arena, the police come into play alongside the local security service as the visible law enforcement agency of the regulating state. Football fans have never just been passive consumers but always stubborn actors who celebrated carnivalesque festivities and temporary states of emergency in the stadium. Like the fan groups, the ceremonies also form a part of the club identity, of which hooligans see themselves as a central part (even if they tend to perform their ceremonies in the "third half").

Stubborn fan cultures: the game in the stands and the fight for the terraces

It is precisely the negative relationship to the police that is of decisive importance. The police as an enemy, expressed in the abbreviation "ACAB" (standing for "All Cops Are Bastards"), testifies not only to anti-elitism but, above all, to a rabid anti-statism, which has traditionally been significantly prevalent in youth and subcultures. Considering that clubs are often a substitute for family or a transfigured community of destiny and that the peer group meets in the stadium, the vehemence with which the fights take place in football stadiums can be explained. This is where one's own public space is claimed as one's own and defended against the encroachment of state institutions. A special atmosphere is created in the stadium itself: A "we" group (home team) plays against a "they" group (visiting team); tradition, honour, performance, hierarchy and identification with one's own club are conjured up here. On-site fan ceremonies are held and a *game is celebrated in the stands* in addition to the *game on the pitch*: Week after week, *away supporters* commute through the republic, in the stadium a *mob* forms a *wall* or even carries out a *storming of the pitch*. Archaic-seeming rituals are also cultivated in the stadium, ranging from the public mockery of the opponent to the ritual of stealing flags, the latter evoking relics from tribal societies. Similar to the rites of local, traditional folk festivals, a premodern echo can still be perceived here: When, for example, an Ultra group has to disband after the violent theft of its flags or insignia. Although archaic symbolism is seemingly only cited here in fan-cultural practice, the temporary transformation of the game and self-dramatisation into violence illustrates the partial dissolution of civilisational norms. Bill Buford's thick description sums it up like this:

> I recall attending the Den at Millwall, the single toilet facility overflowing, and my feet slapping around in the urine that came pouring down the concrete steps of the terrace, the crush so great that I had to clinch my toes to keep my shoes from being pulled off, horrified by the prospect of my woollen socks soaking up this cascading pungent liquid still warm and steaming in the cold air. The conditions are appalling but essential: it is understood that anything more civilized would diffuse the experience.
>
> (Buford, 1991, pp. 167f)

Football in Europe or Latin America, probably more than any other sport, offers the experience of temporarily merging with a crowd, and all stadium-goers know the affects that are released in these moments – affects that can also turn violent. It is precisely in their violent confrontations that certain fan cultures resemble clans in the sense of classical social anthropology, forming a system of a community of meaning and protection. This phenomenon can be observed in several fan scenes and is by no means limited to the subculture under investigation here. However, "hooligan" is still mostly used without hesitation in the media debates, in which hardly any distinction is made between the specifics of an Ultra culture or of hooliganism, and without any knowledge of the concrete groups as a central synonym for football fans who are ready to use violence. In the media, "hooligan" is still not a precise term but an empty signifier that is applied to a variety of phenomena inside and outside the stadium. Stuart Hall et al. pointed out in their 1978 analysis of the media construction of hooliganism in England that the public image of the hooligan is not based on experience but is conveyed by the media and exaggerated within the framework of a "moral panic" (also see Piskurek, 2018).

According to the interpretation of hooliganism, the Ultras, who sometimes compete with each other, are often regarded as elitist in the sense of "aloof". It is true that this fan subculture is sometimes just as unfamiliar with violence as the self-elevated claim to be an elite in fan culture. However, the Ultra fan culture – which is one of the largest youth cultures in Germany – is far more diverse and sometimes almost cosmopolitan in its use of samba drums, flags and dances. Ultras have worked through the history of persecuted Jewish club personalities in clubs such as FC Nuremberg or FC Bayern Munich, sometimes with spectacular actions and meticulous research. Decidedly anti-fascist ultra-groups have made a name for themselves in several German cities, for example, "Aachen Ultras" (ACU99) at Alemannia

Aachen and "*Kohorte* (Cohort)" at MSV Duisburg (Ruf, 2013). The hooligans' attack on sections of the Ultras was aimed at enforcing right-wing hegemony in the battle for the terraces. Hooligans see themselves as the guardians of the tradition of "real" football, which has been sold out to TV stations by clubs and associations. The declaration of hostility against commercialised football unites them with the Ultras, whose telegenic performances have themselves become an elementary part of "modern" football.

Hooligans, on the other hand, act rather stoically as the old school of the fan scene. They invoke "old values", such as nation, masculinity, discipline or club loyalty; the opponent is belittled in an aggressive manner. This change of style in the stadium had tangible consequences in the truest sense of the word: Especially in North Rhine-Westphalian cities, such as Aachen, Duisburg, Dortmund or Düsseldorf, hooligans fought a partly politicised fan culture battle with anti-racist ultra-groups whose choreographies, samba drums and Latin American-inspired chants were considered a violation of classic support. The hooligans' attacks on the dissenters were aimed at restoring the old norm that hooligans are supposed to have dominance in the terraces. At the same time, these assaults served to assert hegemony in the battle for the terraces (Beitzel, 2017). This rigid interpretation of sporting competition, fixated on struggle, opposition and the will to win, makes hooliganism compatible with the ideologism of the extreme right.

Their interpretation of football as a combat sport is in this respect significant in a political sense, even if hooliganism is no longer a mass phenomenon in professional football. According to the annual report of the ZIS for the period 2018/19, "approx. 240 people (approx. 2.5 per cent) of the "perpetrators of violence in sport" recorded nationwide can be attributed to the right-wing motivated sector and approx. 130 people (approx. 1.4 per cent) to the left-wing motivated sector" (ZIS, 2019, p.12). However, since the ZIS figures are a so-called "bright-field analysis", which cannot indicate the number of unreported cases of violence, the actual degree of danger (e.g. through assaults in the unsupervised stadium environment or through verbal provocations) is not recorded nor are the political dynamics of hooliganism, for example, in the fight against the "Islamisation" of the Federal Republic.

HogeSa and the aftermath: Hools as protectors of public order

It became clear at the Cologne demonstration of the HogeSa on 26 October 2014, with around 4500 participants, that the politicised environment of the right-wing open subculture is much larger and that the spheres of football and politics cannot be separated. The "temporary fighting community" (Pilz, 2007) of the HogeSa is of great importance to the new protest movement from the right. October 2014 marked the first appearance of hooligans as determined and high-profile political actors in the Federal Republic of Germany. At that time, the demonstrators succeeded in overturning a police van on Breslauer Platz near Cologne's main railway station. For the scene, this was a spectacular symbolic image of the fallen state power, which the hooligans wanted to take as their own for a high-profile moment. The coverage that lasted for days reinforced the momentum of the demonstration many times over. The new political radius of German hooliganism is, thus, crucial. It was the media response to the first HogeSa rally in Cologne in October 2014 that caused the "evening walks" of Pegida in Dresden to grow. About a week before the Cologne riots, Pegida initially brought about 350 participants onto the streets on 20 October 2014 – a number that was to multiply rapidly after the media response to the first HogeSa march. It is not only in the Pegida environment that hooligans continue to perform duties as security guards or organisers (Sundermeyer, 2015). They, in turn, stage themselves as guardians of order in vigilante-like formations or in the context of street protests, fighting for the restoration of the state's monopoly on the use of force, which has allegedly been dissolved across the board. However, the recourse to the slogan "ACAB" alone shows that it is not actually the endangered state order and its police guardians that are being supported. Hooligans are following their own claim to power as guardians of a self-imposed order.

Street protests by extreme right-wing hooligans, who explicitly pursue political goals in the wake of HogeSa, have been recorded recently in the federal state of North Rhine-Westphalia (to which Cologne also belongs). Since then, this milieu has sometimes been able to organise demonstrations in the triple-digit range and – as has already been noted – has also been leading the staging of numerous protests against the federal government's pandemic measures since 2020. There is a twofold strategy behind the scandalisation of core issues such as internal security and sexualised violence: On the one hand, the agitation is directed against (Muslim) migrants and, on the other hand, against what the right-wing calls the "Merkel system" and its "establishment". The slogans about the "great exchange", i.e. the allegedly deliberately controlled displacement of the "old German" population by immigrants, encountered at the demonstrations are also found almost congruently in parts of the *Alternative für Deutschland* (Alternative for Germany: AfD) party, whose leading figure and honorary chairman Alexander Gauland formulated a telling motif not dissimilar to hooliganism after entering the German Bundestag in 2017. "We will hunt them down", Gauland said, referring to the so-called old parties (Spiegel Politik, 2017). Both formations – hooligans and the AfD – are linked by anti-elitism. Gauland's rabble-rousing, suggesting folksiness, contrasts in a peculiar way with his educated bourgeois biography and his CV as a part of the West German state class for decades. However, it points to a remarkable analogy: Hooligans and populists, despite their self-presentation as a protective force, in fact, see themselves as a disruptive factor of the prevailing public order and claim to voice the truths that are supposedly suppressed because they are inconvenient for the "establishment" in politics and the press. In view of this rhetoric, it is no coincidence that historian Markus Linden (2021) cites AfD politician Gauland as a representative of a "revolutionary conservatism" in Germany. Linden characterises this de facto radical right-wing current as follows:

> Anti-liberalism and supposed anti-totalitarianism, nationalism and religious fundamentalism are stirred into ideological mixtures in revolutionary conservatism, each in different components, which propagate a hooligan-like [sic!, RG] acting out of "resistance" in defiance of democratic norms. Based on an alleged "will of the people", it is permissible to bring the system into line from above and call for a reconquista from below. In case of doubt, it is self-defence against liberal violence.
>
> (Linden, 2021, p.70)

Vigilante groups and "mixed scenes"

Hooligans, according to this logic, see themselves as enforcers of the imagined "will of the people", their "resistance" is articulated in public places. Hooligan formations have, thus, been appearing outside the stadium for a few years not only in martial arts but also increasingly as so-called "vigilantes". The *Steeler Jungs* from Essen-Steele, for example, the Cologne *Begleitschutz* or the *Internationale Kölsche Mitte* have become known nationwide. The security authorities speak of "mixed scenes", in which right-wing extremists, hooligans and rockers, Reich citizens and conspiracy ideologues or "angry citizens" gather. Demonstrations are then held together with classic neo-Nazis. In September 2019, for example, a meeting of around 27 "citizens' militias" with over 700 participants was held in Mönchengladbach in North Rhine-Westphalia. The call was for "resistance" against "the state criminals" and the "unjust regime" of the Federal Republic (Burger, 2020). The *Bruderschaft Deutschland*, which was founded in 2016 in Düsseldorf, the capital of North Rhine-Westphalia, and which was the target of a state investigation at the end of March 2020, is often present at such rallies (Dammers and Mannheim, 2020). This "brotherhood", mainly characterised by extreme right-wing hooligans, also participated in the marches of, for example, the *Steeler Jungs* in September 2019. One of the main organisers of these protests was HogeSa co-founder Dominik Roeseler, who is now a representative of the group *Mönchengladbach steht auf* (Mönchengladbach rises up). His example also illustrates the instrumental relationship of the extreme right associated with the hooligan scene to local politics: In 2014, Roeseler was elected to the Mönchengladbach city council as a candidate for the extreme right minor party Pro NRW. However, he apparently hardly ever showed up for the meetings (Klarmann, 2019). His favourite place is not the local parliament but the street. The most diverse factions of the right are networked here. Functionaries of the far-right parties NPD and Die Rechte were also present at the protests in Mönchengladbach in September 2019. This shows that a cross-scene co-operation, which wants to occupy the streets flexibly and quickly with its demo tactics after a short-term mobilisation via the internet, has been established for a long time not only in East Germany. The paradox of a law-and-order group that is on record as being, in principle, hostile to the state and, thus, only supposedly law-and-order seems to require little clarification within its own ranks.

A "raw bourgeoisie" of hooligans?

It can often be seen at demonstrations of the scene how the hooligans flirt with the political-media attribution as dangerous antipodes to – allegedly effeminate and decadent – bourgeois society. The "Nobody likes us ... " is just as valid here as the "Ahu!" from the film *300*, which freely interprets the Persian Wars, as a battle cry. Although the hooligans can hardly be said to have a "bourgeois" habitus defined by values such as composure or civil conflict resolution, it is, nevertheless, noticeable that disputes in their ranks are fought out according to a competitive logic of "us vs. them". This prompts an examination of the hidden relationships between hooliganism and parts of the bourgeoisie and its ideology. In 2012 – a few years before the founding of the AfD and the public appearance of HogeSa or Pegida – the sociologist Wilhelm Heitmeyer spoke of a "decultivation of the bourgeoisie" as part of his long-term study of the "German conditions" and coined the term "*rohe Bürgerlichkeit* (raw bourgeoisie)" in the course of this discussion (Heitmeyer, 2012, p.35). This habitus, Heitmeyer diagnosed, "shows itself not least in the devaluation of weak groups. Civilised, tolerant, differentiated attitudes once found in higher income groups seem to be turning into uncivilised, intolerant – even brutalised ones" (Heitmeyer, 2012, p.35). Although the subculture of hooligans can hardly be categorised as "higher income groups" or by civil manners, ideological alliances become visible that are also expressed in the politics of the hooligans – the protection of one's own social "privileges" derived from national affiliation against the refugees demonised as invaders is just as much a part of this as the "cultural defensive attitude" that Heitmeyer diagnoses especially with regard to Islamophobia. Despite all the differences between a conventional bourgeois habitus and the quasi-plebeian style of the hooligans, the strength fetish or the philosophy of the social Darwinian interpretation of *survival of the fittest* are propagated precisely in this subculture *qua* cult of violence. Despite their branded clothing, the hooligans have always been boisterous, raw and ambivalent: With their penchant for physical confrontations, they are, so to speak, the "proletarian" counter-image to the bourgeois order, whose creation or maintenance they, on the other hand, demand. Although their media (self-)stylisation makes them look like modern plebeians, hooliganism has a decidedly bourgeois dimension in that sense. The protests against the alleged "Islamisation" of Western cities are about defending "our women" – in this case, they are obviously asserting a title of ownership to their "own" women that follows a genuinely bourgeois logic as well. In this sense, the hooligans are, symbolically speaking, the shadow of the decultivated bourgeoisie. The fact that, according to this attitude, women are degraded to passive objects to be defended by German men from foreign access does not seem to be a problem in the self-perception of the scene, which presents itself as a gathering of women's rights activists without any hint of self-irony. What is cultivated is an atavistic world-view. Be it with the HogeSa or in the case of the FLA, men present themselves here as fighters and warriors. The hooligans apparently identify their own revenants in their central enemy image, the violent Salafists, who define themselves through similar masculinity rituals and move illegitimately into enemy territory, i.e. the large western cities. Therefore, the hooligans do not only pursue power politics in the stadium but also in the German city centres, where they see their claims endangered by Salafists who are ready to use violence. As defenders of the (German or English) territory, they wage a struggle for possession and, according to their self-image, feel themselves to be the true guardians of public order, i.e. as a militant substitute elite against a weak state, which, for example, was indeed unable to stop the sexualised violence at Cologne Central Station on New Year's Eve 2015/16 (NRW, 2016).

There are intersections with politicians such as Alexander Gauland, who are arch-bourgeois by origin, in the hooligans' declaration of hostility against the government. However, despite these convergences between the subculture of the hooligans and the "brutalised" bourgeoisie as stated by Heitmeyer (2012), the differences should not be underestimated. A possessive individualism is specific to the bourgeoisie, which ignores its social preconditions (e.g. property rights, contractual power, wage labour, appropriation of surplus value). In the case of the hooligans, the claim is derived more from a fetishised communal cult that puts the collective before the individual with scene slogans such as "The family holds together". Raw bourgeoisie and hooliganism, thus, move towards each other from a specific starting point: Parts of the bourgeoisie discover the supposed Islamisation of the Occident as the central threat. This obviously gives them the licence to behave in a non-bourgeois way against the enemies of the bourgeoisie and, therefore, to enter into alliances with the non-bourgeois – the Pegida in Dresden or the extreme right-wing demonstrations in Chemnitz are examples of this, as the following section will show yet again.

Hooligans as ventriloquists of public anger

The discursive shifts in the Federal Republic also affect the hooligan subculture. In the context of the demonstrations that followed a killing allegedly committed by refugees in Chemnitz at the end of August 2018, *Kategorie C* wrote a song entitled "*Chemnitz ist überall* (Chemnitz is everywhere)". The lyrics, quoted here only in excerpts, document the world-view outlined above:

> *Es ist diese Politik, die uns zwingt/dass KC dieses Lied nun singt/Unpolitisch kann in dieser Zeit/Wahrlich niemand mehr sein* (It's these politics that forces us/That KC now sings this song/No one can be apolitical in this time/Truly no one can be).

And furthermore:

> *Weiße Frauen sind Freiwild geworden/Auf der Straße regieren die fremden Horden/Kein Schutz, keine Hilfe durch 'Vater Staat'/Protestiert mit uns gegen diesen Verrat!* (White women have become fair game/The foreign

> hordes rule the streets/No protection, no help from 'Father State'/Join us in protesting against this betrayal!).

In the song, the counting rhymes continue to call for "resistance" against the "invaders", the "media" are accused of manipulation and lies, as are leading sections of the state class.[7]

Lines like these are not to be underestimated, as they propagate a weighty change in one's own self-portrayal: Hooligans should be politically active especially now and move away from their old passive attitude. It comes as little surprise that *Kategorie C* appeared in front of the neo-Nazis in Ostritz in 2018. Long existing ideological complicity which had previously been intended to be concealed became fully apparent here. It is also striking that themes are being negotiated here in a very explicit way, as they also circulate in the AfD: The deliberately stoked fear of an "invasion" by refugees, who are dubbed "knife migrants", is part of the agenda-setting in the party and the scene. "When the state can no longer protect citizens, people take to the streets and protect themselves. Today it is the citizen's duty to stop the deadly 'knife migration'!", tweeted Markus Frohnmaier, an AfD member of parliament (Gilbert, 2019). It is, therefore, hardly surprising that the AfD also demonstrated together with hooligans and other factions of the extreme right in Chemnitz in 2018. The term "*Wutbürger* (angry citizen)", which was formulated for the political culture of the Federal Republic of Germany, thus, finds its lurid intensification, its suppressed face, in the hooligan. Hooligans, however, see themselves less as angry citizens and more as citizens of defence. They are German fighters and the protest elite of those population groups for whom the "land of the diverse" – as former Federal President Joachim Gauck called the Federal Republic, which has long existed as an immigration society, in 2015 – is literally "too colourful". Hooligans do not see themselves as outsiders in this polarised situation but as true resisters against the lamented "rule of injustice", which, regarding migration policy, significantly enough, was also lamented in 2016 by the current German Minister of the Interior Horst Seehofer – who at the time of his statement was Prime Minister of the German state of Bavaria. The hooligans, endowed with additional legitimacy by such quotations from politicians in charge of the state, want to offer "our women" the necessary "protection from those seeking protection/asylum", to quote the relevant jargon at demonstrations. This jargon can now be found in the Bundestag as well as in the tabloid media. And since hooligans focus mainly on issues with a trigger effect (Islamism, crime committed by foreigners), they have a certain mobilising power and media impact.

That the hooligans' resentment is directed against the "elites", the "establishment" or the prevailing order is ultimately not surprising. The punchline is in what form the hooligans present their hostility to the elites – namely, as a radical and raw version of the *vox populi*. In this respect, ventriloquist of public anger here does not necessarily mean the claim to form a counter-elite aiming at immediate power replacement at the top of the state. Hooligans do not undergo cadre training or cram in their theoretical work. Instead, hooligans direct their performance as protest actors against "the elites", claiming above all – and this is the crucial point – a pure, uncorrupted representation of the people. Against this background, the explicit politicisation of the hooligan also becomes understandable: The raw, non-sublimated, uncorrupted representation that the hooligans articulate in the stadium and on the street is not only directed vertically against "those up there" and their uniformed security apparatuses but also "horizontally" against Salafists who want Islamist-based social control. In concrete terms, this means that the mobilisation directed towards the "top" and against "foreigners" and aimed at exclusion draws its energy from the fact that the "people" are played off against the "elite" in hooliganism with almost instinctive certainty. And while the formally responsible elite fails or lies according to this reading of the circumstances, the hooligans pursue a policy of self-empowerment – with the goal of a true rule of the people, of which they imagine themselves to be the forerunners and spokespersons.

Conclusion and outlook

There is every reason to abandon the fairy tale of the apolitical hooligan. In the fight for the terraces from 2011 onwards, hooligans mainly took offence at the often colourful, cosmopolitan Ultra culture, which in many places expressed itself explicitly in an anti-racist way. In Germany, however, the sphere of action goes far beyond the stadium. Under the slogan "Merkel must go", which was directed against the German chancellor, hooligans were often part of a protest movement against the state authorities of the Federal Republic, seeing themselves as the people or anti-elite. The importance of this self-image and the associated claim to represent the interests of the people should not be underestimated, especially in the east of the Federal Republic, where many of the protagonists of the protests against the "Merkel regime" have biographical experience of a change of system after 1989. And the overthrow of the system is also the goal here.

In this respect, hooliganism has shown great dynamism since 2014, and not only in Germany. A new type of hooliganism has emerged internationally, not only in Russia and Eastern Europe, which became visible during the riots at the Euro 2016: young, rather ascetically oriented hooligans of a modern type, who also find a new sphere of action in the context of Mixed Martial Arts events (Claus, 2017). Their disciplinary and fitness postulates are often to be seen as part of an armament for civil war. Currently, not all but important parts of the hooligan subculture act as a militant wing of political protests of the extreme right, which is thematically volatile and sometimes ideologically diffuse, but has its most constant feature in Germany in its open contempt for the state. The reference to elite critique remains ambivalent: An elite and state contempt is expressed in the polemic against the state class, which is flanked by a self-exaltation, less as another elite but rather as the "true" and leading "voice of the people". In view of the anticipated crises, not only in the wake of the Corona pandemic, an important militant political actor for the future is also visible here.

Translated from German by Philip Saunders

Notes

1 Listen to the original on the homepage of *Kategorie C*: http://hungrige-woelfe.eu/?audio=kneipentour.
2 Listen to the original on the homepage of *Kategorie C*: http://hungrige-woelfe.eu/?audio=live-in-deutschland.
3 https://www.merriam-webster.com/dictionary/hooligan.
4 https://www.duden.de/rechtschreibung/Hooligan.
5 https://www.etymonline.com/word/hooligan.
6 Listen to the original on the homepage of *Kategorie C*: http://hungrige-woelfe.eu/?audio=pure-emotion.
7 The original can be found on the homepage of *Kategorie C*: http://hungrige-woelfe.eu/?audio=kategorie-c-chemnitz-ist-ueberall.

Bibliography

Allen, C., 2019. The Football Lads Alliance and Democratic Football Lad's Alliance: an insight into the dynamism and diversification of Britain's counter-jihad movement. *Social Movement Studies*, 18(5), pp. 639–646, DOI: 10.1080/14742837.2019.1590694

Alshater, S., 2018. Gewaltaffine Neonazis trafen sich am Wochenende beim "Kampf der Nibelungen" in Ostritz. *Belltower News* [online] 15 October. https://www.belltower.news/gewaltaffine-neonazis-trafen-sich-am-wochenende-beim-kampf-der-nibelungen-in-ostritz-76933/ [Accessed 22 February 2021].

Beitzel, P., 2017. Ausweitung der Kampfzone vom Stadion auf die Straße und zurück. Deutsche Hooligans im (Fan-)Kulturkampf. In: R. Gebhardt, ed. *Fäuste, Fahnen, Fankulturen. Die Rückkehr der Hooligans auf der Straße und im Stadion*. Cologne: Papy Rossa Verlag, pp. 13–47.

Buford, B., 1991/2001. *Among the Thugs. Face to Face with English Football Violence*. London: Arrow Books.

Burger, R., 2020. Sie planten Bürgerkriegsszenarien. *Frankfurter Allgemeine Zeitung* [online] 16 February. https://www.faz.net/aktuell/politik/inland/terrorzelle-plante-buergerkriegsszenarien-16637358.html [Accessed 22 February 2021].

Claus, R., 2017. *Hooligans: Eine Welt zwischen Fußball, Gewalt und Politik*. Göttingen: Verlag Die Werkstatt.

Claus, R., 2020. *Ihr Kampf. Wie Europas extreme Rechte für den Umsturz trainiert*. Göttingen: Verlag Die Werkstatt

Dammers, T. and Mannheim, F., 2020. Durchsuchungen bei der rechtextremen Bruderschaft Deutschland. *WDR: Westdeutscher Rundfunk Köln* [online] 1 April. https://www1.wdr.de/nachrichten/landespolitik/razzia-rechtsextremist-duesseldorf-100.html [Accessed 22 February 2021].

Erhard, F., Leistner, A., and Mennicke, A., 2019. "Soldiers for freedom, nation and blood" – Der Wandel von Anerkennungsordnungen kollektiver Gewaltausübung durch Fußballhooligans im Zuge der *GIDA-Bewegungen. *FuG – Zeitschrift für Fußball und Gesellschaft*, 1, pp. 46–68.

Freeden, M., 1998. Is nationalism a distinct ideology? *Political Studies*, 46, pp. 748–765.

Gebhardt, R., 2015. Die Mär vom unpolitischen Hooligan. *Blätter für deutsche und internationale Politik*, 1, pp. 9–12.

Gebhardt, R., 2017. Blutgrätsche, Kampfschwein, Hooligan. Wildes Nachdenken über Fußball und Gewalt. In: R. Gebhardt, ed. *Fäuste, Fahnen, Fankulturen. Die Rückkehr der Hooligans auf der Straße und im Stadion*. Cologne: PapyRossa Verlag, pp. 133–159.

Gilbert, M., 2019. Von Parasiten und Messermännern. *Süddeutsche Zeitung* [online] 13 September. https://www.sueddeutsche.de/politik/politische-sprache-von-parasiten-und-messermaennern-1.4600177 [Accessed 22 February 2021].

Giulianotti, R. and Armstrong, G., 2002. Avenues of contestation. Football hooligans running and ruling urban spaces. *Social Anthropology*, 10(2), pp. 211–238.

Grüne, H., ed. 2013. *Wenn Spieltag ist. Fußballfans in der Bundesliga*. Göttingen: Verlag Die Werkstatt.

Hall, S., Critcher, C., Jefferson, T., Clarke, J., and Roberts, B., 1978. *Policing the Crisis. Mugging, the State, and Law and Order*. London/Basingstoke: Macmillan.

Heitmeyer, W., ed. 2012. *Deutsche Zustände*. Berlin: Suhrkamp Verlag.

Ingham, R., 1978. *Football Hooliganism. The Wider Context*. London: Inter-Action-Imprint.

Interpool.tv, 2018. Inside HogeSa. Ein Film von Fred Kowasch. *Interpool.tv* [online]. http://www.interpool.tv/inside-hogesa-doku.html [Accessed 22 February 2021].

Klarmann, M., 2019. Braungefärbtes "Wutbürger"-Spektrum. *bnr.de* [online] 9 September. https://www.bnr.de/artikel/aktuelle-meldungen/braungef-rbtes-wutb-rger-spektrum [Accessed 22 February 2021].

Linden, M., 2021. Revolutionärer Konservatismus. Der rechte Angriff auf Freiheit und Demokratie. *Blätter für deutsche und internationale Politik*, 1, pp.62–72.

Mudde, C., 2004. The populist zeitgeist. *Government and Opposition*, 4, pp.541–563.

NRW: Die Landesregierung Nordrhein-Westfalen, 2016. Silvesternacht 2015 in Köln. *Wir in NRW Das Landesportal* [online]. https://www.land.nrw/de/silvesternacht-koeln-landesregierung-traegt-konsequent-zur-transparenten-aufarbeitung-der-ereignisse [Accessed 22 February 2021].

Pilz, G.A., 2007. Feindbild Polizei. *Der Tagesspiegel* [online] 11 March. https://www.tagesspiegel.de/sport/feindbild-polizei/821206.html (Accessed 3 March 2021).

Piskurek, C., 2018. Mediated thugs: re-reading Stuart Hall's work on football hooliganism. *Coils of the Serpent*, 3, pp.88–101.

Ruf, C., 2013. *Kurvenrebellen. Die Ultras. Einblicke in eine widersprüchliche Szene*. Göttingen: Die Werkstatt.

Sonnad, N., 2018. How soccer helped an ethnic slur take over Europe. *Quartz* [online] 15 June. https://qz.com/1306921/world-cup-2018-hooligans-is-an-ethnic-slur-in-history/ [Accessed 22 February 2021].

Spiegel Politik, 2017. "Wir werden Frau Merkel jagen." *Spiegel Politik* [online] 24 September. https://www.spiegel.de/politik/deutschland/afd-alexander-gauland-wir-werden-frau-merkel-jagen-a-1169598.html [Accessed 22 February 2021].

Spiegel Sport, 2000. Hooligans. "Wir sind keine Verbrecher". *Spiegel Sport* [online] 25 0. https://www.spiegel.de/sport/fussball/hooligans-wir-sind-keine-verbrecher-a-70610.html [Accessed 22 February 2021].

Sundermeyer, O., 2015. Hooligans als Schutztruppe. *Frankfurter Allgemeine Zeitung* [online] 2 February. https://www.faz.net/aktuell/sport/fussball/pegida-und-der-fussball-hooligans-als-schutztruppe-13403341.html [Accessed 22 February 2021].

Virchow, F. and Häusler, A., 2020. Pandemie-Leugnung und extreme Rechte. *core nrw – Netzwerk für Extremismusforschung in Nordrhein-Westfalen* [online]. https://www.bicc.de/uploads/tx_bicctools/CoRE_Kurzgutachten3_2020.pdf [Accessed 3 February 2021].

Weiß, V., 2018. Neonazis, Hools und ihre bürgerlichen Freunde. *Zeit Online* [online] 25 March. https://www.zeit.de/kultur/2018-08/rechtsextremismus-chemnitz-ausschreitungen-rechte-gewalt?utm_referrer=https%3A%2F%2Fwww.google.de%2F [Accessed 22 February 2021].

Wellgraf, S., 2017. Policing the crisis. Zur Diskriminierung von Hooligans. *Kriminologisches Journal*, 3, pp.220–242.

Williams, J., Dunning, E., and Murphy, P.J., 1984. *Hooligans Abroad. The Behavior and Control of English Fans in Continental Europe*. London: Routledge & Kegan Paul.

ZIS: Zentrale Informationsstelle Sporteinsätze, 2019. Jahresbericht Fußball Saison 2018/19. Polizei Nordrhein-Westfalen. *Landesamt für Zentral Polizeiliche Dienste* [online], 15 October. https://lzpd.polizei.nrw/sites/default/files/2019-10/ZIS-Jahresbericht%202018-2019.pdf [Accessed 24 January 2021].

13 Nazi-Barbies*

Performing ultra-femininity against the "Feminist Elite" in the Alt-Right movement

Diana Weis

Introduction: what is ultra-femininity?

In the public mind the so-called Alt-Right[1] is mostly regarded as a hate-filled "manosphere" that is defined in large parts by its misogyny and its anti-feminist, anti-woman rhetoric (Nagle, 2017; Sauer, 2020). It is deemed a movement of angry white men to such an extent that in the aftermath of the attempted insurrection in January 2021, the fact that the violent mob storming the Capitol building was scattered with female faces made international headlines (Givhan, 2021; Shaw, 2021; Thomas, 2021). This unwillingness to recognise the role of white women in spreading racism and (neo-)fascism has its historical precursors in the 20th century and has been addressed by a number of feminist scholars (Davis, 1983; Lorde, 1984; Kompisch, 2008). Women in white supremacist movements, including the Alt-Right, often also receive less media and scholarly attention because they are less likely to assume active leadership roles. Instead, they typically self-identify as "shield maidens" that help to soften and normalise white supremacy (Love, 2020). During the Trump administration, Alt-Right imagery increasingly spilled into mainstream media in the US and many other countries. With it came a specific type of female self-stylisation, broadly defined as a bundle of beautification practices, that I refer to as *ultra-femininity*. Although this look is in itself not new, but has been familiar in conservative circles for some time, particularly in the US, I will argue that it is now strategised in a new way. What differentiates the Alt-Right from previous white supremacist movements is its rootedness in digital culture, especially social media. Utilising new aesthetic forms, such as memes, earlier and more effectively than their political counterparts, has given the movement a decided advantage in what is now referred to as a "online culture war" (Nagle, 2017; Hornuff, 2020). In internet culture,

* The term "Nazi-Barbie" used in the title was coined in 2015 on online message boards frequented by members of the Antifa and their sympathizers. It originally referred to the US-American Influencer and White Supremacy Activist Brittany Pettibone, but has since been used more widely to describe a certain type of young female social media personalities connected to the Alt-Right movement. Although I am acutely aware of the critique surrounding the trivialization of the label "Nazi" in contemporary political discourse, I do feel it is a fittingly crass description of the women examined in this paper as it reflects both the use of pop cultural iconography and their ideological agendas.

DOI: 10.4324/9781003141150-17

digital images play an essential part in shaping social discourse – and it is especially young women that have made use of, and in many cases benefited from, these new possibilities of self-representation by creating highly influential aesthetic standards and motifs (Kohout, 2019; Weis, 2020).

Based on approaches from fashion theory and cultural studies, this chapter examines mainstream media coverage, social media posts and refences in popular culture as well as scholarly literature, especially feminist theory, and conducts a discourse analysis of ultra-femininity and related female beautification practices in order to examine the "cultural rules" attached to these practices. The chapter seeks to answer the question how stereotypes associated with ultra-femininity are being used politically – both to captivate and to antagonise. Next to the gendered meanings obviously associated with ultra-femininity, the role of class criticism as a founding narrative of the Alt-Right movement – in this case, the positioning of ultra-femininity as a form of anti-elitist class resistance – needs to be examined as well. What part do "progressive" left-wing and especially feminist approaches to female beauty and beautification play in this? The chapter focuses particularly on a few controversial moments in popular representations of fashion and politics and on "metapolitical"[2] visual strategies within the Alt-Right, highlighting four themes: depictions of right-wing women and the "Fake Melania" meme in liberal-leaning media, a brief look at the history of western female "beautification" and its class implications since the 19th century, a discussion of changing feminist and post-feminist perspectives on "beautification" and, finally, the role of ultra-femininity in pop cultural strategies of the Alt-Right.

From the outset, it should be clear that femininity is a highly loaded concept not without problems or conceptual inconsistencies in feminist theory. Second-wave feminist Susan Brownmiller, for example, described it as "a rigid code of appearance and behavior defined by do's and don't-do's" (Brownmiller, 1984, p.9). In the context of concrete practices of beautification, the subject becomes even more thorny: As female beauty standards are understood as patriarchally imposed, any efforts made to comply with those norms are regarded as irreconcilable with female autonomy. Seen from that perspective, femininity – understood as a cultural performance – in itself is a betrayal of the cause of feminism (Kauer, 2009, p.33). Even though, as I will argue in more detail below, feminist theorists of the third and fourth waves have been more welcoming to the idea of productive female self-imaging through beautification techniques (Davis, 1995; Degele, 2008; Kohout, 2019), the term femininity has remained problematic if only due to its inherent gender binarism. It is usually circumscribed as a symbolic place or practice, defining the bottom position in a binary gender hierarchy (Butler, 1990; Schippers, 2007). However, understanding femininity as a social construction within a power relation does not mean that the symbols chosen to represent it are semantically arbitrary. Approaches in cultural theory and sociology to the construction of beauty norms have argued convincingly that the "materiality of the body", is inseparable from its symbolic properties and must be understood within the context of power relations – especially those relating to class, race and gender (Fleig, 2000; Koppetsch, 2000).

Enhanced women: from "rightwing women all dress the same" to "Fake Melania"

During the Trump presidency liberal-leaning media were outspoken in exposing and expressing their disgust for a certain brand of conventionally attractive, pointedly groomed femaleness:

> But American rightwing women all dress exactly the same, which is to say, mainstream feminine – dresses, not trousers; heels, not flats; no interesting cuts, just body-skimming, cleavage-hinting, not-scaring-the-horses tedium.
>
> (Freeman, 2017)

> You know the look: Hair that is long, layered and blown to salon perfection; makeup that covers the face in foundation, paints the eyes with subtle smokey shadows, and adds coat upon coat of mascara.
>
> (Del Russo, 2017)

> It was a good reminder for Americans about how the Trump administration likes its women: hair done, makeup piled on, and lying through their teeth.
>
> (Valenti, 2017)

One could argue that the women described here share a kind of Wittgensteinian *Familienähnlichkeit* (family resemblance) in that their visual likeness points to further semantic similarities. It can be noted that their self-presentations lay the focus on traditional markers of femininity (long hair, use of make-up, dresses and high heels), while the perceived "sameness" hints at a dismissal of personal individuality in favour of a collective identity. A recurring point of criticism lies in the fact that these women's bodies have been artificially *enhanced* in a multitude of ways. Next to their "piled-on" make-up and professionally "done" hair, acquiring "the look" requires an arsenal of further beautification techniques such as lash and nail extensions, dental work, as well as injections with Botox and fillers. The bodies in question are constructed not only socially, culturally and politically, but also material constructs in a very literal way. Serving as objects of study to a femaleness performed through a series of embodied acts, by "what is put on, invariably, under constraint, daily and incessantly, with anxiety and pleasure" (Butler, 1988, p.530). One could argue that this is true for the vast majority of women in media today regardless of their political affiliation. But what makes fashion and beauty culture such useful tools in setting symbolic boundaries is precisely that its aesthetic markers are nuanced and refined: the tone of lipstick, the height of the heel, the dosage of Botox injections.

In October 2020, just a few weeks shy of the highly anticipated US election, photographs depicting the then-presidential couple boarding the Marine One helicopter in Nashville, Tennessee, gave new fuel to a conspiracy theory commonly referred to as "Fake Melania". As with previous instances, its purveyors cited perceived inconsistencies in Melania's facial appearance as evidence of her being replaced by a body double in public appearances. Several in part contradictory approaches were offered to explain why the "real" Melania was unwilling or unable to be by her husband's side. The more benign interpretations framed the operation as a cover-up for publicly unacceptable behaviour on MT's side, such as secretly separating from her spouse or undergoing cosmetic surgical procedures requiring extensive healing time. However, it was the darker and more bizarre theories that captured the imagination of the audiences and took the concept of *Fake Melania* to another level. They claimed that MT had either been lobotomised into complete submission, murdered and replaced by a look-alike robot or perhaps never existed at all (Benjamin, 2020).

This approach coincides with the aesthetic critique of the "artificial look" of women in the Trump administration in liberal media. Even before the rise of "Fake Melania", MT had been relentlessly ridiculed for her appearance, which was perceived as overly groomed and strangely emotionless. Normative beauty, demureness and sophistication – qualities women are usually praised for in popular culture – oddly seemed to work in MT's disadvantage.

> I am almost sure that Mrs. Trump is not a robot, unlike the women in the famous novel [The Stepford Wives]. I say this despite her sculpted face and the generally 1950s Playboy Bunny appearance that seems to defy human aging. I am still sure that beneath the coaching and stilted speeches, she is a human being.
>
> (Landry, 2016)

Shortly after the inauguration signs reading "Melania, blink twice if you need help" appeared at Anti-Trump rallies, mockingly insinuating that she was held captive, possibly drugged and unable to communicate otherwise (Weaver, 2017). In June 2017 British comedian Tracey Ullmann launched a sketch on her TV show *Tracey Breaks the News* which showed a robotic Melania being remote-controlled from a secret Russian command centre.[3] Around the same time, the catchword *Melaniabot* gained popularity on Twitter (Bakke, 2020, p.147). In 2019 *The View*, a popular daytime US-talk show aimed at a predominantly female audience, dedicated its *Hot Topic* segment, typically dealing with pressing current political or social issues, to *Fake Melania*. Interestingly, the show's gaggle of female hosts adopted an ambivalent approach to the theory: While initially debunking it as "crazy", they later admitted to finding the narrative unsettlingly believable.[4] While it is certainly legitimate to express bewilderment at the ways in which MT chose to fill her position during her husband's term of office, entertaining the idea that the former FLOTUS might, in fact, not be human can be seen as an extreme form of *Othering*. Although conspiracy theories such as *Fake Melania* may appear quirky and exaggerated, they serve as stabilising factors implying social categorisation, often by building on pre-existing stereotypes (Uscinski and Parent, 2014; Klein et al., 2015). Started as a half-joke on social media and quickly picked up by mainstream media (Hyde, 2017), the *Fake Melania* conspiracy theory was mostly referred to in a humorous manner, as an intriguingly absurd emergence of contemporary social media culture. But, as the Trump administration has impressively proved, repetition creates reality, or a semblance thereof (Polage, 2012).

The public discourse surrounding Melania focuses heavily on two qualities: the way she looks and the fact that she is Eastern European.[5] In her analysis of MT's perception in US-American culture, Wiedlack ascribes her Otherness to her Slovenian descent. She argues that even though MT's whiteness and beauty have the potential to blend into elite society, her public sexualisation as well as her classification as a victim of her husband's toxic masculinity assign her to the position of "Eastern European under-classness" (Wiedlack, 2019). Undoubtedly, MT's "Eastern Europeanness" largely fuelled the negative stereotypes ascribed to her public persona. In the specific context of *The View* though, her national origin appears less relevant. The talk show has been praised for its "identifiably" diverse cast of female hosts representing different ethnic and social backgrounds, body types, ages and sexual orientations (FitzSimons, 2019). Despite the show's commitment to inclusiveness, the discussion of the *Fake Melania* theory made it quite clear that the hosts did not consider MT to be "one of them". At the same time, the trope of MT as a robotic, mind-controlled victim serves to absolve her of any willing participation in her husband's white nationalist politics, confirming the utopian feminist idea that women are morally superior to men and less prone to violent or exclusionary behaviour (Gilligan, 1977).

But if MT's *Otherness* is not based on her political views, then what is it that's so wrong about her? What sets *The View* apart from other daytime TV shows, according to *The New York Times*, is its representation of women who are "smart and accomplished", defying the usual "bubbleheads-'R'-us approach to women's talk shows" (James, 1997). To put it bluntly, MT is not considered "smart" enough to serve as identification object for *The View's* audience, she belongs to the "bubbleheads". Taking into consideration that female college-educated democratic voters make up the bulk of *The View's* audience (Schaal, 2020; Flint, 2016) it becomes apparent that the issue taken with MT is closely tied to class-based femininity. Like the majority of Trump's female supporters, she has no college degree (Zachary, Merrill, and Wolfe, 2020). In that view, the social privileges she enjoys were not "honestly acquired" through institutionalised education, but through the time, effort and sacrifices invested into cultivating her body.

Beautification and class shaming

The subtleties of "good" taste have long been utilised by the elites to keep the lower classes at bay. Much research has gone into the ways the taste of the elites presents a discreet yet daunting barrier, hampering the social advancement of those not born into the educated classes (Bourdieu, 1986; Gebauer, 1982; Veblen, 2007). In contrast to his usual overconfident demeanour, former president Trump is reportedly highly sensitive when it comes to the outward perception of his electoral base. A White House staffer revealed that after watching footage of the failed insurrection, "Trump expressed disgust on aesthetic grounds over how 'low class' his supporters looked. [...] He doesn't like low-class things" (Nuzzi, 2021). Yet DT, who famously ordered gold curtains for the Oval Office, has been continuously mocked by the press for the "vulgarity" and "garishness" of his taste (Budds, 2016; Menking, 2020), making him a textbook case of *Nouveau Riche*, whose abundant wealth cannot make up for his lack of cultural capital.

Much of the same malice can be observed in the discourse surrounding female beautification practices more broadly. A tendency to "overdo" things is typically ascribed to working -class and "low"-class women, with heavy make-up and big hair being some of the most widely recognised *white trash* signifiers (Skeggs, 2000, p.141). A similar notion is expressed in the aesthetic critique of the "Trump women" discussed earlier: Through expressions such as "piled-on make-up" "bottle-blonde" or "layer upon layer of mascara", they are placed within the *Nouveau Riche* taste spectrum. These ascriptions can thus be read as a simple yet effective way to discredit social upward mobility. It also, less explicitly, divides the "upper class" into fractions distinguishable by their aesthetic preferences.

The devaluation of "painted women" has a long-standing tradition in bourgeois culture (Peiss, 1998; Ramsbrock, 2011) that is about more than *Nouveau Riche* status. Decorative cosmetic practices were not only considered low class in economic terms, but also widely associated with unacceptably lose ("low") sexual morals, an "abject whorishness" (Penny, 2011, p.6) further barring the women who used them from entering relevant social circles. Common stereotypes of "painted women" such as the Vaudeville girl, jezebel or femme fatale can be read as euphemisms for sex-workers. In the popular *Snatch Game* segment of *RuPaul's Drag Race Season 10*, the winning performance was an impersonation of MT, which portrayed her as a robot dropping sexual innuendos hinting at her husband's love for "Russian hookers" – a term which apparently was meant to include her Slovenian origins. Portraying MT to be, at the same time, non-human *and* a sex worker cites and continues this tradition of devaluing women who come from outside specific elite circles.[6]

Beautification from feminist and post-feminist perspectives

Feminist critiques of conventional femininity have their own rationality and genealogy, but they are not entirely separate from these dynamics either. In her classic book *Femininity*, second-wave feminist Susan Brownmiller (1984) identified female beautification practices enforced by the cosmetic industry as one of the core conflicts threatening the women's movement from within. Interestingly, the trope of willfully creating a false female self by putting on an "impersonal cosmetic mask" (Brownmiller, 1984, p.187) or turning the body into "cultural plastic" (Bordo, 2003, pp.246f.) picks up the familiar perspective that "painted women" cannot be trusted as they are not who they appear to be. Influential feminists of this generation agreed that not conforming to, or even opposing normative notions of female attractiveness are powerful forms of political resistance against an oppressive objectifying system – while at the same time complaining about being stereotyped as old-fashioned, unattractive, bleak and joyless (Daly, 1990; Bordo, 2003; Brownmiller, 1984). This feminist critique of feminine beauty culture is not free of class-based attributions, for instance when explaining some women's eagerness to comply with mainstream beauty norms by ascribing to them a status of social and economic powerlessness, relying on male sexual attention in order to sustain themselves (Adams, 1997; Davis, 2008). Institutionalised education and the rejection of roles ascribed to traditional femininity are regarded as the only truly acceptable forms of female self-expression. By expressing a general disregard for fashion and beauty culture, feminist scholars run the risk of alienating women who partake in and enjoy these practices, making this an easy access point for Alt-Right activists and others to reinforce anti-feminist stereotypes as position them as a form of anti-elitism.

In contrast, however, the third wave of feminism embraced the concept of female agency, honouring individual choices of self-definition as empowering and creative, while labelling second-wave feminists as judgemental of other women and anti-sexual in general (Davis, 1995; Snyder, 2008). The current fourth wave of feminism, shaped by digital culture, appears to continuously oscillate between those two extreme positions: Some voices celebrate the use of make-up and fashion as relevant political strategies regardless of gender (Kohout, 2019) and others warn against the pressures to perform a "robotic capitalist eroticism" (Penny, 2010, p.10) that sells femininity as a bodily property only attainable though consumption (Chae, 2019). Regardless of this opposition, the display of hegemonic white heterosexual femininity is mostly regarded as problematic, even toxic, for its tendency to appropriate and "erase" the experiences of all women not defined by that narrow margin (Butler, 2013; Daniels, 2015; Moon and Holling, 2020). A constricted perspective that focuses on the experiences of white, educated, middle-class women, has increasingly come into criticism (Daniels, 2015; Moon and Holling, 2020). In the current conjuncture, many female authors across a wide political spectrum explicitly defend themselves against class shaming in association with beautification practices. Journalist Cigdem Toprak, for example, criticised the white liberal elite in Germany for denying her intellectual capacity because of her heavily made-up eyes that made her look like "a girl from the streets" (Toprak, 2020). This emerging debate is strongly linked to the perception of migrant women and women of colour who claim specific techniques as markers of their cultural heritage. Author Jacinta Nandi even writes that the moral rejection of beautified women by the white German educated middle classes as "impure" reveals a "racist desire for the Nazi period" (Nandi, 2018, p.45).[7]

However, these dichotomies – even though still powerful within white supremacist visual cultures – can be applied to the politics of femininity in today's Alt-Right movement only to a limited degree. The image of the naturally fresh-faced, linen-clad, braid-swinging "shield-maiden", associated with pre-digital extreme right communities could not be more different from the appearance of today's female Alt-Right influencers. Instead, they opt for a look of a prototypical white "cookie-cutter beauty" in line with the standards of ultra-femininity described above that gained increasing popularity due to the combined rise of digital communication techniques and the cosmetic enhancement industry (Kuczynski, 2006, pp.110f.).

The adoption of this specific feminine style by the Alt-Right can partly be explained by the market rules of the digital playing field, where successful self-staging calls for a different set of presentation strategies. In addition, the idea of attractiveness as a personal achievement, attainable through participation in consumer culture, is one of the founding myths of modern US-American capitalism (Peiss, 1998). Or, as Karl Marx put it, the firm belief in the power of money to destroy ugliness (Marx, 1988, p.219).

Pop culture and the Alt-Right

When approaching the issue of why women are attracted to a movement like the Alt-Right that expresses blatant disregard for gender equality, post-feminism or "emancipation fatigue" (Dietze, 2020) is considered the most obvious explanation. Most theorists agree that the "new momism" culture of neo-maternity[8] also plays a big part (Mattheis, 2018; Hallstein, 2010). But as Angela McRobbie has pointed out, fashion and beautification are areas that do not address women primarily as mothers or caregivers, but rather as a "a gay young thing out for a good time" (McRobbie, 1991, p.145). Therefore, beauty culture within the Alt-Right should be examined separately from its glorification of motherhood.

While the third and fourth waves of feminism disagreed with some basic assumptions of classic second-wave feminism, Alt-Right post-feminism aims to displace feminism altogether, making anti-feminism a more appropriate term to describe the movement. The opinion that emancipation is an aberration, "a norm dictated by the elite", is widespread among Alt-Right women (Dietze, 2020, p.151). The negative stereotype of the "ugly, jack-booted, feminazi psycho lesbian" (Freeman, 2017), a woman so unattractive by conventional standards that she is unable to find a male partner, has been used to discredit the women's movement from the start and is frequently referred to in feminist literature as a bitter joke (Brownmiller, 1984; Daly, 1990; Bordo, 1998). This alone would make embracing ultra-femininity self-evident as a strategy for Alt-Right women. They actively use the "unsexiness" of feminism as an argument, especially when catering to younger women, and update it to current vocabularies. In January of 2020 Alt-Right influencer Brittany Pettibone, whose social media persona earned her the moniker "Nazi-Barbie", posted an advice video titled "Men won't date 'woke' women"[9] on

YouTube that has since received more than 600,000 views. In their quest to erase feminism, influencers like her seek to be understood not as backwards but as "modern women" who have overcome the "false consciousness" dictated by pro-feminist mainstream social and cultural discourse. Even when they are not actively fulfilling the role of the mother (yet), they see their "destiny as women" firmly situated alongside a white male in a heterosexual relationship (Mattheis, 2018). The feminist critique of patriarchal "toxic" masculinity is seen as "an undeserved castration and expulsion of the sexiness from the heterosexual relationship." (Dietze, 2020, p.151).

Still, the adoption of ultra-femininity by radical right-wing women should not be assessed solely as man-pleasing behaviour. Building on reactions to the devaluations described above, there is also an element of class resistance – or, in a more strategic sense, of exploiting class resentments – to it. This paradoxically benefits from increasingly blurry symbolic class boundaries. In 1986 Bourdieu found that French working-class women were less likely to work in fields that "most strictly demand conformity to the dominant norms of beauty" and were therefore typically less inclined "to invest time and effort, sacrifices and money in cultivating their bodies". He went on to say that it is the women of the *petit bourgeoisie* "who have sufficient interest in the market in which physical properties can function as capital" (Bourdieu, 1986, p.206). Since that time, the perceived attainability of beauty norms as well as the possibility of heightened visibility have considerably risen, not least through digital culture, especially social media, making the cultivation of the body appear worthwhile regardless of professions and class status. In a recent study of female supporters of the Tea Party, one woman stated her anger at "feminazis" for class shaming her and her peers as "ignorant, backward, redneck losers. They think we are racist, sexist, homophobic and maybe fat." (Hochschild, 2016, p. 23). For rural right-wing women like her, focusing on beautification can then also be seen as visual resistance against ascribed low-class ugliness.

However, ultra-femininity cannot be understood simply through the opposition between "elitist" and "working-class" taste, and through aligning "elite" with "feminist", which is a reading suggested by Alt-Right metapolitics. For one, it is no secret that even though the Alt-Right cultivates the image of a grass-roots movement, many of its leading figures themselves belong to the "elite" they allegedly aim to destroy. Furthermore, in many cases, the look of ultra-femininity reproduces a "high maintenance" style of beauty that could be observed in the *Nouveau Riche* fraction of the economic upper class long before its utilisation in the Alt-Right. As stand-in for sociologically more precise terms, the *Nouveau Riche* can heuristically be seen as a social stratum with money and influence whose members feels to some extent humiliated by their exclusion from more "cultured", educated, officially credentialed circles. Seen from this perspective, the self-assertive expression of ultra-femininity can also be understood as a further "emancipation of taste" among the *Nouveau Riche*, much like Trump's gold curtains, against the aesthetic standards imposed by the "liberal elite" – which, however, in this case is made to encompass other fractions of the predominantly white upper middle class and the much more diverse and cross-class feminist movement.

Strategies of aesthetic subversion that are typical for subcultures play an important role in that context. In September 2020 British sportswear brand Fred Perry announced the decision to stop selling its iconic Black/Yellow/Yellow twin tipped polo shirt in the US and Canada[10]. The statement came as a reaction to a growing number of media reports worldwide portraying the self-proclaimed *Proud Boys*, a white supremacist, neo-fascist, male-only organisation engaged in targeted violence against left-wing activists, LGBTQI+ and people of colour, that had adopted the polo shirt as an unofficial uniform. This unwelcomed ideological charging of an item of clothing that had been a beloved subcultural staple since the late 1970s is only the latest chapter in an ongoing discourse surrounding the "appropriation" of pop cultural style codes by Alt-Right activists. A few years earlier the figure of the "Nipster" – a compound of the terms "Nazi" and "Hipster" – received some media as well as scholarly attention. The adoption of a dress code favoured by an urban, art-savvy elite sporting undercut hairstyles, skinny jeans and Converse All-Stars trainers was interpreted as a deliberate strategy to appear less threatening and simulate participation in mainstream culture (Rogers, 2014; Gaugele, 2018). However, this reading overlooks that fact that racism and misogyny had always played an integral, if controversial, part in Hipster culture itself (Current and Tillotson, 2018; Dubrofsky and Wood, 2014).[11] In fact, classic Hipster style as it appeared in the noughties of the 21st century was characterised by a preference for what John Leland labelled "caucasian kitsch": Trucker Hats, Wife Beaters and Pabst Blue Ribbon beer (Leland, 2005, p.353). Hipster culture has been frequently accused of "class tourism", college-educated middle-class offspring fashionably posing as "White Trash" (Schiermer, 2013; Weis, 2017). White male hipsters not only loved to "ironically" adopt the fashion style of white low-class country folk, but also their supposed racist and misogynist views that were in stark opposition to the "politically correct" consensus of their peers. Perhaps surprisingly, many white women went along with this under the assumption that the resurrection of sexist stereotypes could now be seen as amusing as gender equality had allegedly already been achieved (Douglas, 2010).

Much like the Alt-Right movement, hipsterism was predominantly perceived as a culture of white men. While there were plenty of female Hipster icons, their style was not so much perceived as subcultural or intellectually stimulating, but as "the familiar 'female' knowledge of how to look" (Tortorini, 2010, p.123). Angela McRobbie has argued that the types of images available to girls and women are limited – in subcultural settings as well as in mainstream society, with a softer, sexually permissive approach generally appearing more acceptable than aggressive and transgressive behaviours (McRobbie and Garber, 1976). Mattheis's study of female Alt-Right protagonists points out a specific obstacle that women in far-right movements face: "How does one act as a warrior of the movement [...] without emasculating men?" (Mattheis, 2018, p.137). While the heightened visibility of Alt-Right actors in mainstream media brought attention to some memorably bizarre male figures such as the "QAnon Shaman" (Sheppard, 2021), the women seem to attract sufficient attention by their femaleness alone (Givhan, 2021; Shaw, 2021; Thomas, 2021). Yet there is an element of the strategic "semiotic confusion" associated with male "Nipster" styles to the look of ultra-femininity as well. In an article on the "TradWife" movement, women proudly proclaiming to be utterly fulfilled by living the life of a wife and mother without any professional aspirations, *New York Times* journalist Annie Kelly writes that they deliberately construct a "hyperfeminine aesthetic" in order to "mask the authoritarianism of their ideology" (Kelly, 2018).

This strategic use of the aesthetic is illustrated by the case of US-Alt-Right influencer Pettibone, who first gained social media fame as a self-proclaimed "pizzagate expert" (Maly, 2020). She relocated to Austria in 2018, where she staged a fairytale wedding to the far-right white nationalist "Identitarian" movement leader and "Nipster" posterboy Martin Sellner. Since becoming a wife, Pettibone's online rhetoric has shifted away from outright political activism and the spreading of anti-Democrats conspiracy theories, to cultural or "metapolitical" issues such as beauty and dating advice. As a social influencer she adopts a self-presentation technique that uses strategic intimacy to appeal to her followers, mostly white young women (Marwick, 2015). In 2018 she published her first book titled "What Makes us Girls" offering a low-threshold access to evolutionary-biologist, culturalistic, anti-feminist ideas and attitudes. The book suggests that beautification and the expression of femininity are acts of self-worth, even if it doesn't offer any make-up or fashion tips. On the cover, Pettibone poses for a glamour headshot, hair and make-up perfectly polished, wearing a demure black dress with lace inserts. In a YouTube video from February 2021 titled "The war on feminine beauty", she argues that women have an "organic desire" that makes them want to look beautiful but that some "deliberately promote ugliness in order to sabotage other women" [12].

Conclusion: ultra-femininity and (white) power

What makes the praising of hegemonic femininity so effective to the Alt-Right agenda is not just its stark visual opposition to the "dowdy feminist" stereotypes, but also that western beauty norms inherently reproduce whiteness as a desirable standard (Davis, 2008; Haiken, 1997). Openly racist motifs and memes that are popular in Alt-Right online circles suggest that the terms "beautiful" and "ugly" serve as thinly veiled euphemisms for white and non-white.

Applying beautification as a conscious strategy means to position oneself within a social field and making a deliberate choice whom to attract and whom to repel (Degele, 2008, p.71). Within the framework of race, gender and class, the Alt-Right women's performance of ultra-femininity deliberately promotes ideas of white supremacy and hegemonic gender roles while challenging norms of beauty and taste that predominate among other elites. Stereotypes associated with ultra-femininity are utilised in two ways: for the purpose of othering women with a different social background or subset of values as well as to create a powerful counter-aesthetic of upward social mobility that is accessible through participation in consumer culture.

Beautification and the creation of glamourous images are techniques commonly applied by ambitious emergent groups and individuals in order to take over or re-invent privileges from elites, thus threatening hegemonic power (Gundle, 2008, p.19). A similar strategy has been adopted by the Alt-Right, whose representational approach to young white women has shifted over the last decade from mostly allegorical motifs to "beautifully photographed icons" consistent with contemporary influencer culture (Dietze, 2020, p.160). This strategy also makes use of the concept of *empowerment* which is highly popular in fourth wave digital cultures where it is commonly used to describe images that provide positive visibility to people enduring exclusion. Exactly *how* female Alt-Right actors are disadvantaged by elitist culture remains an open question. As white heterosexual women they belong to a privileged group most frequently represented in mainstream media. It can be argued that strategies of symbolic class resistance were influential in shaping and furthering ultra-femininity as a form of visual warfare by and within the Alt-Right movement. At the same time, many of the female figures associated with the movement do not actually come from lower or working-class backgrounds. Instead, they have appropriated the concept of empowerment by creating an anti-feminist narrative claiming that a war is being waged against feminine beauty by liberal culture pushing "ugly positive feminists as the beauty ideal" (Mattheis, 2018, pp.141f.).

When examining the power relations within the Alt-Right movement, as well as in the Trump administration, the display of ultra-femininity might serve another, subconscious, purpose as well. In her classic psychoanalytic essay "Womanliness As A Maquerade" (1929) Joan Riviere finds that women in power positions apply practices and behaviours associated with hegemonic femininity to appear less threatening and avoid punishment from the male leader/father figure in an oedipal entanglement.

Ultra-femininity can serve to appease some, but it certainly infuriates others. Seen from this perspective, it is highly effective as an image strategy to visualise the political and cultural opposition of "beautiful" white femininity and the "abject", diverse aesthetics of contemporary feminism. However, ultra-femininity seems less well suited as a tool to redistribute political power – within the Alt-Right movement or society at large – because it embraces, and constantly reproduces the very same limitations that spawned it.

Notes

1 The Alt-Right, short for "Alternative Right" here specifically refers to a new wave of web-based white segregationist subcultures that have gained strength and visibility during Donald Trump's presidential term (Nagle, 2017; 12).

2 The term *metapolitics*, as it is used here, refers to practices addressing, and potentially disrupting, existing political discourse as well as the dominant power structures in a given public sphere (Zienkowski, 2019). The Alt-Right movement uses

metapolitics to strategically frame topics such as family values, gender relations and beauty norms as ideological locations for political debates.

3 Tracey Breaks the News (2017) *Melania Trump Robot*. Available onlone: youtube.com/watch?v=b6NqscIsidQ [accessed 02/25/2021].

4 The View (2019). *Fake Melania Conspiracies Return*. Available online: youtube.com/watch?v=xOPeOUaREVc [accessed 02/25/2021].

5 On this aspect, also see the chapter by Breda Luthar in this volume.

6 RuPaul's Drag Race Season 10, *Snatch Game*. Available online: youtube.com/watch?v=iKl4EXFQTYc [accessed 02/25/2021].

7 As fashion historian and theorist Barbara Vinken has pointed out, ideas of "natural" and "artificial" beauty are deeply rooted in racialized beliefs, with the ideal of the unspoiled nordic women held high in national-socialist Germany, condemning the seductive refinement of "Jewish-oriental" women, cunningly using make-up, perfume and fashion to their personal advancement (Vinken, 2013, pp.152f.).

8 A movement of women, sometimes also referred to as "TradWifes" (traditional wives), that regard choosing full-time motherhood over a career as the ultimate form of female self-actualization.

9 youtube.com/watch?v=SbiWrOMhMho [accessed 02/25/2021].

10 help.fredperry.com/hc/en-us/articles/360013674918-Proud-Boys-Statement [accessed 02/25/2020].

11 It is no coincidence that the co-founder of *Vice* magazine, dubbed the "hipster's bible", Gavin McInnes went on to found the *Proud Boys*. As Mark Greif points out in his analysis of hipster culture, "the markers of hipster ethnicity were straightforward. They were coded 'suburban white'" (Greif, 2010, p. 146).

12 youtube.com/watch?v=R_rhL3eFeuE [accessed 02/25/2021].

Bibliography

Adams, A., 1997. Molding women's bodies: the surgeon as sculptor. In: S. Wildon and M. Laennec, eds. *Bodily Discursions. Genders, Representations, Technologies*. Albany: State University of New York Press, pp. 59–80.

Bakke, G., 2020. *The Likeness. Semblance and Self in Slovene Society*. Oakland: University of California Press.

Bempeza, S., 2020. Aesthetic practices of the New Right – fake and post-truth as a challenge for transgressive art and cultural practices. *Medienimpulse*, 8(2). journals.univie.ac.at/index.php/mp/article/view/3499 [Accessed 25 February 2020].

Benjamin, P., 2020. A brief history of fake Melania. 2020 is so whack that a First Lady body-double doesn't seem so crazy. *Dazed* [online], 28 October. dazeddigital.com/politics/article/50907/1/a-brief-history-of-fake-melania-trump-conspiracy [Accessed 25 February 2020].

Bordo, S., 1998 Braveheart, babe, and the contemporary body. In: E. Parens, ed. *Enhancing Human Traits: Ethical and Social Implications*. Washington, D.C.: Georgetown University Press, pp. 189–220.

Bordo, S., 2003. *Unbearable Weight: Feminism, Western Culture and the Body*. Berkeley, Los Angeles: University of California Press.

Bourdieu, P., 1986. *Distinction: A Social Critique of the Judgement of Taste*. Cambridge, MA: Harvard University Press.

Brownmiller, S., 1984. *Femininity*. New York: Simon & Schuster.

Budds, D., 2016. How Trump made architecture (and cities) worse. *Fast Company* [online] 8 November. fastcompany.com/3065394/how-trump-made-architecture-and-cities-worse [Accessed 25 February 2020].

Butler, J., 1988. Performative acts and gender constitution: an essay in phenomenology and feminist theory. *Theatre Journal*, 40(4), pp.519–531.

Butler, J., 1990. *Gender Trouble. Feminism and the Subversion of Identity*. New York: Routledge.

Butler, J., 2013. For white girls only? Postfeminism and the politics of inclusion. *Feminist Formations*, 25(1), pp. 35–58.

Chae, J., 2019. YouTube makeup tutorials reinforce postfeminist beliefs through social comparison. *Media Psychology*. tandfonline.com/doi/full/10.1080/15213269.2019.1679187 [Accessed 25 February 2020].

Chait, J., 2020. "White women voted for Trump" is the worst election trope. *New York Intelligencer* [online] 1 December. nymag.com/intelligencer/article/did-white-women-vote-for-trump-no.html [Accessed 25 February 2020].

Current, C. and Tillotson, E., 2018. Hipster racism and sexism in charity date auctions: individualism, privilege blindness and irony in the academy. *Gender and Education*, 30(4), pp.467–476.

Daly, M., 1990. *Gyn/Ecology. The Metaethics of Radical Feminism*. Boston: Beacon Press.

Daniels, J., 2015. The trouble with white feminism: whiteness, digital feminism and the intersectional internet. *SSRN* [online] 25 February papers.ssrn.com/sol3/papers.cfm?abstract_id=2569369 [Accessed 25 February 2020].

Davis, A., 1983. *Women, Race & Class*. New York: Vintage.

Davis, K., 1995. *Reshaping the Female Body. The Dilemma of Cosmetic Surgery*. London/New York: Routledge.

Davis, K., 2008. Surgical passing – Das Unbehagen an Michael Jacksons Nase. In: P. Villa, ed. *Schön normal. Manipulationen am Körper als Technologien des Selbst*. Bielefeld: transkript, pp. 41–66.

Degele, N., 2008. Normale Exklusivitäten – Schönheitshandeln, Schmerznormalisieren, Körper inszenieren. In: P. Villa, ed. *Schön normal. Manipulationen am Körper als Technologien des Selbst*. Bielefeld: transkript, pp. 67–84

Del Russo, M., 2017. All the president's clones: why Trump's favorite women look exactly the same. *Yahoo!life* [online] 13 June. https://www.yahoo.com/lifestyle/president-clones-why-trump-favorite-170000231.html [Accessed 25 February 2020].

Dietze, G., 2016. Ethnosexismus. *Movements*, 2, pp.157–165.

Dietze, G., 2020. Why are women attracted to right-wing populism? Sexual exceptionalism, emancipation fatigue, and new materialism. In: G. Dietze and J. Roth, eds. *Right-Wing Populism and Gender. European Perspectives and Beyond*. Bielefeld: transcript, pp. 147–166.

Douglas, S., 2010. *Enlightened Sexism: The Seductive Message That Feminism's Work Is Done*. New York: Times Books.

Dubrofsky, R. and Wood, M., 2014. Posting racism and sexism: authenticity, agency and self-reflexivity in social media. *Communication and Critical/Cultural Studies*, 11(3), pp.282–287.

FitzSimons, A., 2019. How "The View" became the most important political TV show in America. *New York Times* [online] 22 May. nytimes.com/2019/05/22/magazine/the-view-politics-tv.html [Accessed 25 February 2020].

Fleig, A., 2000. Körper-Inszenierungen: Begriff, Geschichte, kulturelle Praxis. In: E. Fischer-Lichte and A. Fleig, eds. *Körper-Inszenierungen. Präsenz und kultureller Wandel*. Tübingen: Attempto, pp. 7–18.

Flint, J., 2016. The TV shows whose audiences have the highest median income. *The Wall Street Journal* [online] 18 November. wsj.com/articles/the-tv-shows-whose-audiences-have-the-highest-median-income-1479493870 [Accessed 25 February 2020].

Freeman, H., 2017. Why do all the women on Fox News look and dress alike? Republicans prefer blondes. *The Guardian* [online] 20 February. theguardian.com/fashion/2017/feb/20/women-fox-news-dress-alike-republicans-blondes-pundits-ann-coulter-kellyanne-conway-rightwingers [Accessed 25 February 2020].

Gaugele, E., 2018. Fashion & Faction: Zur Kritik der neuen Pop-Rechten. *POP. Kultur & Kritik*, 12, pp.90–96.

Gebauer, G., 1982. Ausdruck und Einbildung. Zur symbolischen Funktion des Körpers. In: D. Kamper and C. Wulf, eds. *Zur Wiederkehr des Körpers*. Frankfurt am Main: Suhrkamp, pp. 313–328.

Gilligan, C., 1977. In a different voice: women's conceptions of self and of morality. *Harvard Educational Review*, 47(4), pp.481–517.

Givhan, R., 2021. Women of the would-be insurrection. *The Washington Post* [online] 12 January. washingtonpost.com/nation/2021/01/12/women-would-be-insurrection/ [Accessed February 2020].

Greif, M., 2010. Epitaph for the white hipster. In. M. Greif, K. Ross and D. Tortorici, eds. *What Was the Hipster? A Sociological Investigation* New York: n+1 Foundation, pp. 136–167.

Gundle, S., 2008. *Glamour. A History*. Oxford: Oxford University Press.

Haiken, E., 1997. *Venus Envy. A History of Cosmetic Surgery*. Baltimore: The John Hopkins University Press.

Hallstein, D., 2010. *White Feminists and Contemporary Maternity: Purging Matrophobia*. New York: Macmillan.

Hochschild, A., 2016. *Strangers in Their Own Land: Anger and Mourning on the American Right*. New York: The New York Press.

Hornuff, D., 2020. *Hassbilder. Gewalt posten, Erniedrigung liken, Feindschaft teilen*. Berlin: Wagenbach.

Hyde, M., 2017. Who started the "Melania Trump body double" conspiracy theory? Look no further. *The Guardian* [online] 19 October. theguardian.com/lifeandstyle/lostinshowbiz/2017/oct/19/who-started-melania-trump-body-double-conspiracy-theory [Accessed 25 February 2020].

James, C., 1997. Feet on the ground, heads without bubbles. *New York Times* [online] 21 August. nytimes.com/1997/08/21/arts/feet-on-the-ground-heads-without-bubbles.html [Accessed 25 February 2020].

Kauer, K., 2009. *Popfeminismus! Fragezeichen! Eine Einführung*. Berlin Frank & Timme.

Kelly, A., 2018. The housewives of white supremacy. *New York Times* [online] 1 June. nytimes.com/2018/06/01/opinion/sunday/tradwives-women-alt-right.html [Accessed 25 February 2020].

Klein, O., Van der Linden, N., Pantazi, M. and Kissine, M. 2015. Behind the screen conspirators: paranoid social cognition in an online age. In: M. Bilewicz, A. Cichocka and W. Soral, eds. *The Psychology of Conspiracy*. Routledge/Taylor & Francis Group, pp. 162–182.

Kohout, A., 2019. *Netzfeminismus. Strategien weiblicher Bildpolitik*. Berlin: Wagenbach.

Kohout, A., 2020. Feministische Bildpraktiken in den Sozialen Medien: Empowerment vs. Provokation. *Pop-Zeitschrift* [online] 6 January. pop-zeitschrift.de/2020/01/06/feministische-bildpraktiken-in-den-sozialen-medien-empowerment-vs-provokationvon-annekathrin-kohout6-1-2020/ [Accessed 25 February 2020].

Kompisch, K., 2008. *Täterinnen. Frauen im Nationalsozialismus*. Cologne/Weimar/Vienna: Böhlau.

Koppetsch, C., 2000. Körper und Status. Zur Soziologie der körperlichen Attraktivität. In: C. Koppetsch, ed. *Körper und Status. Zur Soziologie der körperlichen Attraktivität*, Konstanz: Universitätsverlag Konstanz, pp. 7–16.
Kuczynski, A., 2006. *Beauty Junkies. Inside Our $15 Billion Obsession with Cosmetic Surgery*. New York: The Doubleday Broadway Publishing Group.
Landry, P., 2016. Melania Trump is what a Stepford wife looks like in 2016. *Huffington Post* [online] 21 July. huffingtonpost.ca/peter-landry/melania-trump-stepford-wife_b_11087636.html [Accessed 25 February 2020].
Leland, J., 2005. *Hip: The History*. New York: Harper Collins.
Lorde, A., 1984. *Sister Outsider. Essays and Speeches*. Trumannsburg, NY: Crossing.
Love, N., 2020. Shield maidens, fashy femmes, and tradwives: feminism, patriarchy, and right-wing populism. *Frontiers in Sociology*, 5. frontiersin.org/articles/10.3389/fsoc.2020.619572/full [Accessed 25 February 2020].
Maassen, S., 2008. Bio-ästhetische Gouvernementalität – Schönheitschirurgie als Biopolitik. In: P. Villa, ed. *schön normal. Manipulationen am Körper als Technologien des Selbst*. Bielefeld: transkript, pp. 99–118.
Maly, I., 2020. Metapolitical New Right influencers: the case of Brittany Pettibone. *Social Sciences*, 9(7), p. 113. mdpi.com/2076-0760/9/7/113 [Accessed 25 February 2020].
Marwick, A., 2015. You may know me from YouTube: (micro)-celebrity in social media. In: P. Marshall and S. Redmond, eds. *A Companion to Celebrity*. Hoboken: John Wiley & Sons, pp.333–350.
Marx, K., 1988 [1844]. *Ökonomisch-philosophische Manuskripte*. Leipzig: Dietz Verlag.
Mattheis, A., 2018. Shieldmaidens of whiteness: (Alt) maternalism and women recruiting for the Far/Alt-Right. *Journal for Deradicalization*, 7, pp.128–161.
McRobbie, A. and Garber, J., 1976. Girls in subcultures. In: S. Hall and T. Jefferson, eds. *Resistance Through Rituals. Youth Subcultures in Post-War Britain*. London: Hutchinson, pp. 209–222.
McRobbie, A., 1991. The culture of working-class girls. In: A. McRobbie, ed. *Feminism and Youth Culture. From Jackie to Seventeen*. London: Palgrave, pp. 36–60.
Menking, W., 2020 Sad! How Donald Trump transformed New York without any regard for design quality. *The Arhcitect's Newspaper* [online] 8 April. archpaper.com/2020/11/donald-trump-architecture/ [Accessed 25 February 2020].
Menninghaus, W., 2003. *Das Versprechen der Schönheit*. Frankfurt am Main: Suhrkamp.
Moon, D. and Holling, M., 2020. "White supremacy in heels": (white) feminism, white supremacy, and discursive violence. *Communication and Critical Studies*, 17, pp. 253–260.
Mulvey, L., 1989. *Visual and Other Pleasures*. Bloomington/Indianapolis: Indiana University Press.
Nagle, A., 2017. *Kill All Normies. Online Culture Wars from 4chan and Tumblr to Trump and the Alt-Right*. Croydon: zero books.
Nandi, J., 2018. Die Beauty-Lüge. *konkret*, 3(18), pp. 44–46.
Nuzzi, O., 2021. Senior Trump official: we were wrong, He's a "fascist". *New York Intelligencer* [online] 8 January. nymag.com/intelligencer/2021/01/capitol-riot-senior-trump-official-calls-him-a-fascist.html? [Accessed 25 February 2020].
Palmer, J., 2020. Survey: American voters still divided by race and education. *Courthouse News Service* [online] 2 June. courthousenews.com/american-voters-still-divided-by-race-and-education/ [Accessed 25 February 2020].

Peiss, K., 1998. *Hope in a Jar. The Making of America's Beauty Culture*. Philadelphia: The University of Philadelphia Press.

Penny, L., 2010. *Meat Market: Female Flesh under Capitalism*. Croydon: zero books.

Penny, L., 2011. *Meat Market: Female Flesh under Capitalism*. Winchester: Zero Books.

Polage, D., 2012. Making up history: false memories of fake news stories. *Europe's Journal of Psychology*, 8(2), pp.245–250.

Ramsbrock, A., 2011. *Korrigierte Körper. Eine Geschichte künstlicher Schönheit in der Moderne*. Göttingen: Wallstein Verlag.

Riviere, J., 1929. Womanliness as a masquerade. *The International Journal of Psychoanalysis*, 10, pp.303–313.

Rogers, T., 2014. Heil hipster: the young neo-Nazis trying to put a stylish face on hate. *Rolling Stone* [online] 23 June. rollingstone.com/culture/culture-news/heil-hipster-the-young-neo-nazis-trying-to-put-a-stylish-face-on-hate-64736/ [Accessed 25 February 2020].

Rohde-Dachser, C., 2009. Im Dienste der Schönheit: Schönheit und Schönheitschirurgie unter psychoanalytischer Perspektive. In: M. Kettner, ed. *Wunscherfüllende Medizin*. Frankfurt am Main: Campus, pp. 209–227.

Sauer, B., 2020. Authoritarian right-wing populism as masculinist identity politics. The role of affects. In: G. Dietze and J. Roth, eds. *Right-Wing Populism and Gender. European Perspectives and Beyond*. Bielefeld: transcript, pp. 23–40.

Schaal, E., 2020. "The View": do democrats make up a majority of the show's audience? *Showbiz CheatSheet* [online] 25 February. cheatsheet.com/entertainment/the-view-do-democrats-make-up-a-majority-of-the-shows-audience.html/ [Accessed 25 February 2020].

Schiermer, B., 2013. Late-modern hipsters: new tendencies in popular culture. *Acta Sociologica*, 57(2), pp.167–181

Schippers, M., 2007. Recovering the feminine other: masculinity, femininity and gender hegemony. *Theoretical Sociology*, 36(85), pp.85–102.

Shaw, S., 2021. The women of the insurrection. *Ms* [online] 9 February. msmagazine.com/2021/02/09/trump-women-capitol-insurrection-riot-funding/ [Accessed 24 February 2020].

Sheppard, E., 2021. Pro-Trump capitol rioters like the "QAnon Shaman" looked ridiculous – by design. *NBC News* [online] 13 January. nbcnews.com/think/opinion/pro-trump-capitol-rioters-qanon-shaman-looked-ridiculous-design-ncna1254010 [Accessed 25 February 2020].

Skeggs, B., 2000. The appearance of class: challenges in gay space. In: S Munt, ed. *Cultural Studies and the Working Class*. London/New York: Cassell, pp.129–150.

Snyder, R., 2008. What is third-wave feminism? A new directions essay. *Signs*, 34(1), pp.175–196.

Thomas, J., 2021. Capitol mob wasn't just angry men – there were angry women as well. *Honolulu Civil Beat* [online] 31 January. civilbeat.org/2021/01/capitol-mob-wasnt-just-angry-men-there-were-angry-women-as-well/ [Accessed 25 February 2020].

Toprak, C., 2020. Rassismus in Deutschland: Ihr Linken seid genauso ausgrenzend wie die Rechten! *Der Tagesspiegel*, 18 October, p. 6.

Tortorini, D., 2010. You know it when you see it. In: M. Greif, K. Ross and D. Tortorici, eds. *What Was The Hipster? A Sociological Investigation*, New York: n+1 Foundation, pp. 122–135.

Uscinski, J. and Parent, J., 2014. *American Conspiracy Theories*. Oxford/New York: Oxford University Press.

Valenti, J., 2017. This is the only kind of woman it's okay to be in Trump's America. *Marie Claire* [online] 25 July. marieclaire.com/politics/a28389/types-of-trump-women/ [Accessed 25 February 2020].

Veblen, T., 2007 [1899]. *Theory of the Leisure Class*. Oxford/New York: Oxford University Press.

Vinken, B., 2013. *Angezogen. Das Geheimnis der Mode*. Stuttgart: Klett-Cotta.

Weaver, A., 2017. Melania, are you okay? *Athena Talks* [online] 24 January. medium.com/athena-talks/melania-are-you-okay-5e7eadef2c5 [Accessed 25 February 2020].

Weis, D., 2017. Parallax society. In: O. Blumhardt and A. Drinkuth, eds. *Traces. Fashion & Migration*. Berlin: Distanz, pp. 37–39.

Weis, D., 2020. *Modebilder. Abschied vom Real Life*. Berlin: Wagenbach.

White, M., 2018. Beauty as an "act of political warfare": feminist makeup tutorials and masquerades on YouTube. *Women's Studies Quarterly*, 46(1/2), pp.139–156.

Wiedlack, K., 2019. In/visibly different: Melania Trump and the othering of Eastern European women in US culture. *Feminist Media Studies*, 19(8), pp.1063–1078.

Zachary, B., Merrill, C. and Wolfe, D., 2020. How voters shifted during four years of Trump. *CNN* [online] 14 December. edition.cnn.com/interactive/2020/11/politics/election-analysis-exit-polls-2016-2020/ [Accessed 25 February 2020].

Zienkowski, J., 2019. Politics and the political in critical discourse studies: state of the art and a call for an intensified focus on the metapolitical dimension of discursive practice. *Critical Discourse Studies*, 16(2), pp.131–48.

"AGAINST THE ELITES!"

POP CULTURE AND ITS POLITICS

PART V

BREDA LUTHAR, SONJA EISMANN, MORITZ EGE AND JOHANNES SPRINGER, ATLANTA INA BEYER

14 Celebrity and the displacement of class

The folkloristic ordinariness of Melania Trump

Breda Luthar

Before you applaud, remember: she comes from Slovenia, a country too poor to afford irony.[1]

Introduction

There is a constant tendency for society to observe itself, both at the individual and collective level, through the elites. Celebrity discourse and personification more generally occupy an increasingly significant role in the process. While much of the analytical focus tends to be placed on global celebrities, the production of local celebrities has, in the previous twenty years, contributed strongly to drawing the boundaries between deserving elites, national achievers and corrupt elites, that subvert and prevent the shared unity and mythical constitution of a "national community of sameness" (Bauman, 2001) in many countries. The discourse of celebrification in traditional and so-called new or presentational media is one of the central practical and discursive fields and regimes of subjectivation where the public-private split and the "economisation of the social" (Bröckling, 2016, p.xiv) are normalised. Manifestations of celebritisation are driven by separate but interacting moulding forces of mediatisation, personalisation and commodification and are to be understood as long-term structural developments or meta-processes.[2] As a commodity and representational regime of individuality, celebrity depends on conceptualisations of self-transformation, the primacy of the individual, a culture of self-creation and individual entrepreneurial subjectivity that are central to the present neo-liberal conjuncture. As argued by Marshall (2006, 2014), celebrities are "hyper-versions that express the potential and possibilities of the individual" (2006, p.4) in the conditions of liberal capitalism. Practices of celebrification represent not only means of thinking about the individual, tied to the discourse of individual achievement and failure, but are also tied to definitions of collectivity and nationality/ethnicity/race, femininity/masculinity and class.

This chapter addresses the popular representation of Melania Trump, the Slovenian wife of now former United States President Donald Trump, in the Slovene media and the entanglement of a specific celebrity discourse with regressive populism.[3] Mudde and Kaltwasser (2017) argue that populists

DOI: 10.4324/9781003141150-19

often condemn not only the political establishment but also the economic, cultural and media elite, which they present as one homogenous, corrupt group, "the elite", that are working against the national interests, as traitors to the nation. I focus particularly on the way the representational regime of celebrity as a social and cultural formation, in this case, establishes a dualism of good or meritocratic elites, on the one hand, and undeserving elites, on the other. The discursive work on the apparent deservedness of elite status is closely linked to the mobilisation of ordinariness and, in our particular case, to the practice of "nationing" (Turner, 2018, pp.64–74).

Melanija Knavs (later Melania Knauss and then Trump) was born in a small industrial town in eastern Slovenia, where her mother worked in a textile factory for children's clothes and her father owned a dealership for second-hand cars for a while. When she started modelling, she moved to Ljubljana and, after a short one-year episode as a student at the Faculty of Architecture, transferred to Milan and, later, in her late twenties, to New York. There, she met Donald Trump, real estate magnate and tabloid celebrity – apparently at New York Fashion Week. Trump was still married to his second wife at the time and a perfect character for New York tabloids (rich, vulgar, a womaniser) and courted them constantly in order to keep his name in the media and build his brand. Throughout the end of the 1990s, Trump and Knavs regularly appeared together at events and became a staple mention in New York media society pages. They married in 2005 and she was the US first lady from 2016 to 2020. With this story and the narratives surrounding it in mind, I address the representation of M. Trump in Slovene media as an anchor for a range of national ambivalences associated with the post-socialist state and its position on the periphery of global hierarchies and the role of culture in the formation of a neo-liberal imaginary, including a post-feminist structure of feeling.

Celebrity is an industrial formation and a practice of representation that plays a central role in understanding the operation of political and economic power, the notion of class, of elites and particularly ordinariness/normality as a political and moral category. I propose here to understand celebrity as an epistemic object: an economic and cultural phenomenon, a commodity imbued with politics and not the personal position of an individual (Reckwitz, 2017, p.158). My key questions here are how the intimate and the biographical life of M. Trump has been invented and mobilised, what kind of representational repertoires and patterns are employed and how they structure the moral economy of class in this local context. More specifically, what, in the case of M. Trump, is the impact of the representational regime on the discursive construction of elites and the interdependency of class, gender and race/ethnicity as three basic categories of social distinction? And, finally, how does the characterisation of selected elites as ordinary and their success as deserved contribute to the mobilisation of the anti-elitist structure of feeling tied to the nativist invocation of the people in the context of dispossession and capitalist transformation in "post-socialism"? I hope that through particularising the object of research by taking the case of M. Trump, I will be able, as Bourdieu and Wacquant claim, "to perceive it as a particular case and to generalise it, to discover, through the application of general questions, the invariant properties that it conceals under the appearance of singularity" (1992, p.234).

Celebrity, media and the moral economy of class in post-socialism

The analysis focuses on the media narratives on M. Trump in the Slovene national media one month before and two months after the American elections in 2016. It is based on a discursive reading of a sample of 596 articles published during this period, all in local print media or their digital platforms. The selection includes every article that mentioned M. Trump published between 15 October and 15 November 2016 (the last week of the campaign and a few days after the election) and from 20 January to 20 March 2017 (after the inauguration, when M. Trump became first lady), i e. in two stages, altogether three months. Stories on M. Trump were not confined to tabloid media but were found across the media spectrum and spilled into international politics news or business sections of mainstream media, as well as, for example, fashion journalism and travel feature stories. The media representation of M. Trump crossed media genres and sites – tabloid and political journalism were blended; the publicity around her inhabited a cross-section of media discourses in a broader context of the culture of commercial media, from feature stories in quality dailies and their online platforms to tabloids and women's journals.

Celebrity news in all its generic richness is endlessly repetitious and renewable, perfect for visualised media culture and the constant production of "events". Despite the ubiquity of social media platforms as news sources and platforms for celebrity discourse, numerous empirical analyses argue that traditional and social media use each other as sources, creating a spiral of media sources, where contents and topics transfer from one media into another. "New" and "old" media, therefore, create a double spiral (e.g. between tweets referring to traditional news media and vice versa, traditional media referring to tweets) as co-ordinated ritual practice. This is the case here as well. The landscape of traditional media in Slovenia is highly commercialised due to privatisation processes, neo-liberal restructuring, a tiny local market and the growth of digital media.[4] Similar to elsewhere, the ubiquity of celebrity culture exemplifies a transformation of the cultural role of the media (Turner, 2010, p.18) and their self-interested role in the celebrity-commodity. Boundary making in celebrity discourse (e.g. elites vs. the people, bad elites vs. good elites, West vs. East) is part of wider practices of social ordering to which it contributes on a daily basis. It also supports media institutions as a representational authority or privileged access points to a "mediated centre" of society (Couldry, 2012, p.79). This strengthens their role in the battle for the authority to speak for the ordinary people, i.e. for their legitimacy, relevance and economic survival (symbolic and economic value). The production of M. Trump as a celebrity, a fragmented national spectacle, should, thus, be understood against the backdrop of the commodification of media and changing notions of news. The celebrity genres that are relevant here befit the weak media institutions and are particularly suitable for a fast news cycle and short form of news that can be presented as novel and produced with fewer resources. In brief, they are the epitome of entertainment news values and the industrial production of news as a commodity.

It is necessary for interpreting the news material in this analysis to consider the specific political and discursive space in which the local celebrification of M. Trump is embedded. Larger transformations in "post-socialism" (such as the transition to capitalism, with privatisation, marketisation and commodification, and a general redistribution of public wealth) generate violations of widely felt moral entitlements. Post-socialism in Slovenia and elsewhere is not to be considered only as an economic and political transformation but, first and foremost, a semantic transition that includes breaks and continuities. It is at the heart of the transformation and affects the formation of contemporary subjectivity. The present conjuncture across many countries is characterised by an intensive attempt to move the consensus concerning the definition of family, women's reproductive rights and gender politics in general, and the privatisation of public services, tax policy, *inter alia*. In that context, I use the notion of post-socialism reluctantly and only as a descriptive term that has emerged from a particular historical conjuncture.[5] While it is helpful for referring to these situations, it has problematic implications and orientalising tendencies as it privileges a territorial imagination and obstructs a possible reimagining of socialism beyond state socialism with constant ritualistic anti-socialist invocations that contribute to the de-legitimation of any alternative left-wing politics as "communist" (Owczarzak, 2009; Müller, 2019, p.534).[6]

The orientalising tendencies of the notion of post-socialism are contemporary articulations of "Easternism" as an older symbolic geography. This term offers a better conceptual lens for understanding the celebrity discourse in this particular case, as its symbolism rests on the power asymmetries and the (political) economy. In Easternism, the East, similar to the Orient or the Balkan, exists as a discursive configuration that frames and enables the continuing economic and political peripherality of Eastern Europe. This periphery as a historical construct and social formation is, thus, evoked here as an analytical frame, not an object of analysis. Local mediated celebrity discourse operates within this field of Easternism or East Europeanism (Ballinger, 2017). As Don Kalb observes, "the dependent states of Eastern Europe, with their thoroughly comprador capitalism 'in transition', command at best some 30 percent of the wealth of Western Europe" (2011, p.8). The effects of global neo-liberalism are, therefore, more serious here. In this context, the legitimation myth of finally "catching up with the West" and "returning to Europe" is of key importance. I contend that a neo-liberal cultural imaginary of mainstream popular culture provides "a semiotic frame for construing the world" and actively "contributes to its construction" (Jessop, 2010, p.342). Culture has, thus, a central role in the formation of the social and economic. The practice of celebrification and celebrity discourse is an important "justification narrative" or one of the cultural "mechanisms of consent" (Hall et al., 2013, p.207) for a "post-socialist" neo-liberal consensus that frames people's aspirations and moral economy of class.[7]

The particularly local neo-liberalism and the combination of global and local structural adjustments imposed by the European Union (Kalb, 2011, 2012; Hočevar, 2020) have terminated the mythological promise of emancipation associated with the nation state based on ethnicity that flourished in the 1980s and 1990s. Class analysis and even the use of the notion of class generally vanished after the 1980s – not only from the media and popular culture but also, to a great extent, from social sciences and political discourse, where it was primarily replaced by the notion of exclusion (Boltanski and Chiapello, 2005). Mouffe (2005, p.62), in the context of her critique of the adaptation of social democracy to a neo-liberal terrain, has argued convincingly that society became viewed in that context as composed of middle classes; the only exception is a small elite of the very rich and those "excluded". The concept of class has, however, resurfaced in the academic and parts of journalistic discourse in the last ten years. The notion of class particularly in Central and Eastern Europe was marginalised in political and intellectual discourse against the backdrop of neo-liberal doxa and the delegitimisation of a socialist project. Moreover, the privatisation and the radical redistribution of "public property" was not interpreted as class politics but as a "return to civilisational normal". Boris Buden (2017, p.349), for example, argues that one of the core components of contemporary capitalism is the radical break between the communist past and the eternal present of liberal democracy.

The current public discourse on equality/inequality is, in this context, characterised by the conflicting entanglement of neo-liberalism and communitarian egalitarianism. Local public discourse on class, particularly journalism, mainstream popular media and political discourse on social differences, is replete with contradictory expressions of populism. Class differences are individualised and articulated through popular pseudo-psychology and, by and large, treated as a moral problem. This discourse combines neo-liberal philanthropism, on the one hand, and communal egalitarianism, on the other, while its structural and political aspects are suppressed. This repression of class is related to the "nationalisation" of public life in post-socialism more broadly, where people are routinely addressed as a bounded, solidary national group in which internal differences should not matter and where a "mythical fullness […] i.e. fully reconciled society for which we search in vain" (Laclau, 2005, p.119) is the main goal.

The "communist" past of socialist Yugoslavia is constantly evoked in a revisionist historical framework as a symbolic reference point for the time before the mythical founding moment, before the full realisation of the nation. Ideological and cultural distinctions within the "national body" are considered unnatural and destructive. The repression of class is, thus, rearticulated through the nationalisation of public discourse in post-socialism, where everyday banal or ordinary nationalist talk makes for a naturalised *mise-en-scène* of public life. The systematic class politics of accumulation by dispossession and capitalist transformation, thus, came to be understood as a problem of morality: Its criminal excesses were framed as individualised criminality and personal "tycoonism", "theft" or "corruption" (Kalb, 2018, p.307), while poverty was discussed through the notion of exclusion and a morality framework that relies on and produces "orgies of feelings" (Anker, 2014). Against the backdrop of the moral framing of class, the construction of ordinariness as part of the national community is today the central framing device for the mystification of social stratification and the assignment of gender hierarchies within celebrity representations. It contributes, in the final instance, to the naturalisation of the massive redistribution of public property and the tasks of social reproduction.

She was an ordinary, simple girl: the extraordinary ordinariness of M. Trump and the practice of levelling

The "demotic turn" and the performance of ordinariness (Turner, 2009) in its different disguises are a mode of self-presentation for today's elite, from royals to business elite or politicians.[8] M. Trump has never performed a "down to earth" persona, did not give any interviews for Slovenian media and has, through her lawyers, prohibited any commercial use of her name and branding connected with her name or person. However, the celebrity persona of M. Trump is a commodity and commercial property in itself that was further commodified by and for local media. M. Trump's celebrity has been reprocessed and reinvented. These representations were framed strongly in terms of a discourse of familiarity and ordinariness rather than individualism and extraordinary uniqueness. In order to recognise the moral deservingness of M. Trump for her extraordinary position, she had to be firstly identified as ordinary.

Melania was very diligent and hardworking. She got along very well with her fellow students. (Former teacher in *Svet24*, 10 November 2016, p.4)

She was an ordinary, simple girl. (Jeklin, 2016, p.11)[9]

Her normality and domestication are established through stories about her childhood in a small Slovene town, referring to the ordinariness of her lower-middle-class youth, her good work first at school and then hard work as a model, evoking the physical labour of ordinary people. A backstage narrative was produced with the constant references to her background; numerous media stories interviewed former teachers, school friends, fashion photographers who worked with her or a person who met her once at a party in Ljubljana and who would conventionally confirm her modest commonness as a young girl. This reinvokes shared points of commonality and reference points that she shares with all ordinary Slovenes. The legitimacy and symbolic value of her position was, thus, created against the backdrop of her ordinariness and attributed not only to her achievement in the marriage market and her successful makeover, but simply to her unremarkable ordinary self. (The second step was, as we will see on the next pages, the ethnicisation of her ordinariness.) This mediated ritual practice marks her off from other constructed groups (such as possibly undeserving elites) and implies a common, national structure of feeling, a national community devoid of class distinctions, where the deserving elites and "the people" are both conceived as morally untainted.[10] Deservingness is, similar to celebrity itself, an epistemic position, not the objective characteristic of an individual and, as such, an empty signifier. M. Trump represents an ordinary woman in an antagonistic position towards the local elites. She is, thus, able to speak on behalf of other ordinary women and her visibility and wealth are established as wholly legitimate against the backdrop of her ordinariness.

The ordinariness and the notion of ordinary people do not possess a true essence and are always relational and historically contested. They are assigned, performed or addressed within specific contexts for particular purposes. Williams (1983, pp.225–226) refers to "the ordinary people" as an indication of a "generalised body of Others", in this case, those who are not in power. Langhamer (2018, p.21) argued in her study on the notion of ordinariness in 1950s and 1960s Britain that this ordinariness is a social, affective, moral, consumerist and, above all, political category: Ordinariness does not mean average on some statistical basis (Sacks, 1984) but rather the way someone constitutes themself or, as in our case, is constituted and reinvented by the media. A key question is what selection of values, styles and practices gives meaning to the claim to be ordinary in a particular historical conjuncture. Because of the relational and contested nature of ordinariness, the key task is to establish the discursive regimes against the backdrop of which the subject position of "being ordinary" can be legitimately occupied. In order for M. Trump be able to claim ordinariness, a specific meaning and set of cultural and political values had to be attributed to her ordinariness. How, then, is ordinariness constructed and guaranteed and what is the political work that ordinariness is called upon to perform?

First of all, M. Trump's normality was foregrounded through her local background and ethnicity. The performance of her Sloveneness was a crucial condition of her primordial ordinariness.

"One of our own", a Sevnica girl in the White House […]. (Kovač, 2016, p.3)

Still the daughter of Sevnica […]. (Živčec, 2016, p.24).

And Melania, a comely girl born on the Sevnica side of the Alps […]. (Turk, 2016, p.19)

The reiteration of locality ("Sevnica girl", "the daughter of Sevnica", "a comely girl born on the Sevnica side of Alps […]") contributes to the construction of her ordinariness through a biographical narrative which domesticates and humanises her. M. Trump's ordinariness is rarely stated explicitly but always implied by referring to her former ordinary hardworking and humble self and to her ethnicity. They have ensured her a place in the world of ordinary people and, at the same time, a role of a nationally important achiever. Her "extraordinary ordinariness" (Littler, 2018, p.121) is constructed through, firstly, a biographical narrative that is reiterated countless times and includes the transformation from an ordinary girl to the wife of a millionaire speculator, from a lower-middle-class girl in her provincial hometown, where everything revolved around the children's clothing factory where her mother worked, to the first lady. Her biographical narrative is a narrative of success with two stereotypical mythological elements: life in a small provincial town and a lucky break (rags-to-riches story). Her inclusion into the realm of ordinary women humanises and necessarily depoliticises her persona and Trump politics in general. It can be seen as a media ritual that serves as a tool in ongoing ideological contestations over class, nation and gender.

"Nationing" the ordinary: nation-as-family metaphor and a creation of a folkloristic periphery

There are always background assumptions and a hidden common ground in the genre of ordinary life stories and life experiences behind the extraordinary lives of celebrities. M. Trump has to be domesticated in order to pass for ordinary. Her ethnic familiarity is a source of cultural intimacy and an indispensable element of her legitimacy. Or, as argued by Edensor, "[…] it must be domesticated, localised and personalised" (2002, p.92) to claim a sense of belonging to national identity as a large and vague entity. *Schadenfreude*, i.e. the expression of dislike and disidentification towards celebrities and audiences' pleasure over their misfortune, is an important aspect of celebrity culture (Cross and Littler, 2010). Just like *Schadenfreude*, but in more positive tones, this domestication by ethnicisation enacts a process of levelling whereby value is inscribed in the other in line with a radical reassessment of its supposed intrinsic value. We are not encouraged to be full of envy or resentment or to challenge her position but invited to enjoy her success and visibility in an "interpassive" way. Her normality is, thus, not based on universal criteria but insistently foregrounded on the criteria of ethnicity.

Americans adore her, they are enthusiastic about her and are not bothered by her accent as they understand that she is an immigrant to America, they hold her to be beautiful, sophisticated and successful (Sabadin, 2016, p.2).

N. Pinoza, who was in contact with Melania, convinced us that Melania is still a kind and warm person, although she may have borrowed her speech from her more intelligent colleague (we are not applauding this, of course) […] (Bajt, 2016, p.31).

Audiences of local media were addressed or interpreted as national subjects to embrace different affective dispositions, such as admiration for "one of our own", an ordinary girl who made it in the big world. Creating such symbolic boundaries and dichotomies (e.g. people vs. elite) is a mode of public signification that is not limited to populist performance. However, a key question is who can belong and who is excluded from such an essentialised understanding of "our people". Regressive/ethno-nationalist celebrity discourse, in this case, limits the right to be included to ethnic belonging. M. Trump's moral legitimacy is derived simply from her ethnicity that establishes primordial solidarity. The process of "nationing" enables the production of ordinariness identified on ethnic grounds. Ethnicity and nationality are here naturalised, essentialised, understood as the natural order of things and invested with moral values that elevate the ethnic–national over all other social groupings. Donald Trump's election victory in 2016 was represented in the local media within the framework of "Trump won, America got a Slovenian bridegroom", "[…] our bridegroom Donald Trump" or the "the Slovenian son-in-law D. T.".

Donald, Trump, a Slovenian bridegroom […] (Štefančič, 2016, p.24)

The first Slovenian son-in-law was elected on 8 November […] (Zore, 2016, p.24)

The word bridegroom (Slovenian: *ženin*) is a folkloristic archaism and represents a cliched code common to the speaker and the audience, characterised by the consciousness of the past states of the language while establishing the possibility of a cultural intimacy based on a regressive reification of history. The use of the language of kinship, family membership and nation contribute to establishing a community of ordinariness and sameness. Furthermore, the treatment of M. Trump's marriage and her husband's presidency as a family affair establishes cultural and, therefore, political homogeneity and national unity discursively – without reference to class.

The meritocratic trope and M. Trump's regressive post-feminist femininity

Gender is a formative dimension of nationalisms. Nira Yuval-Davies (1997) warns against a gender-blind theorisation of nationalism and the class system as the construction of nationhood usually involves a specific notion of "manhood" and "womanhood". Nationalisms are, thus, from the outset, constituted in gender power, with women typically constructed as the symbolic bearers of the nation. The family tropes noted above (a Slovene bridegroom, the daughter of Sevnica) and the iconography of the family are well-established cultural projections from patriarchal family life to nation (McClintock, 1995, p.358). By depicting class hierarchies in familial terms, these metaphors construct social differences as a category of nature and are not structured along class lines. The honour and recognition of a nation in this common-sense rhetoric of banal and everyday nationalist talk was, as in a classical patriarchy, closely linked to her female virtues.[11] As argued by Sarah Banet-Weiser (2018a, p.11), popular feminism (post-feminism as articulated in popular media) that thrives in popular culture and popular journalism tends to deny the socio-economic and cultural structures that are shaping our lives. In a related sense, the claims to M. Trump's ordinariness were underpinned not only by ethnicity but also by a particular notion of entrepreneurial yet traditional femininity based on natural gender differences. The proclamation of ordinariness did gender and class work. Gender and her post-feminist femininity rooted in the notion of natural sex differences are here bound to and mobilised into a "classless" nationalist project.

The celebrity of M. Trump is shaped in the local context by a set of entangled and contradictory discourses on femininity – post-feminist femininity (her entrepreneurial and makeover abilities) and traditional femininity (self-reserved, timid, supportive) – that are in tension but support each other. According to Rosalind Gill (2007, 2017), post-feminism should be understood as a distinctive kind of gendered neo-liberalism, a sensibility that is made up of a number of interrelated themes (e.g. self-discipline, individualism, choice, empowerment and a revival of ideas about natural sexual difference). It is a structure of feeling that is expressed and reproduced through a new vocabulary, where personal happiness, self-care and work–family balance as a normative frame and ideal are replacing rights, social justice and a critique of patriarchal structures (Gill in Banet-Weiser, Gill and Rottenberg, 2020, p.5; also see Banet-Weiser, 2012, 2018b). Mainstream celebrity is, thus, restricted to those who can display very specific types of femininity. M. Trump's media persona was, in a related sense, entrenched within a "neoliberal justice narrative" (Littler, 2018, p.68) that prescribes competitive individualism as the remedy for inequality and patriarchal structures but, at the same time, revives the notion of essentialist gender differences and virtues of traditional femininity.

Along the lines of post-feminist sensibility, M. Trump's visibility was interpreted as her meritocratic achievement, the proof of her innocently apolitical entrepreneurial ability and hard work. Feature stories in popular media and women's magazines devoted to her shrewd "female way" of catching a rich real estate mogul and media personality are particularly prominent here. The episode of their first meeting at a party at New York Fashion Week where she supposedly insisted on taking his telephone number without giving him hers and let him wait for her call to arouse desire was reiterated endlessly. A journalist for a tabloid newspaper who visited M. Trump in Florida after her wedding described M. Trump as "[...] extremely nice, warm, sophisticated with charisma typical of people who have high goals in life [...]" (*Svet24*, 2016, p.4). Her entrepreneurial ability and merit lies, therefore, in her victorious triumph on the marriage market and successful makeover from a young provincial catalogue model to a global beauty and sophisticated fashion icon:

> [...] This was an inspiration for our programme called 'A town of beautiful girls', that we have offered to different target groups for the last couple of months. It is still too early to talk about a success of a programme, but I believe success will come sooner or later. [...].
>
> (a representative of a regional tourist board, Stanković, 2016, p.5)

As a calculating, enterprising self, she found her inner talent, made a venture of her life, worked hard and maximised her human capital (her body and female cunningness), and marketed herself in the right way to become what she wanted be.

The image of traditional female domesticity, on the other hand, was continuously referred to in the interviews and short statements from, for example, former teachers, neighbours, school friends and inhabitants of her home town, and invited experts from fashion photographers to politicians or international relations specialists announcing a new era of visibility for Slovenia.

She speaks about politics, but at home, never in public (Klarič, 2016, p.4).

Those who know her say that she will be excellent in her position as a first lady because she is quiet, dignified and very beautiful. She will not impose her views on her husband (*Svet24*, 2016, p.4).

Finally, it is not unimportant if she whispers advice or an opinion about something to her husband at breakfast or morning coffee. That is always binding for a husband. Maybe even about the dispute over the Bay of Piran, as said by our Croatian friends (Mirošič, 2016, p.3)

M. Trump's ordinariness is manufactured and sustained against the backdrop of essentialist notions of femininity based on traditional popular values. The traditional notion of gender is a key moment in the construction of her ethnic–national ordinariness through the biographical narrative: She is quiet but stern; she never challenges her husband in public; she does not speak too much; she has political opinions but expresses them in private; her political influence is, therefore, the soft feminine power of a loyal spouse; she adores her son, to whom she reportedly taught the Slovene language; and is, above all, a devoted mother. Her motherhood is referenced as part of the strategy of personification and a widely used narrative that suggests the caring aspect of her character. Displaying female strength without the performance of female fragility would disqualify her as a woman. The meritocratic biographical story of her individual achievement and "makeover", thus, appears side by side with the image of traditional domesticity, with its naturalised location of men in the public sphere and women in the private sphere of consumer agency. Both aspects of her celebrity persona suppress the difference between "her people" and good elites like her and downgrade the political nature of her role by treating it as question of a family affair. The discourse on M. Trump is to be understood against the backdrop of shared norms of communal conventions that represent what it takes to be "our (wo) man" and, by doing so, restating these notions: humble and hardworking, therefore, probably quietly smarter than she looks (according to her biographers), but also beautiful, sophisticated and with classical highbrow taste. Her body is, thus, the source of her value.

By staging M. Trump's entrepreneurial, yet, traditional femininity and ordinariness, class is present but is reduced to matters of consumer choice. Lifestyle markers of difference, the visual signs of her meritocratic success, such as clothes, jewellery and interior design, are domesticated as matters of personal good taste and an expression of her innate talents.

> […] quiet, nice, sophisticated, diligent, hard-working and with classical taste (she radiated in a pink Gucci pussy-bow blouse, she shined in a long white dress by the French designer Mugler, in a red coat by Ralph Lauren, even in trousers she looks more than excellent) […]
>
> (see *Avenija*, 2016, p. 7)

In spite of the occasional focus on her choice of designer clothes in women's magazines and supposedly understated elegant taste, class differences are presented as non-existent while "ordinariness" is reiterated. The difference between M. Trump's life and that of the readers is acknowledged but class differences are translated into her distinctive lifestyle, which is enabled by her meritocratic success. Diana Kendall (2005, pp.30–35) identifies this overlooking of differences in representations of the lifestyle and opportunity of the rich and famous as the consensus framing ("we are all alike"). Designer clothes, hairstyle and the type of Gucci pussy-bow blouse worn by M. Trump at a Republican Convention only represent a subtle recognition of differences in taste or lifestyle but do not subvert the egalitarian narrative. Her taste and her appearance is, thus, an aspect of her successful self-management and hard work on the self.

This section has suggested a coexistence of neoconservative values and a discourse of post-feminist self-management and choice – referring to not only M. Trump's devoted motherhood and female humbleness but also her body and beauty as female capital and a resoluteness and clever entrepreneurship in search of a rich husband. Traditional parochial femininity is, therefore, entangled with the post-feminist discourse of individualism and individual empowerment, where one's life is directed by the apolitical ideology of personal choice and self-determination. This type of feminism has remarkable visibility in the media, especially through "[...] psychologizing discourse and promotion of female confidence, work on the self, positive thinking, self-esteem and self-love" (Gill, 2017, p.617). While the making of ordinariness in popular media does not eradicate class differences, it makes them appear non-threatening, just and meritocratic, attributing it to moral deservingness and/or talent. Meritocracy discourse is a key ideological term in the reproduction of neo-liberal culture and, according to Littler (2018, p.15), an under-theorised ideological engine of late capitalism. It represents the attempt to absorb the language of equality into entrepreneurial self-fashioning, -regulation and -care that can create individual empowerment that does not deal with the wider structural causes of economic or gender inequality. Celebrification is just one example of ongoing cultural work through which the post-feminist structure of feeling and patriarchy, both in lockstep with neo-liberalism, are articulated (Adamson, 2016). In that broader context, M. Trump's marriage is the proof of her entrepreneurial ability and self-fashioning, as a "generalisable model of profitable self-production" (Hearn, 2008, p.208). Her beauty and sexuality were her assets and her hard work to trade them in is considered as her meritocratic talent or a field-specific capital that paid off in the patriarchal marriage market.

Easternism in action

The subject position of "ordinary Melania" is defined against the backdrop of a set of other subject positions, such as local and global elites. Celebrity culture has a highly formal structure that permits and contains extemporisation by a cast of recognisable social types and standard narratives. The notion of ordinariness cannot be defined without referring to something the "ordinary" reiterates and against which it can be measured – someone exceptional and inauthentic, i.e. social elites in general. With Mudde and Kaltwasser's (2017) comments on populists' construction of a homogenous, corrupt group, "the elite" (see above), in mind, it can be noted that celebrity discourse contributes to a populist logic by promoting, in this particular case, the image of an organic society – of an economic, cultural and social harmony of ordinary people and meritocratic deserving national elites (the nationally important achievers) against morally corrupt or incompetent elites. Popular journalism and the business media in Slovenia are full of fallen CEOs that used to be treated at some point as national treasures and entrepreneurial heroes of post-socialist transformation and capital accumulation (Littler, 2007). Stories of deserving and undeserving elites are reverse sides of the same coin.

Eastern Europe is, against the backdrop of the ancestral discourse of the success of "one of our own", evoked here through colonising tropes that rest on the notion of the sexual availability of the exotic other in a "backward", powerless country that lacks standing and recognition on the global stage. M. Trump was constructed as an ordinary girl who has made it and a national treasure of female virtues. At the same time, however, she was reduced to national "human capital" and market metrics. Countless interviewees, experts and ordinary people have contemplated unique chances for the monetisation of her role as a first lady for the "national brand". According to numerous interviewees, she is heralding no less than a new era of Slovenia's global visibility. A new era of geopolitical relations of Slovenia and the possibility of disrupting the geopolitical status quo were announced. This particular case of self-orientalisation is established through the ongoing lamentation in the media that this unique opportunity to attract American tourists, gain more national recognition and political influence on a regional and global political stage, and affect the redistribution of symbolic visibility will not be exploited properly by incompetent local political elites.

Never before was it so easy to achieve a promotion of Slovenia. […] but obviously our officials sit too deeply in their armchairs to understand this (Lahovnik, 2016, p.4).

We Slovenes don't know how to make something out of this opportunity. If our officials were more resourceful, if nothing else, they could make mugs or T-shirts with her image on and sell these (Pelko, 2016, p.34).

There is a question: How will Slovenia be able to capitalise the fact that a fellow Slovenian countrywoman came to the White House as a first lady? (Malovrh, 2016, p.4).

This an opportunity for Slovenia to take a step into history (Božič, 2016, p.2).

The figure of undeserving elites that is rejected in an emotionally loaded way is – similar to the figure of "ordinary people" – an empty signifier. It stands, alternately, for politicians, public officials in general, bureaucrats, intellectuals or the state in general, in brief, everybody in public office with the exception of "outsider elites", and contributes to the creation of a frontier within the social in Laclau's sense (2005). This critique of elites stands firmly within the imaginary of regressive populism, where anti-elitism lends itself to supporting the disillusion with collectivism and everything political, while the conception of the honest, ordinary "people" stays firmly within a nativist imaginary. It is, however, also embedded into a post-colonial dichotomic categorisation of the East as the West's discursive other, as ahistorical, unable to change and the antithesis of modern entrepreneurial competence ("[...] this is the opportunity that has to be seized [...] it is typical for Slovenes to always look for the mistakes, to spit on everything [...]" Mirošič, 2016, p.3).

Celebrity discourse, therefore, functions here fully within the typically regressive logic that rests on the ordinary people vs. undeserving local elites' dualism. It also draws on local repertoires of rules in which gendered cultural identities are played out within this national context, as shown above. Representing M. Trump within the completely depoliticised framework of controlled idealised post-feminist virtues should, therefore, not be regarded as being in opposition to the staging of virtuous commonness and ordinariness of traditional domesticity. On the contrary, they both contribute to the affirmation of a common national sociality, where ordinary people are in cultural harmony with deserving elites.

Reporting on M. Trump in mainstream liberal western media, on the other hand, often implied (also in a classic Easternist sense) that her role as American first lady was illegitimate by virtue of her low level of cultural and social capital and her fashion practice lacking the attributes of discerning taste. Her undistinctive use of high heels, the lack of political presence and the conventional forms of staged authenticity, her failure to manage the distance between exceptionality, ordinariness, distance and intimacy in the usual ways, the apparent absence of spontaneity and of easy-going entitlement – all of these were converted into positional signifiers of her outsider status. Her fashion choices of mainstream high-end prêt-à-porter brands were seen as too formal and, in fact, spectacles of ostentatious display. She typically displayed a knowledge of what to buy and wear, but not what to wear when and how to embody and perform first lady public persona (Smith Maguire, 2019). In Bourdieusian terms, we could say that her style devalued itself by the very intention of distinction (Sacks, 1984, p.249). As an Eastern European trophy wife, M. Trump was lacking the understatement, sobriety and measure of an American political spouse. This was never more obvious than after the Biden inauguration when the media were reflecting on the past four years. *The Guardian*, for example, applauded the serene, inconspicuous and all-American image of the new first lady, Jill Biden, and Vice President Kamala Harris and praised their fashion choice of independent American brands for the event. The author reminds us how "staggeringly expensive and pointedly non-American" M. Trump's clothes had always been (Cartner-Morley, 2021). The British tabloid *The Sun*, on the same occasion, reprimanded M. Trump for her lack of class (Kelly, 2021). M. Trump's nationality implies that her status represents a mimicry of an image of a lady she cannot fully assume. She mimics the style of the class above her and plays the social scripts of class difference inadequately. The media discourse, thus, normalised the authority of orientalist discourse and policed the shifting symbolic and class boundaries, thereby preserving the hegemonic practice of othering discursively. With a few exceptions, however, this superiority remained within the discourse of civility as an instrument of prestige and power.[12] It did, however, imply the categorical undesirability of women "like her" transgressing class and national boundaries.

Conclusion

Larger transformations, such as the privatisation, marketisation, radical commodification and a general redistribution of public wealth, that are subsumed under the term "post-socialism", give rise to a violation of moral entitlement felt by many. Popular discourses on class (either in celebrity narratives or the majority of media genres, such as human interest-oriented current affairs programmes, sit-com or reality TV) tend to reduce questions of class to a moral problem and elite status (and wealth) to the myth of meritocracy. This also helps to preserve communitarian concepts of cultural belonging symbolically by imagining the morally pure people against immoral and undeserving elites. The Slovene media representations of M. Trump should, therefore, be viewed as linked to contemporary political and economic formations. Gender and the entanglement of traditional and post-feminist femininity are bound to and mobilised into a "classless" nationalist project with a regressive conception of ordinary people along ethnic lines. Her "becoming ordinary" through ethnicisation enacts a process of levelling, on the one hand, but maintains the distance in wealth and celebrity through the specific merit of a deserved success, on the other hand.

Popular culture and particularly the commodified celebrity discourse perform cultural work that contributes significantly to the regressive populist imaginary as a composition of traditional themes, such as family, motherhood, self-reliance, personal success and other elements of a "repertoire of anti-collectivism" (Hall, 1979, p.17). It generally represents a cultural preoccupation with individualism and the obfuscation of the structural and systemic. As such, it has important implications for the analysis of the current moment and the moral economy of class in the current populist conjuncture. Celebrity as a genre of representation, discursive practice and discursive effect, then, is constitutive of hegemonic struggles and class dynamics. It has important consequences for how the individual self, the collective and the relationship between the two are performed culturally and defined in a particular historical conjuncture. Since talk of the subject always implies a regime of subjectification, a description of the subject is also always a prescription: Representing M. Trump within the completely depoliticised framework of meritocratic achievement should, therefore, not be regarded as being in opposition to the staging of her virtuous commonness and ordinariness.

On the contrary, these share the same discursive/performative effects, as they both contribute to the affirmation of a common national sociality where ordinary people are opposed to false elites but in cultural harmony with deserving ones.

Notes

1 Bill Maher, American stand-up comedian and TV host (Real Time with Bill Maher, HBO 2016).
2 In situating celebrity culturally and socially, I follow here Driessen's (2013) distinction between celebrification and celebritisation. The former captures the transformation of ordinary people and public figures into celebrities, i.e. changes at the individual level. The latter, however (celebritisation), is the term for a meta-process that grasps the changing nature of celebrity and the societal and cultural embedding of celebrity in contemporary societies.
3 On the binary construction of people and elites in different populisms, see Morgan's (2020) cultural sociological critique of dominant definitions of populisms and his argument that populism should be understood as a form of public signification rather than an ideology and, therefore, as a form of cultural work.
4 The right-wing Slovenian Social Democratic Party has been instrumental in undermining quality media in the last 20 years. At the time of writing, the media are under heavy pressure from the ruling SDS party that is collaborating closely with the Hungarian Fidesz (a right-wing national-conservative political party) as part of an effort to establish a regional illiberal right-wing alliance. Hungary is financially supporting the Slovenian right-wing media (the press, television stations and an online tabloid news platform) or has a major share in its media companies by operating through different companies within the circle of Fidesz.
5 As argued by Ege and Gallas (2019, p.92), conjunctural analysis is based on the wager that there are themes around which contradictions coalesce that characterise a distinct conjuncture.
6 Štiks and Horvat (2012, p.39) similarly argue that two main reasons behind the rhetoric of incomplete transition from socialism into liberal democracy and a free-market economy seem to be the avoidance of a full confrontation with the consequences of the transition and the preservation of the discourse and relations of Western dominance vis-à-vis the former socialist states, thus, in constant "need" of tutelage and supervision from the West.
7 For our purposes here, the notion of "moral economy" is understood as a consensual understanding of what constitutes rights and a fair distribution of resources (Thompson, 1971), where moral assumptions as much as actual deprivation can be the source of collective action. It concerns the understanding of the functioning of capitalism and social reproduction in a particular historical moment.
8 See Jo Littler's analysis of self-presentation of the business elite as ordinary, just like us, only richer, and the promotion of an image of "extraordinary ordinariness" (2018, p.121). Also see Repo and Yrjölä (2015) on "middle class monarchy".
9 All media quotations in the chapter are translated by B. Luthar.
10 Regressive populism's main characteristic is, as argued by Müller (2017, p.19), the moralistic imagination of politics that sets "morally pure and fully unified" and, ultimately, completely fictional people against elites.
11 From the perspective of a study of the gendered nature of imperialism and colonialism, McClintock (1995, pp.357–358) argued that since the subordination of woman to man and child to adult was deemed a natural fact, hierarchies within the nation could be depicted in familial terms. This results in understanding social difference as a category of nature.

12 For exceptions to the rule of superior decency, see the Maher quote above or a Diet_prada post on Instagram where a caption in "immigrant" English is added to a well-known photograph of M. Trump planting a tree in the White House rose garden in extremely high heels: "I make grave for husband now" (7 November 2020).

Bibliography

Adamson, M., 2016. Postfeminism, neoliberalism and a "successfully" balanced femininity in celebrity CEO autobiographies. *Gender, Work & Organization*, 24(3), pp.314–327.

Anker, E.R., 2014. *Orgies of Feeling: Melodrama and the Politics of Freedom*. Durham: Duke University Press.

Avenija, 2016. Prva dama stila [The first lady of style]. *Avenija*, 21 October, p. 7

Bajt, M.K., 2016. Melanija in Melanija [Melanija and Melanija]. *Grazia*, 1 November, p. 31.

Ballinger, P., 2017. Whatever happened to Eastern Europe? Revisiting Europe's eastern peripheries. *East European Politics and Societies and Cultures*, 31(1), pp.44–67.

Banet-Weiser, S., 2012. *Authentic™: The Politics of Ambivalence in a Brand Culture*. New York: New York University Press.

Banet-Weiser, S., 2018a. *Empowered: Popular Feminism and Popular Misogyny*. Durham: Duke University Press.

Banet-Weiser, S., 2018b. Postfeminism and popular feminism. *Feminist Media Histories*, 4(2), pp.152–156.

Banet-Weiser, S., Gill, R. and Rottenberg, C., 2020. Postfeminism, popular feminism and neoliberal feminism? *Feminist Theory*, 21(1), pp.3–24.

Bauman, Z., 2001. *Community: Seeking Safety in an Insecure World*. Cambridge: Polity Press.

Boltanski, L. and Chiapello, E., 2005. *The New Spirit of Capitalism*. London: Verso.

Bourdieu, P. and Wacquant, L.J.D., 1992. *An Invitation to Reflexive Sociology*. Chicago: The University of Chicago Press.

Božič, K., 2016. S pragmatizom pridobimo [We gain with pragmatism]. *Večer*, 14 November, p. 2.

Bröckling, U., 2016. *The Entrepreneurial Self. Fabricating a New Type of Subject*. London: Sage.

Buden, B., 2017. Afterword: and so they historicized. In: D. Lugarić, D. Jelača and M. Kolanović, eds. *Cultural Life of Capitalism in Yugoslavia: (Post)Socialism and Its Other*. Cham: Palgrave Macmillan, pp.345–350.

Cartner-Morley, J., 2021. Biden and Harris dress to reassure that normal service is restored. *The Guardian International Edition* [online] 20 July. https://www.theguardian.com/fashion/2021/jan/20/biden-and-harris-dress-to-reassure-that-normal-service-is-restored [Accessed 9 March 2021].

Couldry, N., 2012. *Media, Society, World. Social Theory and Digital Media Practice*. Cambridge: Polity Press.

Cross, S. and Littler, J., 2010. Celebrity and schadenfreude. The cultural economy of fame in freefall. *Cultural Studies*, 24(3), pp.395–417.

Driessen, O., 2013. The celebritization of society and culture: understanding the structural dynamics of celebrity culture. *International Journal of Cultural Studies*, 16(6), pp.641–657.

Edensor, T., 2002. *National Identity, Popular Culture and Everyday Life*. London: Bloomsbury.
Ege, M. and Gallas, A., 2019. The exhaustion of Merkelism: a conjunctural analysis. *New Formations: A Journal of Culture/Theory/Politics*, 96(96), pp.89–131.
Gill, R., 2007. Post-feminism?: new feminist visibilities in postfeminist times. *Feminist Media Studies*, 16(4), pp.610–630.
Gill, R., 2017. Postfeminist media culture. Elements of sensibility. *European Cultural Studies*, 10(2), pp.147–166.
Hall, S., 1979. The great moving right show. *Marxism Today* [online] January 1979, pp. 14–20. http://banmarchive.org.uk/collections/mt/pdf/79_01_hall.pdf [Accessed 3 November 2019].
Hall, S., Critcher, C., Jefferson, T., Clarke, J. and Roberts, B., 2013. *Policing the Crisis: Mugging, the State, and Law and Order*. London. Macmillan.
Hearn, A., 2008. "Meat, mask, burden": probing the contours of the branded "self". *Journal of Consumer Culture*, 8(2), pp.197–217.
Hočevar, M., 2020. Kapitalistična država in kriza liberalne demokracije [The Capitalist State and the Crisis of the Liberal Democracy]. Doctoral thesis [online]. University of Ljubljana, Faculty of Social Sciences. https://repozitorij.uni-lj.si/Dokument.php?id=135333&lang=slv [Accessed 3 February 2021].
Jeklin, B., 2016. Gostilna pri Melaniji [Inn at Melanija's]. *Zarja*, 15 November, pp. 11–15.
Jessop, B., 2010. Cultural political economy and critical policy studies. *Critical Policy Studies*, 3(3–4), pp.336–356.
Kalb, D., 2011. Headlines of nation, subtexts of class: working-class populism and the return of the repressed in neoliberal Europe. In: D. Kalb and G. Halmai, eds. *Headlines of Nation, Subtexts of Class*. Oxford; New York: Berghahn Books, pp.1–36.
Kalb, D., 2012. Thinking about neoliberalism as if the crisis is actually happening. *Social Antropology*, 20(3), pp.318–330.
Kalb, D., 2018. Upscaling illiberalism: class, contradiction, and the rise and rise of the populist right in post-socialist Central Europe. *Fudan Journal of the Humanities and Social Sciences*, 11(3), pp.303–321.
Kelly, L., 2021. Dr Jill Biden is giving Melania Trump a quick lesson in class. *The Sun* [online] 22 January. https://www.thesun.co.uk/news/13824923/jill-biden-melania-trump-quick-lesson-in-class/ [Accessed 9 March 2021].
Kendall, D., 2005. *Framing Class: Media Representations of Wealth and Poverty in America*. Lanham: Rowman & Littlefield.
Klarič, M., 2016. Od male Sevnice do praga Bele hiše [From small Sevnica to From small Sevnica to Doorstep of White House]. *Svet24*, 7 November, pp. 4–5.
Kovač, D., 2016. Od očetovega milijona do predsednika ZDA [From father's million to the President of the USA]. *Dnevnik*, 10 November, pp. 3–4.
Laclau, E., 2005. *On Populist Reason*. London: Verso.
Lahovnik, M., 2016. Dan D za prvo damo [D-day for a first lady]. *Slovenske Novice*, 7 November, p. 4
Langhamer, C., 2018. 'Who the hell are ordinary people?' Ordinariness as a category of historical analysis. *Transactions of the Royal Historical Society*, 28, pp. 175–195.
Littler, J., 2007. Celebrity CEOs and the cultural economy of tabloid intimacy. In: S. Redmond and S. Holmes, eds. *Stardom and Celebrity: A Reader*. London: Sage, pp.230–243.

Littler, J., 2018. *Against Meritocracy. Culture, Power and Myths of Mobility*. London and New York: Routledge.

Malovrh, D., 2016. Kranjska v vesolju, špeh v Beli hiši [Krainer sausage in the space, bacon in the White House]. *Slovenske Novice*, 10 November, p. 4.

Marshall. P.D., 2006. *The Celebrity Culture Reader*. New York and London: Routledge.

Marshall, P.D., 2014 [1997]. *Celebrity and Power: Fame in Contemporary Culture*. Minneapolis: University of Minnesota Press.

McClintock, A., 1995. *Imperial Leather. Race, Gender, and Sexuality in the Colonial Contest*. London: Routledge.

Mirošič, I., 2016. Hišica v preriji ali paraziti na periferiji [Little house on the prairie and the parasites in the periphery]. *PN*, 11 November, p. 3.

Morgan, M., 2020. A cultural sociology of populism. *International Journal of Politics, Culture and Society* https://doi.org/10.1007/s10767-020-09366-4 [Accessed 3 February 2020].

Mouffe, C., 2005. *The Return of the Political*. London: Verso.

Mudde, C. and Kaltwasser, C.R., 2017. *Populism: A Very Short Introduction*. Oxford: Oxford University Press.

Müller, J.W., 2017. *What Is Populism?* London: Penguin.

Müller, M., 2019. Goodbye, post-socialism! *Europe-Asia Studies*, 71(4), pp.533–550.

Owczarzak, J., 2009. Postcolonial studies and post socialism in Eastern Europe. *Focaal*, 53(2009), pp.3–19.

Pelko, T., 2016. Če bi Melanija postala prva dama, bi bil zelo ponosen [If Melania becomes first lady, I will be very proud]. *Svet24*, 8 November, p. 34.

Reckwitz, A., 2017. *The Invention of Creativity: On the Aestheticisation of Society*. Cambridge: Polity Press.

Repo, J. and Yrjölä, R., 2015. "We're all princesses now": sex, class, and neoliberal governmentality in the rise of middle-class monarchy. *European Journal of Cultural Studies*, 18(6), pp.741–760.

Sabadin, D., 2016. Prva ženska ali prva Slovenka [First lady or first Slovene]. *PN*, 8 November, p. 2.

Sacks, H., 1984. On doing "being ordinary". In: J. Maxwell Atkinson and J. Heritage, eds. *Structures of Social Action: Studies in Conversation Analysis*. Cambridge: Cambridge University Press, pp.413–429.

Smith Maguire, J., 2019. Media representations of the nouveaux riches and the cultural constitution of the global middle class. *Cultural Politics*, 15(1), pp.29–47.

Stanković, D., 2016. Osemletnega Andreja so kazensko presedli k sošolki Melaniji [Eight-year-old Andrej was punished and had to sit near schoolmate Melanija]. *Dnevnik*, 10 November, pp. 5–6.

Štiks, I. and Horvat, S., 2012. Welcome to the desert of transition! Post-socialism, the European Union and a new left in the Balkans. *Monthly Review*, 63(10), pp.38–48.

Svet24, 2016. Melanija Trump – iz Sevnice na vrh sveta [Melanija Trump – from Sevnica to the top of the world]. *Svet24*, 10 November, pp. 4–8.

Štefančič, M., 2016. Dan zapisan sramoti [A day of shame]. *Mladina*, 11 November, pp. 24–31.

Thompson, E.P., 1971. The moral economy of the English crowd in the eighteenth century. *Past & Present*, 50(1), pp.76–136.

Turk, R., 2016. Sajens fikšn [Science fiction]. *PN*, 11 November, pp. 19–20.

Turner, G., 2009. *Ordinary People and the Media*. London: Sage.

Turner, G., 2010. Approaching celebrity studies. *Celebrity Studies*, 1(1), pp.11–20.

Turner, G., 2018. Commercialism, the decline of the "nationing" and the status of the media Field. In: G. Rowe, G. Turner and E. Waterton, eds. *Making Culture. Commercialisation, Transnationalism, and the State of "Nationing" in Contemporary Australia*. New York, London: Routledge, pp.64–74.

Williams, R., 1983 [1976]. *Keywords: A Vocabulary of Culture and Society*. London: Flamingo.

Yuval-Davies, N., 1997. *Gender & Nation*. London and New York: Sage.

Zore, J., 2016. Na gradu pričakujem Trumpovo rezidenco [I am expecting Trump to set his residency in a castle]. *Delo*, 10 November, p. 24.

Živčec, D., 2016. Ameriško jutro v Melanijini Sevnici [American morning in Melania's Sevnica]. *Večer*, 10 November, pp. 24–25.

15 Who says who's cool, and how much is it worth?

The convergence of elite luxury fashion with streetwear styles

Sonja Eismann

There are basically two conflicting views of fashion: one that it is elitist, aesthetically progressive, tasteful, of high quality and pricey. The other, that it is a mass consumer good, unimaginative, of poor quality, tasteless, vulgar and cheap. Both exist alongside each other and are intertwined in a paradoxical way: Where the perspective of fashion as an elite good is often accompanied by notions of well-balanced restraint and excess, that of fashion as vulgar paints it as off balance in almost all respects. It is always too ostentatious or too boring, too colourful or too bland, too short or too long, too sexy or too frumpy. One of the most striking analyses of these paradoxical perspectives on fashion came in the form of an exhibition by Judith Clark and her psychoanalyst husband Adam Phillips which scrutinised notions of vulgarity in fashion over a time span of 500 years for her show "The Vulgar" at the Barbican in 2016. By placing items as seemingly different as Grecian-style dresses by Vionnet, Dutch bonnets, Yves Saint-Laurent's Mondrian dress, and garments by Rudi Gernreich and Vivienne Westwood alongside each other, visitors' preconceptions of elegance and obscenity were challenged. Many of the designs that had caused outcries in their respective eras seemed perfectly "tasteful" to present-day spectators. The curators state in an interview with Barbican associate curator Sinéad McCarthy that

> (t)he exhibition's thesis is that nothing is intrinsically vulgar; vulgarity is the consequence of description. The exhibition intentionally resists ascribing vulgarity, but starts from the assumption that we know vulgarity when we see it. It is, however, surprisingly diverse in its references when you look at the use of the word over time: the common or the popular, the vernacular, or associated with imitation – all are included in fashion.
>
> (McCarthy, 2016)

Whereas their affirmative view of vulgarity draws heavily on concepts of the body, sexuality and popular culture as multifaceted, vitalising forces in (high) fashion, this article is more interested in how class and race influence what we perceive as elitist and anti-elitist clothing. While the two antagonistic viewpoints in respect to fashion stated above may be held in general or only

DOI: 10.4324/9781003141150-20

toward certain trends or brands, it cannot be ignored that there is an almost inevitable element of bias in them: Clothing from expensive labels, worn mostly by economically well-off people, is less often condemned as vulgar than cheap fast fashion items worn by the poor, or ostentatious designs by the "nouveaux riches", who are commonly castigated for possessing money but no taste, since the latter cannot be commodified so easily. But what happens when these markers of distinction are turned on their head in a sort of class and race drag?

In this chapter, I would like to propose the example of the recent rise of streetwear inspired by hip-hop and skate culture into the ranks of high fashion to demonstrate how established lines of distinction between popular, even vulgar, and "elite" aesthetics may – or may not? – be blurred. This very phenomenon marks a shift from established paradigms of thinking about fashion hierarchies and notions of high, low and counter culture. Pierre Bourdieu contrasted the tastes in sports of the old elites with "a new form of poor-man's elitism" in the "new fractions" of the bourgeoisie and petit bourgeoisie in his chapter on sports, fashion and class tastes in "Distinction", first published in 1979 (Bourdieu, 1984, pp.219–220):

> On the one hand, there is respect for forms and for forms of respect, manifested in concern for propriety and ritual and in unashamed flaunting of wealth and luxury, and on the other, symbolic subversion of the rituals of bourgeois order by ostentatious poverty, which makes a virtue of necessity, casualness towards forms and impatience with constraints (...).
>
> (Bourdieu, 1984, p.220)

However, today, regarding the case of luxury streetwear, the situation is different in ways that might not have been anticipated in the 1970s. There is no longer a clear distinction between the "flaunting of wealth" and "ostentatious poverty". Wearers of this newly lauded style of clothing – such as hooded sweaters, trackpants or sneakers – are keen to display the "authentic" humbleness of precarious urban living in racialised parts of the city, and simultaneously, their own economic prosperity through the fetishisation of these very identities in popular culture. Parts of the population who were looked down upon and sometimes even feared by the moneyed elites have become role models for a sophisticated, athletic and cutting-edge look – a very expensive one – that distinguishes its wearers as not only in sync with their times but ahead of them. It might even be said that the very milieus whom the economically well-off feared during, for example, the London riots, that were set off by the police killing of Marc Duggan, a young man of Afro-Caribbean and Irish descent on 4 August 2011– predominantly young people, quite a few of them of colour and clad in sportswear, angrily protesting and looting and destroying stores – have now become those to whom they turn for fashion inspiration. By drawing on the example of the rise of streetwear, which until recently used to be associated with poverty, inelegance and even deviance, this article aims to analyse how new concepts of elites in fashion and the appropriation of non-elite vestimentary aesthetics that are often seen as provocations of elite tastes may overturn preconceived notions of class hierarchies but without finally deconstructing them.

Blurring social standing with clothing

Sartorial choices have been a means to display one's personal wealth and status in society since the beginnings of fashion. From the late 14th century, when fashion cycles as we still know them today started to develop, until the epistemological break that constituted the French and the Industrial Revolution, access to fashion was largely limited to social and financial elites. With the advent of mass production and a less rigid class system in the 19th century, more people had a greater set of sartorial choices. As Diana Crane writes in her survey on "Fashion and Its Social Agendas",

> [i]n the late nineteenth century, clothing appears to have had a special significance as one of the first consumer goods to become widely available. Clothing was useful for "blurring" social standing, as a means of breaking away from social constraints and of appearing to have more social or economic resources than was actually the case.
>
> (Crane, 2000, p.67)

At around the same time, after the Second Industrial Revolution, sociologists, such as Thorstein Veblen and Georg Simmel, posited that the display of wealth of the leisure class, as Veblen termed it, inspired people from lower classes to emulate them. In his famous treatise "The Theory of the Leisure Class" (Veblen, 1899), he coined the term "conspicuous consumption", which also included clothing, as a mechanism for showing off unproductive riches with the aim of asserting social and economic power. Simmel further elaborated on what was later to become known as the "trickle-down effect" of fashion by stating that "the fashions of the upper stratum of society are never identical with those of the lower; in fact, they are abandoned by the former as soon as the latter prepares to appropriate them" (1904, p.133).

However, power balances in the game of fashion influences changed distinctly with the formation of popular culture as a governing principle for virtually all areas of life after the end of World War II. On the one hand, by "the 1960s, the fashion industry had begun to produce and distribute more than enough products for everyone to be able to dress fashionably" (Medvedev, 2010, p.646). On the other, it was no longer so much the association with wealth that made clothing styles desirable for an increasing number of consumers but their degree of pop- or subcultural authenticity. Ted Polhemus wrote in the catalogue for the Streetstyle exhibition at the Victoria and Albert Museum in 1994–1995, where the styles of subcultures such as the Teddy Boys, Ravers and B-Boys were portrayed: "Today, as high culture has given way to popular culture, it is the litmus test of 'street credibility' that is crucial. If it won't cut it on the corner, forget it" (1994, p.6). In opposition to the trickle-down model, which to him seemed outdated, he created a "bubble up" theory which he described as follows:

> First there is a genuine streetstyle innovation. This may be featured in a pop music video and streetkids in other cities and countries may pick up on the style. Then, finally – at the end rather than the beginning of the chain – a ritzy version of the original idea makes an appearance as part of a top designer's collection. Instead of trickle-down, *bubble up*.
>
> (Polhemus, 1994, p.10)

Significantly, the first photo that accompanies these introductory words shows two stylish young men of colour standing somewhere in an English city. The caption reads: "West Indian men on a streetcorner in Liverpool in 1949" (Polhemus, 1994, p.6).

"The very best in luxury along with the very best sneakers" – the democratisation of fashion?

Today, in an age where "anyone across the world could imitate a new style instantaneously" due to "the democratization of fashion" (Medvedev, 2010, p.646) as well as the gigantic amounts of clothing that fast fashion retailers produce daily, we see a striking convergence of these two seemingly antagonistic models. Street fashion, in a hip-hop context, started out as the necessity to look "fresh" in a climate of rivalling street gangs in the Bronx of the early 1970s. It implied "a certain capacity to stand out on account of one's style despite the economic conditions of poverty in which one finds oneself" and was associated with sports brands like Adidas, Nike or Reebok (Chiais, 2020, pp.65–66).

After the erosion of distinct subcultures in the new millennium, the term is nowadays almost exclusively associated with the styles of young urban youths of colour who prefer sports-related, extremely casual attire, and it has become the Holy Grail for luxury fashion houses. It has also fuelled the growth of subsidiary business branches. The people behind Highsnobiety, which went from "being a streetwear blog to a media brand and production agency" (Bobila, 2017), not only keep millions of readers informed about the latest trends in streetwear fashion and adjacent fields such as tech or urban arts with their website and biannual print magazine. They also "build campaign ideas that resonate with the next generation of tastemakers" (Bobila, 2017) with their production agency Highsnobiety+, that works with clients such as Gucci, Louis Vuitton or Nike.

> I think what we saw [...] was that you could mix high luxury with what was happening on the streets. If you look at the last 12 years of what we've done, our content has always covered the very best in luxury along with the very best sneakers from Nike and Adidas, as an example

says Managing Director Jeff Carvalho in an interview with fashion website Fashionista (Bobila, 2017). Highsnobiety was founded in 2005 by advertising student David Fischer in Geneva, Switzerland, as a blog to highlight products from the streetwear range that appealed to him personally. It is not surprising that the rise of this once rather intimate blog coincides with the erosion of power of established fashion media. Whereas influential print magazines used to have privileged access to fashion information and events, the democratisation of the digital realm has greatly reduced these advantages. Fashion shows that were once open only to elitist circles from magazines such as Vogue, Harper's Bazaar or Elle, who, thus, knew about new trends well before the general public, can now be filmed by Instagrammers and instantly be shared with whoever has access to the internet. Therefore, it seems logical that not only did fashion media or the reception of fashion became more permeable, even more "democratic", but that fashion itself also started to be more diaphanous. The makers of Highsnobiety have noted the merge of streetwear "with luxury fashion as early as 2011, when Riccardo Tisci debuted a Rottweiler print with his men's Fall 2011 collection" (Bobila, 2017). Alongside the continuing growth of the media brand itself, it is estimated that "the streetwear market, which, by 2015, was valued at about $75 billion [...], and its impact on high-end brands will continue to grow" (Bobila, 2017).

From Dapper Dan to Supreme – imitation, scarcity and the question of ownership

This marks a distinct shift in attitudes of high-end brands towards urban leisure or sportswear, which used to be, when worn outside of sports functions, deemed proletarian or even sub-proletarian. The sporting apparel of the popular and racialised classes, unlike "true" subcultural styles such as Punk or New Wave, used to be rather unrecognisable as a distinct style by high fashion producers and the general public, but was perceived more as a sign of inertia or refusal to dress in "real" clothes. The 2000s and early 2010s saw repeated moral panics regarding the "decadence" of the lower classes, which were stereotypically depicted as forever living in casual sporting clothes, or, specifically in Great Britain, (*white*) "Chavs" who were specifically shamed for looking cheap even in expensive clothing since their "alleged consumption of high-end brand Burberry goods has negatively affected the brand" (Le Grand, 2020, pp.215–216). At the same time, these "poor looks" were exploited for ironic fashion statements, such as, with a much criticised undertone of cultural appropriation, the use of the "China Bag plaid" by Marc Jacobs for Louis Vuitton in 2007. Whereas Karl Lagerfeld infamously (and only allegedly) stated in 2012 that whoever wears track pants has lost control over their lives, six years later, Louis Vuitton appointed Virgil Abloh, an African-American designer who had started out with streetwear, as director of their ready-to-wear men's line, as the first person of African descent in the brand's history. Demna Gvasalia, the Georgian-German designer behind the much hyped Vetements luxury label, quickly became famous for selling unisex casual wear, inspired by street fashion, for high prices – most notably the yellow-red DHL T-shirt for 245 Euro in 2016. Supreme, the legendary skate brand from New York City, "dropped", as it is phrased in streetwear speech, a highly coveted, pricey collaboration with Louis Vuitton in 2017 – 17 years after they had received a cease and desist letter for producing an unauthorised collection featuring the same prestigious French logo.[1] But they were in no way the first to try out this tongue-in-cheek approach to a luxury brand with a street style attitude that imitated but also demystified the signifiers of "elite" exclusivity. One of their most famous forerunners was Daniel Day, aka Dapper Dan, who is now recognised as a central figure in the genesis of street fashion. From the middle of the 1970s onwards, this self-taught tailor from Harlem sold textiles – counterfeited and of his own design – on which he imprinted the logos of fashion houses such as Gucci, Fendi and Louis Vuitton (cf. Chiais, 2020, pp.66–67) to locals and celebrities such as Mike Tyson, Salt'n'Pepa and LL Cool J alike. With his combination of street smarts and the desire to partake in the unattainable, glamorous lifestyles represented by luxury brands (that can also be noticed in the naming of the houses of queer Black and Latinx ballroom subcultures in the New York of the 1980s and 1990s: e.g. House of Balenciaga, House of Gucci, House of Mugler), he laid the foundations for the chic streetwear of today. Dapper Dan had to close his shop on 125th street in 1992 due to lawsuits for counterfeiting from several brands. However, the affair is not over at this point, as Eleonora Chiais recounts:

> Almost thirty years later, in 2017, Gucci's creative director Alessandro Michele in fact presented in his Cruise Collection in Florence a jacket that was practically identical to the one created by Dapper Dan in 1989 for Olympic athlete Diane Dixon. This instantly sparked off a controversy and, under the leadership of Dixon herself, the Italian brand was asked to attribute the creative paternity of the jacket to the Harlem tailor. After an agitated exchange of words, the brand entered into co-operation with Dapper Dan, making him become the protagonist of a brand advertising campaign in the same year and asking him to design a collection to be sold in the Gucci stores in the spring of 2018. Nor did it end there. The brand itself, in fact, chose to help Dan reopen his historic shop, supplying him with the materials and the templates of the original logos for impressing on his garments, permitting him to restore life to his activity.
>
> (Chiais, 2020, pp.67–68)

Whereas in Dapper Dan's times, there was scarcity because of limited means of production in a one-person business, street style is now characterised by artificial shortages created by the producers themselves. The semiotics and the economics of "the drop" – the release of a limited-edition product, frequently a co-operation between two well-known brands, or even artists, which is heavily publicised and sure to be sold out soon, has become one of the most important elements of streetwear consumer culture. Partaking in the world of luxury streetwear as an "elite" consumer requires a lot of disposable income but also the agility and knowledge to be able to secure these scarce goods early enough.

Interestingly, this logic of artificial scarcity has crossed over into one of the most elite fields of culture and commerce, i.e. the art world. The latest hype in digital art, so-called NFTs (non-fungible tokens) that are situated as files on block chains and usually display jpgs or gifs, are announced as "internet drops" that can be acquired by collectors.[2] Beeple aka Mike Winkelmann, the computer artist who made a sensation when an NFT of 5000 of his digital works sold for more than $69 million on March 11, 2021, at Christie's, has already worked with luxury fashion brand Louis Vuitton – artworks of his were featured on their Women's Spring 2019 ready-to-wear collection. Another instance in which cool commodity cultures merge with high fashion.

"You'll be cool when we say you're cool" – streetwear culture and its elites

Former gatekeepers of refined tastes and abundant wealth – editors of fashion magazines, publicists or art connoisseurs – are, in a way, outdone by the product culture in all of these cases. Neither their access to money nor their definition of "distingué" taste puts them in the dominating position of an elite when it comes to the crossover of streetwear and luxury fashion. While in earlier decades, subcultural aesthetics were desirable for many because of their societally antagonistic meanings, but became virtually worthless once they were appropriated by moneyed elites (because this was deemed to strip them of remnants of anti-elite attitudes), streetwear culture is not a subculture in the classic sense, but rather an attitude that is able to incorporate financial success and an embrace by the upper classes into its own concept of cool. Andrew Raisman, CEO of Copdate, a website and app for "copping every drop", puts the power dynamics in streetwear luxury fashion this way: "You want to be cool? You'll be cool when we say you're cool, and that's the only thing that matters" (Milnes, 2018).

Not surprisingly, there is a wide range of criticism levelled against this unlikely alliance in the world of fashion.

> Leading the forefront of this streetwear-ification of luxury are designers like Abloh, Kanye West and Demna Gvasalia, whose designs for both Vetements and Balenciaga can be so mundane, yet so expensive, that anyone who isn't acting like they get it must think it's some kind of joke,

writes journalist Hilary Milnes in her article on the "infiltration" of luxury fashion by streetwear (2018). Something that looks so cheap cannot be so expensive, seems to be her argument. On the other hand, defenders of the values and authenticity of subcultures have another axe to grind:

> The significance of streetwear is now being dictated by the financial interests of high-end fashion brands. In the past, groups that did not identify with the mainstream or elitist lifestyles that ruled the fashion scene, used personal style to pursue their own needs, shaping a style for themselves and their communities as a whole. Whether it were skaters or hip-hop influencers facing issues like marginalization, racism, or classism, streetwear projected the sentiments of such new social groups and subcultures. Looking back, it visually addressed political inequalities and defied conventional ideology. The fact that high-end brands now sell garments directly influenced by streetwear subcultures and style them within their collections next to $3000 handbags, is not congruent at all to what streetwear is all about; luxury customers will never lead a marginalized lifestyle, and may probably never experience real social inequality and hence, cannot fully identify with it. The owners of luxury labels, who are almost exclusively white and rich, are now profiting off of the symbols that were once created to openly denounce them.
>
> (Cavazos, 2017)

But DJ and streetwear influencer Mick Batyske, who is quoted in Milnes' piece, takes another stance on the publicly perceived gulf between the image of streetwear as cheap, informal clothing and the prices attached to it by big labels:

> The price points are astronomical. But when you buy luxury streetwear, you're not paying for the most handcrafted, highest quality piece of garment. You're buying into a subculture. It's something you either are a part of or want to be a part of, and when you think about it, that's what any luxury brand has always been about.
>
> (Milnes, 2018)

What Batyske is alluding to is not only the aura of luxury that customers try to get into by shopping for costly brands but also the fact that the subculture in question is, unlike quite a few others, not adverse to money but, instead, sees it as an authentic validation of its own success. Streetwear is closely associated with young Black men from underprivileged inner-city areas who are, in turn, most commonly associated with hip-hop culture. As a result of the political, social and economic marginalisation of Black communities in the Global North, wealth and its symbolisations in the form of commodities have been regarded by quite a few of them as acts of defiance or rebellion against an oppressive, exploitative system of *white* supremacy that tries to restrict access to its resources. Inside this framework, it may seem logical to elevate streetstyle into the ranks of luxury labels to parallel it with the ascent of the Black community as a whole. In that sense, streetwear producers "on the ground" are not averse to elites or to elitism as such but to the conditions of who, i.e. members of which groups, is allowed or not allowed to enter into elite spheres and permitted to be successful in the first place. On the other hand, what Ian Cavazos states in the quote above about the majority of owners of these labels being *white* and rich still holds true – in many if not most cases, they are the ones profiting from the boom in luxury streetwear.

All of this concerns primarily the side of producers. However, what are people who can afford to buy these clothes and who are very often not part of the community who inspired their designs getting out of the conjunction of these two disparate realms?

Inconspicuous consumption

Shayne Oliver is the founder of independent fashion label Hood By Air and one of a growing number of ground-breaking Black designers combining streetwear with avant-garde elements. He put male bodies into latex pants, high-heel boots, spaghetti tops or tight-fitting bodices with a futuristic street smart slant long before more mainstream designers used traditionally female clothing codes for male apparel. Having closed down his brand for a number of years before reappearing in 2020, he has a rather disillusioned view on the subject:

> Today, everyone wants to look like a rapper from the 90s or 2000s, but they go to the luxury brands in order to dress that way. This kind of streetwear is for older people – for people who have some sort of false nostalgia for a past where they didn't dress like a rapper because it was too urban for their taste, but they want to do it now because the big luxury brands are endorsing the trend and it makes them feel young again when they dress this way. It's a type of stay-young fashion, and it's quite strange.
>
> (Tudor, n.d.)

In his view, then, the customers of luxury streetwear are "older" (than 35, most likely), well-off, and defined by a common desire for youth and someone else's past. Charlene Lau has termed this modern-day predilection for clothing that looks cheap but is not, in a reversal of Veblen's fin-de-siècle thought, "'inconspicuous consumption': flashing lots of cash, but not – at least on a very surface level – looking like you have" it (Bramley, 2017). This makes the act all the more luxurious. What is more, wearing sportswear, which is often synonymous with streetwear, gives the impression of active, able, hardened bodies that are in no way required for the participation in the middle and upper classes of today's digitised neo-liberalist capitalism but which have become, for this very reason, all the more desirable. Further proof of this is the continuing boom in "athleisure", a combination of athletic and leisure apparel, which demonstrates that you have the body to work but the money to relax.

What is more, *white* fashion observers argue that middle- or upper-class consumers of the younger generations – they, too, represent a large market – especially feel the opportunity to embrace ideas of diversity and progressiveness: Ideas, it could be added, that neo-liberalism endorses to make capitalism more complete and all-encompassing. In a performance of race and class drag, they slip into the styles associated with "authentically" poor people of colour and wear their identities as a second, very visible skin, without actually having to come close to the latter's precarious lives and realities. Big fashion houses try to cater to this supposed "millennial mindset" by hiring new artistic directors who appeal to the desire for diversity and authenticity. Or, as David Fischer of Highsnobiety says, "We build campaign ideas that resonate with the next generation of tastemakers" (Bobila, 2017), meaning moneyed members of the millennial and Gen Z age cohort.

That streetwear is considered vulgar but – traditionally – the opposite of ostentatious is another asset in this blurring of distinction lines between the high and the low. However, although established concepts of street and luxury, Black and *white*, or bubble up and trickle down are challenged in this remarkable reversal or even criss-crossing of influences, it remains to be seen whether the latter manages to shatter the system of class and race privilege in the fashion industry in the long run – or if the fashion train keeps on travelling to unexpected locations, as it is, forever paradoxically, expected to do. Virgil Abloh, one of the most prominent figureheads of the luxury streetwear trend, surprised fashion fans by answering the question about the role of streetwear in the 2020s as follows: "I would definitely say it's gonna die, you know?" (Allwood, 2019). Instead, he proposed to go vintage and make use of our archives. The question then is: What will be deemed worthy to be archived in the long run? In the near or far future, will it still be items such as Chanel suits, from established designers such as Yves Saint-Laurent, that are already being collected like prestigious works of art, or will it also be designs by people who will follow in the footsteps of Abloh? Elites are characterised by longevity and tenaciousness.

Notes

1 https://stockx.com/news/supreme-x-louis-vuitton-kim-jones/.
2 https://www.monopol-magazin.de/was-sind-nfts-und-warum-sprechen-alle-gerade-davon.

Bibliography

Allwood, E.H., 2019. Virgil Abloh: Streetwear? It's definitely gonna die. *Dazed* [online] 17 December. https://www.dazeddigital.com/fashion/article/47195/1/virgil-abloh-end-of-2010s-interview-death-of-streetwear [Accessed 3 March 2021].

Bobila, M., 2017. How Highsnobiety went from being a streetwear blog to a media brand and production agency. *Fashionista* [online] 11 April. https://fashionista.com/2017/04/highsnobiety-agency-streetwear [Accessed 16 March 2021].

Bourdieu, P., 1984 [1979]. *Distinction. A Social Critique of the Judgement of Taste.* Translated by Richard Price. Cambridge: Harvard University Press.

Bramley, E.V., 2017. Distressed fashion. Making sense of pre-ripped clothes. *The Guardian* [online] 18 August. https://www.theguardian.com/fashion/2017/aug/18/distressed-fashion-making-sense-of-pre-ripped-clothes [Accessed 16 March 2021].

Cavazos, I., 2017. How high end brands have co-opted streetwear. *Couturesque* [online] 17 August. https://www.couturesquemag.com/single-post/streetwear-politics [Accessed 16 March 2021].

Chiais, E., 2020. From the Bronx to the boutiques. The rise of street style in the fashion industry. In: M. Stefano, ed. *The Culture, Fashion, and Society Notebook 2020*. Milan: Pearson Italia, pp. 55–77.

Crane, D., 2000. *Fashion and Its Social Agendas. Class, Gender, and Identity in Clothing*. Chicago: University of Chicago Press.

McCarthy, S., 2016. Introducing the vulgar. Fashion redefined. *Barbican* [online] 5 October. https://www.barbican.org.uk/read-watch-listen/introducing-the-vulgar-fashion-redefined [Accessed 16 March 2021].

Medvedev, K., 2010. Social class and clothing. In: V. Steele, ed. *The Berg Companion to Fashion*. Oxford: Berg, pp. 645–647.

Milnes, H., 2018. How streetwear infiltrated luxury fashion. *Digiday* [online] 5 March. https://digiday.com/marketing/streetwear-infiltrated-luxury-fashion/ [Accessed 16 March 2021].

Le Grand, E., 2020. The "Chav" as folk devil. In: C. Critcher, J. Hughes, J. Petley and A. Rohloff, eds. *Moral Panics in the Contemporary World*. New York: Bloomsbury, pp. 215–234.

Polhemus, T., 1994. *Streetstyle*. London: Thames and Hudson.

Simmel, G., 1904. Fashion. *International Quarterly*, 10, pp. 130–155.

Tudor, E., n.d. Shayne Oliver: "Streetwear is for old people". *Nowfashion* [online]. https://nowfashion.com/shayne-oliver-streetwear-is-for-old-people-25844 [Accessed 16 March 2021].

Veblen, T., 1899. *The Theory of the Leisure Class. An Economic Study of Institutions*. New York: Macmillan.

16 Against hipsters, left and right

A figure of cultural elitism and social anxiety

Moritz Ege and Johannes Springer

"The workers, not the hipsters": mainstream anti-hipster populism in Germany (2016)

In this chapter, we examine the meanings and functions of the cultural figure of the hipster in the context of political populisms, especially in Germany. For that purpose, we choose an entry point that lies a few years back. After the US election in 2015, Sigmar Gabriel, then head of the German Social Democratic Party (SPD), Vice Chancellor and Minister for Economic Affairs and Energy, published a newspaper opinion piece titled "The workers, not the hipsters – why Social Democrats must change after Trump's victory" (2016). In the article, Gabriel argued that contemporary right-wing populist movements wanted to return to the hierarchical cultural ways of the 1950s. In doing so, they were able to exploit people's disappointment with growing social inequality and the increasing disconnect between political leaders and "the citizens". These, he writes, were the main reasons for the wide-spread hatred towards the elites which had come to the fore in Trump's victory in the US. Against that tendency, around the globe, parties like the social democrats had to, once again, take care of the people for whom that party had been founded 150 years ago. Gabriel declared: "If you lose workers in the rust belt, hipsters in California will not save you" (2016, p.32).

Crucially, however, Gabriel argued that what was needed to "win back" the working class was an increased attention to *cultural* questions, rather than primarily economic ones, such as redistributive policies. With all its left-populist rhetoric, in policy terms, his anti-hipster diagnosis led him to conclusions that were different from, for example, Bernie Sanders' near-successful Democratic Socialist strategy. For politicians like Gabriel, the "the workers, not hipsters" analysis supported a corporatist-centrist political position, similar to the British "Blue Labour" discourse a few years earlier.[1] Social democratic advisor Niels Heisterhagen made similar points, designating the cultural adversary as "hipsters with MacBooks" (2018). He, too, was supporting a "communitarian" corporatist political course focused on the interests and the (apparent) cultural orientations of the core, primarily male industrial workforce in, for example, the automotive industry. Numerous

DOI: 10.4324/9781003141150-21

politicians, social scientists, public intellectuals and pundits echoed similar lines in the following years.[2]

In this chapter, we ask why this figure was given such prominence. What were the cultural and political preconditions that made it plausible for the hipster to take on this role of "the elite"? And what do these figurations tell us about configurations of anti-elite sentiments and their political implications in that conjuncture? These questions concern specific sociocultural and discursive situations. At the same time, cultural historians have shown that the hipster figure has been present throughout the pop-culturalisation of public and political life, beginning in English-speaking countries in the mid-20th century. So has pejorative talk about hipsters, from mild mockery of their pretentions to full-grown hipster hate. Many texts on hipsters mention the figure's negative reputation; a few focus on "hipster hate" directly (Athill, 2018; Erbacher, 2012; Ikrath, 2015; le Grand, 2021; Rabe, 2012; Stahl, 2010; Süß, 2019), without, however, centering the "elite" motif.[3] Anti-hipster sentiments grew in the US during the 2000s and 2010s. In Germany, the hipster figure first gained broader prominence around 2010.[4] It entered mainstream political rhetoric and sociocultural analysis in 2015, in the "Brexit/Trump moment". We argue that this wave of anti-hipster statements documents and contributes to broader shifts in anti-elite attitudes and rhetoric. They are indicative of a view of the social as a *cultural* figuration that has come to dominate the political scene.

In order to trace anti-hipster as anti-elite discourse, to decode its implications and to better understand its conjunctural contexts, we review some of the ways in which figuring hipsters as elites has allowed social observers to criticise cultural distinctions, practices and subjectivities that have emerged from subcultures and pop culture more broadly, situated as they are in capitalist societies with their basic structures – and fine distinctions – of social classes, gender ideologies, racial hierarchies and attributions and other contradictions. Overall, we consider "hipster" a pop-cultural, everyday ("lay") term that is shaped by "classificatory struggles" (Bourdieu, 1984, p.184) and serves a number of communicative functions. It should not be taken for an exhaustive scientific analytic/description.[5] The question therefore isn't whether the hipster really exists or not, or to whom exactly it refers, or whether the term adequately represents the people to whom it points. Rather, the analytical task is to survey the wide range of phenomena to which such labels refer and to better understand what happens when these cultural figures are used by specific actors in specific contexts to make sense of complex and contradictory social, economic, cultural and political processes.

The chapter draws on publicly available sources (newspaper articles, popular music) and observations that we gathered over the years during research projects on sub- and pop culture during field studies, primarily in Berlin, London and Chicago. We will review arguments made in the literature on the topic in recent times, and we will add some media analysis. On a theoretical level, we argue that the case of the hipster illustrates the broader importance of processes of figuration for understanding how social inequalities are culturalised.

One more time: "What was the hipster"

More than ten years ago, Mark Greif, Kathleen Ross and Dayna Tortorici asked: "What was the Hipster?" (Greif, Ross, and Tortorici, 2010). In posing this question in the past tense, they performed a hipster gesture: being one step ahead of their time. In a quasi-parody of social research, these authors – mostly literary scholars, i.e. not social scientists – defined hipsters as a "subcultural formation" (Greif, 2010b, p. viii) that has become prominent in US cities, but also, increasingly, globally, that they roughly dated as belong to the time from 1999 to 2010.[6]

The term "hipster" famously emerged in the 1940s in the subterranean world of bebop jazz, among African Americans, where it referred to the possessors of esoteric, oppositional, or countercultural knowledge. It was popularised by – primarily *white* – jazz afficionados in the 1950s, denoting a person with subcultural attitudes and knowledge. Leland points out that at the time, "to be a hipster was to be labeled a hoodlum, hooligan, faggot, n*-lover, troublemaker, derelict, slut, commie, dropout, freak" (Leland, 2004, p.38). Semantics of a revolutionary, edgy, Faustian figure were shaped in now-canonised essays like Broyard's "A Portrait of the Hipster" and Mailer's "The White Negro". In the following decades, the term "hip" remained in the vocabulary. During the 1990s and 2000s, primarily in the context of "Indie" music culture, the loose meanings associated with the term again began to coalesce around a specific cultural figure. This figure increasingly became the object of commentary and other pieces of social observation in city newspapers, blogs, films and comics. Now, hipster referred to a social type in the post-college phase, i.e. their early/mid-20s, and its aesthetics-centred, (often seemingly voluntary) low-budget lifestyle in gentrifying urban areas. It was mostly used pejoratively or ironically, seldom as self-designation. Here is one of many lists of the typical male hipster at the turn of the millennium, which Greif et al. (following earlier journalistic observers) took to be indicative of an "ironic" nostalgia of "white-ethnic", lower middle-class, suburban aesthetics and sentiments:

> trucker hats, undershirts called 'wifebeaters', worn as outwear; the aesthetic of basement rec-room pornography, flash-lit Polaroids, fake wood paneling; Pabst Blue Ribbon [cheap beer], 'porno' or 'pedophile' mustaches; aviator glasses; Americana T-shirts for church socials, et cetera; tube socks; the late albums of Johnny Cash, produced by Rick Rubin; and tattoos. *Vice* magazine … Alife … American Apparel. (Greif, 2010c, p.9)

There were hipster sub-typologies, such as the proper Indie rock hipster, the bike messenger hipster, the humanities student, the barista, the self-avowed member of the "creative class", the green hipster or the hipster rapper. Some of the signifiers of hipsterdom changed subsequently, as numerous blog and newspaper articles, memes and books of hipster observation remarked.[7] A few years later, for example, such a list would probably have included fixed-gear bikes, iPhones, or "third-wave coffee".

While the well-known narrative we have repeated here is useful enough, it also conceals a certain messiness in how the term was and is being applied: Overall, it is unclear whether "hipsters" are anything like an actual group at all. Greif (2010b, pp.9f) usefully – but also somewhat inconclusively – distinguished between three senses in which the term was being used: first, to designate a subcultural type, though one that had become increasingly dominant. Secondly, the term referred to an aesthetic pattern and corpus, "hipster culture", epitomised at the time by the films of Wes Anderson or, later, Greta Gerwig.[8] In a more general sense, "hipster" was also used to (critically) denote a larger stratum of "hip consumers" who may not embody a subcultural type in a stricter sense of the term or adhere to much of hipster culture, but signal some form of adherence to that aesthetic. Cultural analysts characterise hipsters as the epitome of late-modern cultural conundrums – such as conformist individualists (Schiermer, 2014), anti-consumerist consumers or anti-fashion fashionistas (Rüß, 2021) – but these are meanings that are often also implicit in critical everyday usages of the term.

Even if, by now, there are numerous newer studies of hipsters, including media analyses (MacDowell, 2012; Newman, 2013; McIntyre, 2016; Dorrian, 2020; le Grand, 2020, 2021), analyses of various aspects of hipster culture (among others, the contributions in Steinhoff, 2021) and academic fieldwork on hipsters from geography and sociology (Murray, 2020; Scott, 2017),[9] the basic operating principles of hipsterdom have remained constant. Similar cultural values, attitudes and dynamics are expressed by different signifiers in homological fashion.[10] Nevertheless, there have been a few shifts as well: Aside from concrete matters of consumption and taste, the hipster is now seen more strongly as the embodiment of distinction-oriented people of the new-ish middle classes in cities: "a young, trendy, highly stylized, urban, middle-class person … engaged in occupations or entrepreneurship in the creative industries" (le Grand, 2021, p.33). In that sense, the use of the term has expanded its reach, from a small subcultural typification to a larger-scale, quasi-sociological term.

Anti-elitist hipster critique in pop culture since the 1950s: a typology

Transatlantic pop music history can serve as an indicator for the evolution of anti-elite attitudes, affects and rhetoric connected to the hipster figure. Pop songs often take up newly emergent cultural matters, terms, phenomena and attitudes before more formal commentators do (see already Klapp, 1954; Hall, 1968). Furthermore, as the hipster arose within a musical subculture and expresses, at least historically, countercultural challenges to a cultural and political status quo, pop music and subcultural worlds are particularly important fields of resonance for the establishment, critique and negation of this figure. Using pop music lyrics since the 1960s as a jumping-off point, we argue that there are five basic types of critique of the hipster in pop music discourse: the square, subcultural, anti-consumerist, left-political and pop-sociological critiques. These are connected to a range of anti-elite sentiments and arguments. Their political implications have shifted considerably over the decades.

Square critique

The proponents of what we call the square critique of hipsters argue that hipsters are irresponsible, frivolous, detached from reality and even scandalous in their conspicuous disregard of dominant societal values. The square critique usually reacts to the presence of hipsters – a presence that is seen as challenging, even offensive, to the critic and her/his audience – who often defend "traditional" values and forms of life. From the beginning of this discourse, the square critique is intertwined in culture wars. In Roger Miller's (1964) country song *Squares Make the World Go Round*, for instance, the narrator draws on a dichotomy between the "square little man", who assumes responsibility for politics and society – he should be mayor or governor. This "little man" may not be glamorous, the narrator implies, "but government things can't be made do/by hipsters wearin' rope-soled shoes". Hipsters can't be trusted to get anything done responsibly – just as their light shoes are ill-suited for physical labour or projecting a respectable public persona.[11]

In cases like these, hipsters are thought to refuse to confront the serious nature of life, including political responsibility and hard work. During the 1940s and 1950s, the early decades of the hipster, the social type – as a lived subculture that was increasingly becoming represented in different media – embodied a bohemian critique of conformist ("square") ways of life, and of repressed subjectivity, not least in the realm of sexuality. In ethnographies from the time, such as Ned Polsky's (1967) work on pool hall hustlers or Howard S. Becker's (1951) study of jazz musicians, the hipsters' disdain for the squares becomes palpable. The "square critique" of hipsters therefore has a resentful structure: Hipsters look down on squares, they critique societal norms and those who adhere to them in a conformist fashion. But, hipster-haters protest, so-called "squares" are the more responsible, more realistic and moral people. Questions of lifestyle, class and labour, racial attitudes and sexuality become closely intertwined in hipster critique from the beginning.[12] The actual content of hipsters' critique of societal norms is largely ignored in texts like these.

Chronologically, this is an early form of anti-hipster figuration that is set in the Fordist era. However, it is a recurrent one, as recent examples illustrate: When, for example, German conservative politician Jens Spahn scolds international "elitist hipsters" in Berlin for "shutting themselves off from normal citizens" (2017, p.40), the square critique resonates.

Subcultural (immanent) critique

The second type, which emerges later, comes from a different position. It can be called the *subcultural critique* of hipsters: Its proponents argue, from an anti-hegemonic, anti-establishment position, that hipsters have turned counterculture, subcultural creativity and the radical nature of dissident life choices into a field of trivial distinctions and power games and into a commodity. For an illustration, we turn to a band that comes out of the 1980s

New York City hardcore scene, *Sick of It All*, who called out hipsters ("Fly-by-night scenester, Fly-by-night hipster") in their song *Who Sets the Rules* (1994) as "snobby fools, setting the rules, living in contradictions and lies". Here, hipsters are hypocritical moralists: "Living independent – what a joke, those caught dipping in a trust fund won't go broke – all this rhetoric is so hard to bear when the fool assumes a high and mighty air". Here, the scenester-hipster doesn't necessarily follow a distinct aesthetic code. Rather, the term refers to a position, an orientation and a social-psychological type that is fundamentally opposed to dominant culture and politics only in its self-image. Hipsters embody upper-class privilege, which they conceal ("caught dipping in a trust fund"), and their critics accuse them of classism. Furthermore, hipsters offend because of their orientation towards the merely fashionable *and* their desire to tell others what to do: Being "snobby", they think they are better than others. This attitude is connected, the lyrics imply, to a lack of experiences with the real world, and hence, with a general sense of the hipsters' inauthenticity ("never suffered, never paid the dues, living contradiction, live a lie oh so fashionable"). There is some overlap with the square critique of hipsters' lack of down-to-earth realism. *Sick of It All* show themselves very concerned about the scene's purity, about the sincerity and deeper conviction of its members and they are ready to police its boundaries by calling out inauthentic "fraudsters". Here, the hipsters are the children of the wealthy – the economic elite, in a basic sense – within the scene, and there is a sense that as would-be rule-setters, they are vying for cultural influence – in that subculture, but perhaps also beyond, in a world where a (broadly speaking: Post-Fordist) culturalised economy is taking shape.

In historical terms, this example comes from the years before the onset of the "hipster boom" in the late 1990s and early aughts. The speakers' own social position is that of underprivileged men with *white* working-class backgrounds who *have* paid their dues in life and in the scene over time. Their cultural position is that of purist, authenticist believers in a subculture with its oppositional meanings and values. This critique of hipsters thus has a sociological or economic and a cultural component. To some extent, hipsters are also accused of turning subcultural existence into a fashionable commodity. In more recent times, they would have been accused of "gentrifying" the scene. The anti-consumerist critique of hipsters, however, also goes beyond inner-scene distinctions and reaches into a wider post-subcultural world. Overall, the relevance of subcultural critiques of hipsters has increased in recent decades and can be found in a number of genres. Perhaps paradoxically, it appears to have greater resonance in a post-subcultural world, where subcultural aesthetics are everywhere, but their borders and a sense of commitment to subcultures seem more nebulous.[13]

Anti-consumerist critique

Staying in New York, but moving on a few years and changing musical aesthetics, the early 2000 anti-folk formation – at the time and in the place that

Greif et al. comment upon, when the hipster is also a relatively clear subcultural type – *The Moldy Peaches* (2001) sing: "See the hipsters in the park, hair so styled, clothes so dark. Prefab moulded hamburgers. I don't want a bite of yours." Here, too, hipsters fail at being authentic. Most importantly, however, they lack real individualism and creativity. They are a social type recognisable through their aesthetics, but, again, their existential failures as well: The lyrics indicate a disdain for cookie-cutter subcultural identities which can be consumed "prefab"-style and don't feature the creative, nonconformist DIY-sensibility that genres like anti-folk demand of its adherents.

In contrast to the *SOIA* example, hipsters are considered something like a scene of their own in this case, not just a fraction within a scene, reflecting the broader shift towards an Indie "hipster culture". For *The Moldy Peaches*, however, these hipsters do not seem to embody subcultural "elites". Rather, they represent a consumerist pseudo-avantgarde that – expanding a bit on this theme – embodies the neoliberalisation of youth cultures/subcultures, gentrification and a touristic gaze. The implication is that if hipsterism dominates the local scenery, true subcultures must break off from that world and build other, less consumerist worlds. The song lyrics are generally more poetic, playful and cryptic and also less class-coded than the earlier examples. In terms of sartorial, bodily and also musical aesthetics, however, it seems likely that the authors of *What Was the Hipster* would have considered *The Moldy Peaches* themselves a hipster band. Indeed, the flaneur-like practice of observing and classifying social types ("see the hipsters in the park") can itself be seen as a foundational hipster gesture. While the cultural figure had become recognisable, then, by the early 2000s, its application and boundaries remained contentious.

As the hipster aesthetic and type spread after the early 2000s, in part through consumption practices, the anti-consumerist elements became the dominant pattern of critique in anti-hipster-elite discourse. At the same time, it is important to note that the anti-consumerist and the subcultural critique of hipsters often mix. The anti-consumerist critique can be expressed by people with strong ethico-political commitments to anti-consumerism. But it is also brought forth by people who themselves live somewhat happily within a consumerist world, but have aversions against the specific, purportedly anti-consumerist, symbolically low-brow type of consumer distinction embodied by hipsters. Such critique tends to come from people invested in traditional upper-middle-class lifestyles or high fashion, for example. In its resentment against (young) people with strong moral stances, anti-consumerist critique can also take on similarities with the square critique of hipsters mentioned above.

(Left) political critique

The *(left) political critique of hipsters* is closely related to the subcultural and the anti-consumerist critique. It stresses hipsters' obsession with seemingly trivial matters of style, it laments their political harmlessness or complicity and – in its radical forms – accuses them of treason to a political movement or "the revolution". Whereas the strictly subcultural critique defends subcultures from imposters, the left political critique claims that hipsters have replaced real politics with mere (sub)culture. In that view, hipsters are *not political* enough – even if they believe(d) in a loosely defined cultural revolution or cultural–political vanguardism.

These arguments originally emerged in the process in which "new" left political movements separated from the "old" left in Western countries during the 1950s and 1960s/1970s and contributed to the related split between "hip" countercultures (such as the hippies and punk rockers) and more straightforwardly political movements at the time, such as socialist, communist or anarchist groups, or the dissident labour movement. (Radical) left political critique of subcultures and more specifically hipsters as politically inefficient and counterproductive starts out as a move in the strategy debates within a broadly dissident, oppositional world, but it quickly turned into a wider social diagnostic and a critique of emergent social milieus or fractions of the new middle class.

The left critique of hipsters can also be brought forth as an identity-based critique – as a critique of class-based and racial privileges. In this vein, for instance, Patrice Evans accused *white* hipsters in the US of not questioning their own racial privileges and remaining complicit with structures of white supremacy from which they benefit (Evans, 2010; see also Greif, 2010a), as in gentrification processes in urban settings. Anti-gentrification discourse has often intertwined with a (left) political critique of hipsters (Friedrichs and Groß, 2021; le Grand, 2020; Myambo, 2021).

There also is a more strictly class-oriented, left-*populist* version of this political critique that scolds hipsters for being anti-majority, anti-popular or demophobic. It resonates in arguments like those of centrist social democrat Niels Heisterhagen quoted above. Left critiques of hipsters in this vein thus oscillate between a critique of distinction and consumption, a critique of "elitist" anti-popular and anti-common-people attitudes and a (cultural) defence of the really existing working class. To become a hipster, from this viewpoint, is to turn away from the majority of society that a political movement could be expected to want to represent or "win over". In its implicit or explicit cultural defence of the really-existing popular classes and their ways of life, this version of left political critique can resemble the square critique, combining economic populism with conservative views of racial and gender politics and broader matters of values and culture. If, however, a populist critique is embedded in a radical pro-popular tradition, such as the IWW in the US, the left-populist critique may also oppose racism and nationalism.

Cultural figures: What kind of thing is a hipster, theoretically speaking?

Having established this typology of anti-hipster sentiments and arguments and having highlighted the concomitant forms of anti-elite critiques, as well as some relevant sociocultural contexts, we agree with Steinhoff that hipsters are a "nebulous cultural figuration that needs to be examined in the specific cultural, historical, economic and political context of its (re)production" (Steinhoff, 2021, p.2). But what kind of sociocultural entity are hipsters, ultimately? Is the hipster merely a stereotype prevalent in the media? Or should they also be approached as an actual group? A set of tastes and practices? Questions of this sort are among the oldest in research on subcultures. For present purposes, we want to introduce an analytical vocabulary that takes seriously theories of cultural figures and of figuration. The different meanings of "hipster" – a social type, a subculture, an aesthetic, a social theoretical problematic, or a series of critiques – should not primarily be understood as deriving from clearly distinct different phenomena "out there", but rather from different aspects of *what cultural figures are and do*: They connect and mediate between typifications, representations, social-diagnostic discourses, references to social mechanisms, labelling practices and everyday self-fashioning – an overall process we call cultural figuration, following the work of cultural anthropologists like Mary Weismantel (2001) and John Hartigan (2005).[14]

First, then, like in the early instance of square critique, "hipster" refers to a cultural figure in the sense of *a social type* or a series of related (family-resemblant) social types: "hipster" is an intersubjective mental abstraction of people with certain properties, that is, of people who look, think, behave in certain predictable ways. Some recent researchers – sociologists, anthropologists, geographers – have used "hipster" in this sense: as a descriptive or analytical term for groups whose members share certain traits and, to some extent, an overall identity. Scott, for example, professes to use "hipster as an umbrella term that is indicative of a broad subgroup within the new petite bourgeoisie – those creating cultural micro-enterprises." (Scott, 2017, p.62; see also Murray, 2020) If not considered in relation to broader processes of figuration, this type of usage tends to presuppose a form of *realism* in social theoretical terms: It assumes that these patterns exist and a set of people are "typifiable" in Alfred Schütz's classic sense of that term (Klapp, 1958; Schutz, 1962). Speaking of hipsters as a social type in this way goes a certain way. But it is neither exhaustive nor fully satisfactory. In empirical terms, it is unclear how the "type" should be defined and delineated. Is it really just about micro-entrepreneurs, as Scott claims? This seems very limited. Or do stylistic markers suffice for membership in that category, such as a haircut, bike, taste in beer or coffee? Analytically, a category that relies on such different criteria and presupposes homologies between them seems dubious. Furthermore, despite its uses, this definition tends to ignore the role of media in shaping the ideas people may have of figures like the hipster.

Second, therefore, the hipster can be seen as a figure in the sense of a *discursive construction in various forms of media and genre*. In most cases, what people "know" about hipsters derives from these representations rather than from direct experience. Being an ontologically different kind of object, these constructions do not derive straightforwardly from a faithful representation of the related social type. Often, in the case of the hipster, these meanings are negative. Whereas the hipster-as-social-type view is usually epistemologically realist, this view tends to be constructivist – and, in contexts of social observations, is brought forth in a critical fashion, connected to analyses of those figures' social functions. Analytical terms like "folk devil" (le Grand, 2021; building on Cohen, 2011; Stahl, 2010) or "scapegoat" can be used to point out the cultural work that is done through stereotypical media figuration.

In a third and closely related sense, the hipster serves as a *diagnostic (epistemic) figure* in the context of specific narratives and meta-narratives – in discourses, be they academic in a stricter sense or not. Such figures are made to embody overarching cultural tendencies. By talking and writing about a cultural figure like the hipster, people often aim to formulate a broader diagnosis of "our times", like in Greif et al.'s diagnoses. Understanding the work performed by the hipster as a diagnostic figure requires a more interpretative, even speculative approach that highlights hidden meanings – for example, the implicit messages that are inherent in subcultural styles, such as the vintage/retro orientation of most hipster figurations since the 2000s, a major theme of later cultural analyses.[15]

The meanings associated with the hipster should also be seen in light of a fourth aspect of figuration a: the hipster as the *embodiment of a sociocultural principle or mechanism*. Since its emergence in the United States of the 1940s, "hip" has referred to something like a *modus operandi* of cultural avant-gardism outside of "legitimate" high culture. In that sense, hipsterism can be seen – in the tradition of analyses of subcultural style – as a *generative principle* of aesthetic and existential differentiation from mass society which, over time and in different places, results in different aesthetics, not always necessarily those of the people labelled "hipsters". In that sense, to be hip is to claim a cultural or aesthetically advanced status: For example, New York's New Wave impresario and Warhol companion Glenn O'Brien noted in 1987:

> To be hip [...] has been to be where it's at, pointed towards where it's all going. Hip is the posture of the futuristic elite, who are living today by tomorrow's standards, ideals and ideas. To be hip is to live in the future. Twenty minutes or more. And to be hip is to possess the attributes of the future as they are perceived from where it's at.
>
> (O'Brien 1997, quoted in Rabe, 2012, p.201)

Recent anti-hipster discourse that describes hipsterism primarily as (anti-) consumerist pretension, the self-delusion of early adopters, has its roots in meanings like these. It is also clear that hipsters, in that sense, can hardly be populists – as aesthetic vanguardists and therefore elitists, they are at least in that sense distinct from "ordinary" people.

Fifth, as a cultural figure, the hipster is also constituted through verbal *acts of labelling*. This takes place in all sorts of media and in face-to-face interactions: "you hipster", "hipsters like us", "these hipsters over there". In that sense, "hipster" is not only a type, or a construction, an element in a diagnostic narrative, or a logic, but also a label that people use, more or less tactically and strategically, in everyday interactions, where interpersonal relationships and broader meanings ("empirically situated social identities, often identified through certain nicknames in lay discourse", le Grand, 2021, p.32) mesh. Stressing this aspect has been an important contribution of empirical, mostly ethnographic sociological and anthropological accounts of figuration. Decoding the meanings and the uses of the hipster figure as a sociocultural phenomenon requires us to figure out who calls whom what, when, where, in what context and why. As Ege (2011, 2013) has argued in regard to the "Proll" figure, which in the German context often functions as a counterpart to the hipster, such labelling contributes strongly to "classification struggles" between social groups (Bourdieu, 1984; see also le Grand, 2020). This is enacted to a significant extent within the informality of everyday life.

In each of these aspects of the figuration of hipsters, anti-elitism plays out slightly differently, but they ultimately fuse into one process of anti-elite figuration. The linkage between the hipster *as social type* – the first sense mentioned above – and the "elite" moniker depends on more or less intuitive observations about that type's "typical" position in society and (privileged) social/family background. However, these are far from self-evident, as the wide range of hipster sub-typologies in early writings on hipsters illustrates – a bike messenger or barista, for example, is unlikely to be part of a socioeconomic elite. In *media constructions* of the hipster figure, as we have seen, the association with an elite is established through connotations to cultural and social elitism (such as in song lyrics), but also through explicit references. Political discourses – such as the ones from Germany in the post-2015 conjuncture we cited in the beginning of this article – play an important role here, as articulations between groups and meanings are being established, confirmed or challenged. For example, McIntyre has shown how the right-wing press in the UK and the US associated the hipster figure and its apparent contradictions, such as anti-capitalist consumerism (2016, p.93), with the left-wing Occupy protests in New York in 2011. If the people protesting are hipsters, the articles seemed to imply, their causes cannot really be authentic and worthy. Constructing the Occupy protests as populated by hipsters, and the hipster figure as elitist, hypocritical *and* left-wing, delegitimated a radical and to some extent heterogeneous political movement. In doing so, right-wing media fought off any "homological process of identification through different fields of class relations" (McIntyre, 2016, p.94). Or, to put it slightly differently: It strategically stressed the cultural differentiation and elitism of hipsters in order to portray radical left-wing politics as anti-popular. This type of media blends into social-diagnostic narratives. Here, linkages between anti-hipsterism and anti-elitism are spelled out most directly, especially in narratives where hipsters figure as representatives of a rising/emergent social stratum, of a new upper-middle class that increasingly replaces older "elite" formations from an industrial and socially more conservative age and, crucially, is in a dominant relation towards lower social strata. The fourth sense, that of *a cultural logic or principle* (being "hip" as being ahead of others) presents hipness as structurally elitist, which provokes counter-reactions (Tyler, 2015, p.506).

All of this helps explain why the hipster *as a cultural figure* could become diagnostically useful to so many people, but it remains somewhat general. In the last two sections of this chapter, we move from the typological overview to short, more context-specific spotlights onto examples of anti-hipsterism since the late 2000s. One case is an ethnographic snapshot on a group of teenage hip-hop practitioners in Chicago, the other one is about two rap songs from Berlin and their contexts. While the contextual differences and the types of material (ethnography vs. song lyrics analysis) bar any direct comparison, these spotlights illustrate how aspects of the cultural figure of the hipster play out in recent pop-cultural products and situations. Equally importantly, they document forms of anti-hipster, anti-elite critique that seem quite similar, but differ significantly in their political implications, even if they both express a critique of the cultural dimension of social inequality.

Hipster hate as self-defence? An ethnographic snapshot from the US, 2009

The first case stems from a phase of field research one of the authors of this chapter (Ege) did in Chicago around 2009/10. We therefore start with an ethnographic vignette. It is situated in a youth centre where Ege was doing participant observation over a period of two months. During the summer break, people between 15 and about 19 years of age who were active in hip-hop culture gathered there under the supervision of a few "old heads" of that scene who were paid by the city parks department for that purpose. The group was ethnically diverse, primarily male and almost all of them were native Chicagoans. This was a self-selected, affinity-based group whose members were working on their artistic skills. None of them came from middle or upper-class families. They came from different parts of the city to attend this programme. On intermittent days, they focused on graffiti "stylewriting", breakdance (sometimes also footwork to Chicago juke), and writing rap lyrics and spoken word poetry.

One late afternoon, a cypher for freestyle rapping formed, and people took turns on the mic. True to form, this included insulting competitors' lack of skills, their old-fashioned haircuts, cheap sneakers, fake gangsterism et cetera. Ethnic stereotypes weren't entirely taboo, but they were a touchy subject. One of the young men, Tony, a (*white*) Italian American, generally had a strong presence in the group. He was an accomplished graffiti writer, a very good talker and dresser and moved between different social scenes with ease. Rapping, however, was not his main forte. In the cypher, he became the target of an attack that was unusually sharp. Mark, who was less suave, less eloquent and less "hip" in his usual comportment, but also a much better rapper, counted off the ways in which Tony was a cultural imposter and appropriator who merely played at being into hip hop. This culminated in Mark's calling Tony a "hipster" and a "hipster-rapper" – a label that summed up these accusations, completed by a swipe at his predilection for wearing "tight pants", a main signifier of hipsterdom at the time. Mark's diss was met with loud cheers. Tony seemed hurt. Afterwards, everyone made peace again, but there was a sense that this stuck.

Mark, who is African American, actually also sometimes wore tight pants that summer, being open to fashion trends. Clearly, such stylistic rules were far from written in stone among those young men. The protagonists of this little event remained on friendly terms and in some ways – in relation to hip hop as a shared culture – had much more in common with one another than with most of their age-group peers. It was not as if this act of labelling in a cypher battle laid bare categorical distinctions between them. Rather, the rap cypher was an opportunity to sound out social and political tensions that were usually bracketed when the group met and created its common scene, but, as the episode and the reactions showed, remained in play nevertheless. Later in the week, when the ethnographer asked Mark about the meanings of the term, he explained that hipsters were privileged people who *wanted* to be countercultural – but without committing to radical politics and without questioning themselves and their privileges. So, ultimately, they just remained consumers. Clearly, there was a serious political and cultural background to hipster critique in this case.

Other people the ethnographer met in these months who were political activists, many of them of colour, used the term in a similar way. Doing so, among other things, helped them challenge and (briefly) flip a power relationship: Usually, hipsters came into the neighbourhoods where people like Mark and, to be fair, Tony as well, had grown up. Hipster types, this was the contention, were used to having their way. Their aesthetic and moral claims to being opposed to mainstream society were ultimately pretence because they remained cultural and aesthetic. Their whiteness protected them, or at least so it seemed, from being truly excluded and from experiencing police violence. Furthermore, hipsters were usually quite judgemental about others when it came to matters of style and ethics, in which they considered themselves ahead of everybody else: The hipsters' pejorative, demeaning gaze, where everybody else is stuck in the past, can hurt. Public anti-hipster *discourse* made a difference in this context: Now that the hipster was a recognisable, pop-culturally resonant category, these people were easier to ridicule and critique, be it in the cypher or elsewhere, and it was more likely that such ridicule and critique would have resonance and be recognised. Those who would usually do the classifying, were now being classified themselves.

This, then, was an egalitarian, left-wing critique of the hipster from a minority position that became possible partly because the term had turned into a public cultural figure. It relied on many of the aspects highlighted above but re-formulated them more strongly in terms of ethnic and class identity, which to Mark (and many others) were very closely intertwined with political stances – *white* hipsters were not, as it were, sufficiently "treasonous" to whiteness. Life-world-level figuration like this relies on the different aspects of figuration processes mentioned above, but the type of anti-elite critique that happens on this ethnographic level, "on the ground", cannot be fully deduced from media discourses alone.

Berlin 2012/14: Hipster Hass

The second case comes from the Berlin mass-market hip-hop world of the late 2000s and early 2010s, where the local anti-hipster discourse of the time[16] spawned a few tracks featuring hipster disparagement, among them popular ones like *Sido's* (2012) "Ich will mein Berlin zurück" (*I want my Berlin back* – from the "hipsters" who have taken over) and *Fler's* (2014) *Hipster Hass* (*Hipster Hatred*), respectively (on Fler, see Süß, 2019). *Sido* and *Fler*, both quite successful, were positioned at the time as *white* working-class rappers from (post-)proletarian areas of West Berlin. In his previous work, *Fler* – originally part of a roster of artists on *Aggro Berlin* records that included famous rappers with Arab or Turkish backgrounds – had presented himself as a proud *white*-ghetto nationalist wise to the ways of the "street", a "Deutscha Bad Boy" – signalling his Germanness, for instance, by using archaic-seeming *Fraktur* font on his album covers.

In both songs hipsters are seen as gentrifiers: The people who are in the process of destroying the "real", non-bourgeois city which the rappers embody. Words like gentrification are not used, however, and while political–economic processes of collective displacement underlie the stories, these matters are primarily addressed through antipathies towards specific figures and the ways of life and values they stand for. The main problems with the hipster figure in *Fler's* lyrics are to do with gender, sexuality and the aesthetics of the male body: The chorus of *Hipster Hass* is the sarcastic line "I'd like to be a hipster, but my shoulders are too broad", followed by "so you better run because the way back to Kreuzberg is far" – but really, *Fler* says, hipsters should return to where they are (supposedly) from and where they belong: wealthier, more boring German cities like Munich or Stuttgart. *Fler* raps:

> Behind our backs you're sneakily talking about chavs [*Prolls*], but you're sitting there with your legs crossed like a poof, bitch – You go to Berghain while I can't get into Cookies. Politically incorrect, it must be down to the muscles. Maskulin, we don't wear skinny jeans. I only wear Givenchy and Burberry whereas you're wearing jumpers from bands nobody knows.

The skinny vs. wide jeans trope documents the global circulation of this attribute; in this case, however, the point is that a masculine physique with broad shoulders and muscular thighs is incompatible with tight jeans.[17]

Overall, in this representation, hipsters figure as an undifferentiated group of cultural "others" that inhabits certain areas of the city. These lyrics blend a strong critique of classism that is in some ways reminiscent of the first example: Hipsters dismissively call others "chavs", the night club door policy from which they benefit bars proud working-class men like Fler, too.

More broadly speaking, *Fler's* lyrics and stylised *hexis* are based in a cultural – rather than political or economic – class-consciousness and a caricature-like, over-the-top heterosexual-masculinist politics of bodies and sexuality: The "true" Berlin body politic of broad-shouldered masculine types versus physically feminised, sexually ambiguous hipsters. True to hyperbolic form, the rappers threaten, among other things, a massacre on the *Bread and Butter*, the Berlin Fashion Week – the fashion world being stereotypically associated with attributes such as inauthentic/superficial, female and gay. At the same time, with the references to the high-end brands he and his friends wear, *Fler* highlights their own expensive and avowedly self-confidently *nouveau-riche*-tastes, which they oppose to hipsters' cheap but pretentious second-hand "homeless/hobo-style" (*Pennerstyle*) and obscure, esoteric aesthetic orientation – the implication is that while this may seem advanced to hipsters themselves, it is irrelevant to everybody else. While this stresses that hipsters' cultural power is just pretence and a temporary hype, like the rise of certain rappers, in another sense, hipsters are presented as having become more powerful figures. Fler envisions hipsters as people who control the entry points to whole cultural spheres and have managed to conquer the city, at least temporarily. In the lines quoted above, this is illustrated by nightclub door policy: The club *Cookies* that he can't get into due to his *Muckies* (an endearing term for muscles) was an In-Club for Berlin Mitte from the mid-90s onwards. It was not underground and gay as famous *Berghain* (later), but influential among the heterosexual in-scene, attracting artists, models, people from the fashion, media, creative industries. Like the decisionmakers at the *Cookies* door (more likely women than muscular bouncer-types), hipsters are able to classify and exclude others; "Prolls" or chavs are the object of this. When *Fler* reacts with physical threats, this is presented as a matter of natural self-defence to this assault on the "real" city – just like the association with being "politically incorrect". Here, too, the public circulation of the hipster label and the cultural figure affords a critique of symbolic inequality and dominance, and they allow *Fler* to demonstrate that he sees through these "neo-postmodern, tight-pants wearing aliens".

The references to the distribution of power in society are highly ambiguous in this song: Hipsters are culturally dominant, but they also lack any *real* power – they aren't the real elite, be it artistic or financial, as documented by their middling media jobs and second-hand outfits. In a sense, the lyrics imply that Fler would respect the hipsters more if they held actual power and wealth. In this context, then, the normative orientation of the anti-elite critic is not so much a sense of equality, but a "natural" state in which broad-shouldered, down-to-earth men "still" dominate the city. The music video, which by 2018 had garnered three million views on YouTube, presents *Fler* as an underground figure speaking from a car park, a subway station or cruising the hood at night, whereas the hipster is depicted as a flaneur or art connoisseur. In doing so, anti-hipster tracks like *Fler's* hark back to old rap music tropes of realness and urban charisma, illustrating the difference between the territorialism of "the street" and the less easily visible dominance that capital exerts through cultural and financial gentrification. This class war primarily plays out as a struggle over hegemonic masculinity. The question *Fler* poses is what his well-trained body, the fact that he is good at being a heterosexual man in the sense that he sees it, his seemingly "uncultured" money and his narrative of urban roots, are worth in an urban cultural economy where hipsters seem to dominate not just night club door policy. While Fler and "street" rappers like him do speak "from below" and intend to offend middle-class tastes (including that of academics), to which (left-wing) academics shouldn't react with knee-jerk self-defence, it would be mistaken to see these rappers as populist left-wing critics in the making – *Fler's* work and public persona are in most respects right-leaning, nationalist, authoritarian, individualistic, at times social Darwinist. This includes open sexism and violence against women. The hyperbolic, but nevertheless serious masculinist aspects of Fler's critique of feminised hipsters, embody this authoritarian tendency and signal its general direction. These political implications are an important vector of recent anti-hipster as anti-elite discourse.

Conclusion

In this chapter, we turned to the hipster figure and highlighted its anti-elite dimensions. Our initial questions were how the hipster could come to take on such importance in political diagnoses from around 2016, and what cultural meanings resonated in such critiques. We then briefly introduced some of those meanings by surveying the literature on the hipster and historicised them by presenting a typology of anti-hipster critiques in pop culture and countercultures since the 1950s. Analytically, we suggested that the hipster should be seen as a cultural figure and spelled out some of the implications of that term.

The spotlights in the last two sections of this chapter were not intended to make broad claims about the overall conjuncture in the respective contexts: It would have been possible to find progressive rap artists in Berlin using the hipster figure differently and authoritarian-leaning hip-hop practitioners in Chicago. Nevertheless, building on the history of anti-hipster sentiments and arguments that we reviewed in this chapter, they illustrate important but divergent tendencies in anti-elite discourses that form around 2010, foreshadowing later usages. The first case study illustrates a progressive-egalitarian, if slightly identitarian, direction that such a critique can take: Here, the critique of hipsters is a critique of collective, undeserved privileges of a certain stratum of the urban middle class, of anti-political individualism, a consumerist lack of self-awareness and insufficient commitment to social change. The other case study attacks similar targets, speaking from a (working) class position, but it has a strong conservative-reactionary tendency in at least two senses: in regard to the politics of gender and sexuality and in its relation to broader questions of equality. By joining in anti-hipsterism as an anti-elite discourse, cultural producers and political strategists of various stripes have attempted to activate resentments against the apparent winners of the (urban) "knowledge economy". These strategies have in turn also contributed to the hipster's prominence – up to the point where the term can serve as shorthand for election analyses.

The overall background for this is that over time, hipster critiques have become intertwined with critiques of the rising sociocultural milieus and their values, the "new petty bourgeoisie" of cultural intermediaries, as Pierre Bourdieu (1984) called them. The latter – however they are being defined in a stricter sociological sense, as a petty bourgeoisie, a "new class" or a "creative class" – are often taken to be a crucial formation of contemporary elites. It is important to recall, however, that Bourdieu had described them as a *dominated* fraction within the dominant class, removed from an actual power elite (Davies, 2016; Mills, 1956). While the rise of service and knowledge economies, the increasing dominance of startup culture and related capital factions may have shifted the scales somewhat, Bourdieu's overall observation remains apt. To equate "being against hipsters" with "being against the elite" is to articulate a culturally suggestive and evocative syllogism and to point out an important form of power, but it has its limitations. In many cases, like in Sigmar Gabriel's statements or *Fler's* lyrics, it is connected to conservative cultural and political strategies.

The critique of hipsters is indicative of a broader culturalisation of inequality and inequality discourse. This is not merely an ideological distraction. From a Cultural Studies perspective, it should be stressed that culturalisation relates to the irreducible *culturality* of inequalities, as they have again come to the fore in recent autosociobiographies and in political critiques of "classism". The actual distribution of wealth, and the overall composition of the ruling class, however, can be rendered invisible through an affective-laden focus on easily recognisable cultural figures of the supposed elites: Hipsters primarily appear as an *undeserved* elite then. In such contexts, regressive forms of cultural critique prevail.

It would also be possible to approach the topic from another angle and advocate a completely different, affirmative reading. If, normatively speaking, we see the positive elements in everything that has been denigrated in hipsters, the latter could be seen to stand for cosmopolitanism, playfulness, futurity, the critique of the status quo and a non-fundamentalist relation to gender and sexuality. Furthermore, many hipsters are arguably part of the contemporary "precariat", rather than of actual elites, and their arrogance towards others and their ways of life could at least in part be seen as motivated by the oppressive nature of those ways of life. This was the line taken, for example, by a few groups who more or less ironically gave themselves names like "Hipster Antifa" in Berlin eight or nine years ago and who were – in line with many groups within the German left – happy to remain subcultural rather than striving to become popular. Maybe this represented an innovative solution to dilemmas of the left, but more likely, it was no more than the usual strategy of affirming the less popular side of a fraught dichotomy. If, then, looking at hipsters inspires all sorts of social diagnoses, and looking at hipster hate, as we have done, can help us make sense of the ways in which social tensions are being culturally articulated, being *for* or *against* the hipster probably is not going to help much in finding new strategies for cultural politics. Maybe the hipster's historical role has been to bring us to this point, to embody these contradictions and quandaries, and new figurations are needed to get out of the place in which we seem stuck.

Notes

1 Varieties of which have been taken by prominent left-leaning social scientists like Wolfgang Streeck and Wolfgang Merkel; see for an American right-wing version, Gutfeld, 2014. See also Beyer's chapter in this volume.
2 Conservatives jumped on board that train as well, giving anti-hipster discourse a cultural–nationalist twist. In August 2017, Jens Spahn, then a rising star of the German conservative party's right wing, complained in an interview that there were numerous bars and restaurants in Berlin where the staff spoke only English. Spahn found this distasteful and exclusionary. His main line was that "elitist hipsters are shutting themselves off from normal citizens" (2017). Spahn complained that there were areas in Berlin where a "colorful bubble has emerged where everyone feels conspicuously open to the world, but what is really being lived is a heightened form of elitist-globalised tourism. All those who can't keep up with the easyJet generation have to stay outside. For example, those Germans whose English isn't good enough. And, curiously, those immigrants who took the effort to learn the German language instead of English." While expats were a part of the problem, his ire was mainly directed at German linguistic "self-dwarfication". On the actual complexities of this issue, see Schulte, 2019: 189.
3 Many authors deduce the hipster's prominence from a larger economic and cultural constellation. As Rabe points out, the concept of the hipster as manically in the know, always ahead, implicitly became a norm in a neoliberal, entrepreneurial, post-Fordist digital economy (Rabe, 2012: 202). Dorrian (2020) makes a similar attempt by positioning the hipster in the precarious world of post-financial crisis capitalism: "Millennial hipsters are thus caught in a paradox in which they are cast as elitist if they attempt to signify difference through cultural taste, conversely, if they are to comply with the neoliberal ideologies of post-recession capitalism – which have favoured the individual entrepreneur – they are portrayed as gentrifying subjects who embody the inequalities of contemporary society" (3).
4 This began with newspaper articles (Greif, 2011; Rosen, 2009) and long-form radio features (von Lowtzow, 2009) and culminated in the Suhrkamp translation and extended version of the book by Greif et al., 2012.
5 Terms such as "hipster capitalism" (Scott, 2017) are suggestive but ultimately not particularly useful.
6 While all the contributions to the book stress that "hipster" is a pejorative term that no one unambiguously applies to themselves, Greif also noted shifting connotations: In his view, the word had "been used for insult and abuse" (Greif, 2010b, p. viii), but it was also "gaining a neutral or even positive estimation in the culture" (Greif, 2010b, p. viii).
7 See numerous online videos and booklets (Cassar and McRae, 2016; Moe, 2015; Morris and Hazeley, 2015).
8 Especially among literary and media studies scholars, "hipster culture" denotes a culture or a sensibility (lived and/or represented) that is characterised by specific affective and aesthetic categories, from detachment/cool to quirkiness (MacDowell, 2012; Newman, 2013; Steinhoff, 2021).

9 In 2014, Schiermer could still point out a gleaming

> neglect of the hipster phenomenon on the part of academic sociology. (…) There exists an immense quantity of opinions and observations on the hipster phenomenon made by journalists, bloggers and layman experts of all categories. The entry 'hipster' yields 75 million hits on Google – and thus exceeds the entry 'sociology' (73 million).
>
> (Schiermer, 2014, p. 168)

10 Periodisations abound in writings about the hipster, see, e.g. Greif's (2011) sketch of the major transformation in the aughts from the "white hipster" to the green "primitive hipster" and its entanglement with major American political crisis. For a discussion of Greif's periodisation, see Springer (2011); Springer and Dören (2016).

11 These tendencies are present in most genres. Mostly exempt seem only genres and youth cultures such as disco that appreciate what Thomas Meinecke and Eckhard Schumacher (Meinecke and Schumacher, 2012) positively describe as being central about the hipster, namely the capacity to be a performative snob, aligned with camp and artificiality.

12 Leland (2004) notes: "The church, the law, capital and mass opinion all lined up against hip, as against a disease. Voices of authority took pains to be corny. Athletes, celebrities, politicians, war heroes and civic leaders all presented their rectitude—literally, their squareness—as a bulwark against hip's sinuous slink. People who smoked a joint or loved out of hetero wedlock were labeled dope fiends or sex fiends; rhythm was considered a threat to civilization. Police narco units of the 1950s specialized in tossing jazz musicians." (13) For an account of how different contemporary culture wars play out with reference to hipsters against the background of a libertarian right, see Burns (2022).

13 The discussion about the dissolution of subcultures and the emergence of post-subcultural formations dominates research in the 1990s and 2000s (see Hesmondhalgh, 2005); the emergence of the hipster figure is intertwined in the same processes.

14 On figuration, see also Ege (2013); Ege and Wietschorke (2014).

15 Such as Reynolds, 2011.

16 See Slobodian and Sterling (2013) for an analysis of the local constellation of the sell-off of Berlin's social housing and the reinvention of the city as a destination for tourists, "digital bohemians", "expats", etc. and international capital – while local magazines like *Zitty* declared "American hipsters" and their European epigones to be the central problem. Stahl describes a similar development for Montreal's Mile-End (2010).

17 "Maskulin", aside from the obvious meaning, references one of Fler's rap crews, *Südberlin Maskulin*.

Bibliography

Athill, C. N. (2018). Who Are You Calling a Hackney Twat? Gender and Stigma in Media Representation. *JEA*, 2(1), pp. 46–65.

Becker, H. S. (1951). The Professional Dance Musician and His Audience. *American Journal of Sociology*, 57(2), pp. 136–144.

Bourdieu, P. (1984). *Distinction: A Social Critique of the Judgement of Taste*. London; New York: Routledge.

Broyard, A. (1948). A Portrait of a Hipster. *Partisan Review*, 15(6), pp 721–727.

Burns, N. (2022). New York's Hipster Wars. https://www.newstatesman.com/ideas/2022/05/new-yorks-hipster-wars(Accessed: 28 May 2022)

Cassar, J. and McRae, C. (2016). *How to Spot a Hipster*. Melbourne, Australia: Smith Street Books.

Cohen, S. (2011). *Folk Devils and Moral Panics: The Creation of the Mods and Rockers*. Abingdon, Oxon; New York: Routledge (Routledge classics).

Davies, W. (2016). Elite Power Under Advanced Neoliberalism. *Theory Culture and Society* [Preprint]. Available at: http://research.gold.ac.uk/18744/ (Accessed: 13 March 2017).

Dorrian, R. (2020). Millennial Disentitlement: Greta Gerwig's Post-Recession Hipster Stardom. *Celebrity Studies*, pp. 1–18.

Ege, M. (2011). Carrot-cut Jeans: An Ethnographic Account of Assertiveness, Embarrassment and Ambiguity in the Figuration of Working-class Male Youth Identities in Berlin. In D. Miller and S. Woodward (Eds.), *Global Denim*. Oxford; Providence: Berg, pp. 159–180.

Ege, M. (2013). *'Ein Proll mit Klasse': Mode, Popkultur und soziale Ungleichheiten unter jungen Männern in Berlin*. Frankfurt: Campus Verlag.

Ege, M. and Wietschorke, J. (2014). Figuren und Figurierungen in der empirischen Kulturanalyse. Methodologische Überlegungen am Beispiel der »Wiener Typen« vom 18. bis zum 20. und des Berliner »Prolls« im 21. Jahrhundert. *LiThes. Zeitschrift für Literatur- und Theatersoziologie*, (11), pp. 16–35.

Erbacher, E.C. (2012). Hip or Square? Pop Cultural Negotiations of Hipster Lifestyles between Commodification and Subversion. In: L. Koos (Ed.), *Hidden Cities: Understanding Urban Popcultures*: Brill, pp. 13–22.

Evans, P. (2010). Hip-hop & Hipsterism. Notes on a Philosophy of Us and Them. In *What Was the Hipster? A Sociological Investigation*. New York: n+1 Foundation, pp. 103–111.

Friedrichs, A. and Groß, F. (2021). (Re-) Dressing the Naked City: Hipsters, Urban Creative Culture, and Gentrification in New York City. In H. Steinhoff (Ed.), *Hipster Culture: Transnational and Intersectional Perspectives*. New York: Bloomsbury Academic, pp. 27–46.

Gabriel, S. (2016). Die Arbeiter, nicht die Hipster. *Der Spiegel*, 11 November. Available at: https://www.spiegel.de/politik/die-arbeiter-nicht-die-hipster-a-c4b33eb2-0002-0001-0000-000147863974 (Accessed: 11 September 2022).

Gabriel, S. (2017). Sehnsucht nach Heimat Wie die SPD auf den Rechtspopulismus reagieren muss. *Der Spiegel*, 2017 (51).

le Grand, E. (2020). Moralization and Classification Struggles Over Gentrification and the Hipster Figure in Austerity Britain. *Journal of Urban Affairs*, pp. 1–16. DOI: 10.1080/07352166.2020.1839348.

le Grand, E. (2021). Folk Devils and the Hipster Figure: On Classification Struggles, Social Types and Figures in Moral Panic Research. In M.D. Frederiksen and I. Harboe Knudsen (Eds.), *Modern Folk Devils: Contemporary Constructions of Evil*. Helsinki: Helsinki University Press, pp. 27–46.

Greif, M. (2010a). Epitaph for the White Hipster. In *What Was the Hipster? A Sociological Investigation*. New York: n+1 Foundation, pp. 136–167.

Greif, M. (2010b). Introduction. In M. Greif, K. Ross, and D. Tortorici (Eds.), *What Was the Hipster? A Sociological Investigation*. New York: n+1 Foundation, pp. vii–xv.

Greif, M. (2010c). Positions. In *What Was the Hipster? A Sociological Investigation*. New York: n+1 Foundation, pp. 4–13.

Greif, M. (2011). Was War der Hipster. *Süddeutsche Zeitung*, 7 January, p. 12.

Greif, M., Ross, K. and Tortorici, D. (Eds.) (2010). *What Was the Hipster? A Sociological Investigation*. New York: n+1 Foundation.

Greif, M., Ross, K., Tortorici, D., Geiselberger, H. (Eds.), (2012). *Hipster: eine transatlantische Diskussion*. Berlin: Suhrkamp (Edition Suhrkamp. Sonderdruck).

Gutfeld, G. (2014). *Not Cool. The Hipster Elite and Their War On You*. New York: Crown Forum.

Hall, S. (1968). *The Hippies: An American "Moment"*. Birmingham: CCCS (Stencilled Papers).

Hartigan, J. (2005). Introduction. In *Odd Tribes. Toward a Cultural Analysis of White People*. Durham; London: Duke University Press, pp. 1–32.

Heisterhagen, N. (2018). *Die liberale Illusion. Warum wir einen linken Liberalismus brauchen*. Durchgesehene 2. Auflage. Bonn: Dietz (Dietz Standpunkte).

Heyd, T.; von Mengden, F.; Schneider, B. (Eds.) (2019). *The Sociolinguistic Economy of Berlin. Cosmopolitan Perspectives on Language, Diversity and Social Space*. Boston; Berlin: de Gruyter Mouton (Language and social life, Volume 17).

Hesmondhalgh, D. (2005). Subcultures, Scenes or Tribes? None of the Above. *Journal of Youth Studies*, 8(1), pp. 21–40.

Hubbard, P. (2016). Hipsters on Our High Streets: Consuming the Gentrification Frontier. *Sociological Research Online*, 21 (3).

Ico, Maly (2019). Hipsterification and Capitalism: A Digital Ethnographic Linguistica Landscape Analysis of Ghent. Online: https://research.tilburguniversity.edu/en/publications/hipsterification-and-capitalism-a-digital-ethnographic-linguistic.

Ikrath, P. (2015). *Die Hipster. Trendsetter und Neo-Spießer*. 1st ed. Wien: Promedia Verlag.

Klapp, O.E. (1954). Heroes, Villains and Fools, as Agents of Social Control. *American Sociological Review*, 19(1), pp. 56–62.

Klapp, O.E. (1958). Social Types: Process and Structure. *American Sociological Review*, 23(6), pp. 674–678.

Leland, J. (2004). *Hip. The History*. New York: Harper Collins.

Lin, J. (2021). 10. Boulevard Transition, Hipster Aesthetics, and Anti- Gentrification Struggles in Los Angeles. In: C. Lindner and G. Sandoval (Eds.), *Aesthetics of Gentrification*. Amsterdam: Amsterdam University Press, pp. 199–220.

von Lowtzow, C. (2009). Wo ist vorne? Wie die Hipster die Welt eroberten (Zündfunk Generator). *Bayern* 2, 13.12.2009.

MacDowell, J. (2012). Wes Anderson, Tone and the Quirky Sensibility. *New Review of Film and Television Studies*, 10(1), pp. 6–27.

McIntyre, A.P. (2016). Millennials Protest: Hipsters, Privilege, and Homological Obstruction. In G. Austin (Ed.), *New Uses of Bourdieu in Film and Media Studies*. New York: Berghahn, pp. 89–106.

Mailer, N. (1957). *The White Negro. Superficial Reflections on the Hipster. Advertisements for Myself*. Cambridge und London: Harvard University Press, 1992, pp. 337–358.

Meinecke, T. and Schumacher, E. (2012). Geradeaus Williamsburg. In M. Greif et al. (Eds.), *Hipster: eine transatlantische Diskussion*. Berlin: Suhrkamp (Edition Suhrkamp. Sonderdruck), pp. 171–187.

Mills, C.W. (1956). *The Power Elite*. Oxford: Oxford University Press.

Moe, D. (2015). *Hipster Animals: A Field Guide*. 1st ed. Berkeley: Ten Speed Press.

Morris, J. and Hazeley, J. (2015). *The Ladybird Book of the Hipster*. Penguin UK.

Müller, J.-W. (2016). *What Is Populism?* Philadelphia: University of Pennsylvania Press.

Murray, M.A. (2020). White, Male, and Bartending in Detroit: Masculinity Work in a Hipster Scene. *Journal of Contemporary Ethnography*, 49(4), pp. 456–480.

Myambo, M.T. (2021). Glocal Hispsterification: Hipster-Led Gentrification in New York's, New Delhi's and Johannesburg's Cultural Time Zones. In H. Steinhoff (Ed.), *Hipster Culture: Transnational and Intersectional Perspectives*. New York: Bloomsbury Academic.
Newman, M.Z. (2013). Movies for Hipsters. In G. King, C. Molloy, and Y. Tzioumakis (Eds.), *American Independent Cinema*. Routledge, pp. 71–82.
O'Brien, G. (1997). *Soapbox. Essays, Diatribes, Homilies, and Screeds 1980–1997*. Gent: Imschoot.
Polsky, N. (1967). *Hustlers, Beats, and Others*. Chicago: Aldine.
Rabe, J.-C. (2012). Gegenwärtigkeit als Phantasma. Über den Hass auf Hipster. In M. Greif et al. (Eds.), *Hipster: eine transatlantische Diskussion*. Berlin: Suhrkamp (Edition Suhrkamp. Sonderdruck), pp. 188–204.
Reynolds, S. (2011). *Retromania: Pop Culture's Addiction to its Own Past*. London: Faber & Faber.
Rosen, B. (2009). Blick auf die Szene: German Elite mit Konfusionsfrisuren. *Tagesspiegel*, 24 October. Available at: https://www.tagesspiegel.de/berlin/stadtleben/german-elite-mit-konfusionsfrisuren-6516685.html (Accessed: 13 October 2019).
Rüß, C. (2021). Hipster and (Anti-)Fashion. In H. Steinhoff (Ed.), *Hipster Culture: Transnational and Intersectional Perspectives*. New York: Bloomsbury Academic, pp. 105–128.
Schiermer, B. (2014). Late-modern Hipsters: New Tendencies in Popular Culture. *Acta Sociologica*, 57(2), pp. 167–181.
Schutz, A. (1962). *Collected Papers I. The Problem of Social Reality*. The Hague: Martinus Nijhoff (Phaenomenologica).
Schulte, L. (2019). Stancetaking and Local Identity Construction among German-American Bilinguals in Berlin. In: T. Heyd, F. von Mengden and B. Schneider (Eds.), *The Sociolinguistic Economy of Berlin. Cosmopolitan Perspectives on Language, Diversity and Social Space*. Boston, Berlin: de Gruyter Mouton (Language and Social Life, Volume 17), pp. 167–194.
Scott, M. (2017). "Hipster Capitalism" in the Age of Austerity? Polanyi Meets Bourdieu's New Petite Bourgeoisie. *Cultural Sociology*, 11(1), pp. 60–76.
Slobodian, Q.; Sterling, M. (2013). Sacking Berlin: How Hipsters, Expats, Yummies, and Smartphones Ruined a City. *The Baffler*, 23, pp. 138–146.
Spahn, J. (2017). Berliner Cafés: Sprechen Sie doch deutsch!. *Zeit Online*, 23 July. Available at: https://www.zeit.de/2017/35/berlin-cafes-hipster-englisch-sprache-jens-spahn (Accessed: 13 October 2019).
Springer, J. (2011). Der primitive Hipster im Wald. *skug – Journal für Musik*, 84, pp. 54–55.
Springer, J.; Dören, T. (2016). Einleitung. In: J. Springer and Thomas Dören (Eds.), *Draußen. Zum neuen Naturbezug in der Popkultur der Gegenwart*. Bielefeld: transcript Verlag (Kultur- und Medientheorie), pp. 7–24.
Stahl, G. (2010). The Mile-End Hipster. Montreal's Modern Day Folk Devil. In: B. Binder, M. Ege, A. Schwanhäußer and J. Wietschorke (Eds.), *Orte – Situationen – Atmosphären*. Frankfurt am Main: Campus Verlag, pp. 321–329.
Steinhoff, H. (Ed.) (2021). *Hipster Culture: Transnational and Intersectional Perspectives*. New York: Bloomsbury Academic.
Süß, H. (2019). "Ich wär' auch gern ein Hipster, doch mein Kreuz ist zu breit" – Die Ausdifferenzierung der HipHop-Szene und die Neuverhandlung von Männlichkeit. In T. Böder et al. (Eds.), *Stilbildungen und Zugehörigkeit: Materialität und Medialität in Jugendszenen*. Wiesbaden: Springer Fachmedien (Erlebniswelten), pp. 23–44.
Tyler, I. (2015). Classificatory Struggles: Class, Culture and Inequality in Neoliberal Times. *The Sociological Review*, 63(2), pp. 493–511.

Weismantel, M. (2001). *Cholas and Pishtacos: Stories of Race and Sex in the Andes*. Chicago, IL: University of Chicago Press (Women in Culture and Society).

Music video

Fler, 2014. Hipster Hass. YouTube. https://www.youtube.com/watch?v=xCQwrOVxA74

Songs

Freddy, Quinn, 1966. Wir.
Merle, Haggard, 1969. Okie from Muskogee.
Roger, Miller, 1964. Squares Make the World Go Round.
Sick of It All, 1994. Who Sets the Rules.
Sido, 2012. Ich will mein Berlin zurück.
The Moldy Peaches, 2001. These Burgers.

17 The ghost of Europe is shifting shape

How the film *Folkbildningsterror* intervenes in left debates around class vs. identity politics

Atlanta Ina Beyer

Right-wing populists and authoritarians have been on the upswing internationally in recent years. Increased racism and the scapegoating of refugees, anti-feminist and "anti-gender" ideologies, as well as the return of conservative familisms, foster their rise. The current conjuncture is – again – shaped by an increasingly authoritarian neo-liberal capitalism. The classic analysis of the shift towards an increasingly authoritarian populism in 1970s Britain by Stuart Hall and John Clarke, *inter alia*, reminds us that the subsequent birth and rollout of neo-liberalism (Thatcherism) were linked to an aggressive culture war and a number of moral panics (e.g. Hall et al., 2013; Hall, 2014). At present, Christa Wichterich notes a "cultural battle raging over the power to define sexuality, gender and family as well as sexual and reproductive rights" (2019, p.103). In this politics of identity from the right, feminists and a so-called genderism are accused of disintegrating the "natural or allegedly God-given social and (patriarchal) gender order", and "to destroy social and cultural identities" (Wichterich, 2019, p.103). The "totalitarian-sounding term" genderism, Wichterich argues, is constructed as a hate object, and is intended "to arouse fantasies of fear and moral panic" (2019, p.103. In this logic, liberal

> elites in politics and the media have succeeded in anchoring gender and gender mainstreaming in the canon of 'political correctness', which is now being forced upon the majority of the population. Feminism is an elite project and only speaks for a minority. By contrast, right-wing populist forces claim to represent a silent majority.[1]
>
> (Wichterich, 2019, p.103)

Wichterich, thus, describes just the latest edition of the same old right-wing tactics of divide and conquer. However, the charge of "elitist" feminist and queer identity politics cannot only be found on the right but also in parts of contemporary debates on the left and liberal spectrum. Nancy Fraser, for example, links the rise of the right in the US to a crisis in "progressive neoliberalism" (2017). She describes the latter as "an alliance of mainstream currents of new social movements (feminism, anti-racism, multiculturalism, and LGBTQ rights), on the one side, and high-end 'symbolic' and service-based

DOI: 10.4324/9781003141150-22

business sectors (Wall Street, Silicon Valley, and Hollywood), on the other" (Fraser, 2017).[2]

According to Fraser, Trump voters in the US did not just vote against neo-liberalism but against progressive neo-liberalism and its entanglements of corporate and "minority" politics. Didier Eribon's *Returning to Reims* points to the situation in France. In the book, he ponders why sections of the French working class – traditionally close to the French Communist Party PCF – have increasingly voted for Le Pen's right-wing *Front National* (now *Rassemblement National*). He partly blames the left itself for having moved away from the working class. When a German translation was published in 2016, heated – and ongoing – debates arose about why the left had "lost" the working classes. Sizeable parts of the German left, such as the now defunct *Aufstehen* movement around Left party politician Sahra Wagenknecht, academics Wolfgang Streeck and Andreas Nölke (2017), or dramaturge Bernd Stegemann (2018), blamed a supposed focus on identity/minority politics and representation for this loss (Nölke, 2017, pp.88f., 228ff.; Stegemann, 2018, pp.10, 102f.). The distinction between "cosmopolitanism" and "communitarianism" (Merkel, 2016) is used to discursively merge left-liberal politics – assumed to be the dominant form of left (party and cultural) politics – with a neo-liberal "cosmopolitan elite" (Nölke, 2017, p.42). Feminist, anti-racist and queer politics (also understood as uniformly left-liberal) are accused of being complicit with neo-liberal policies.

This discourse, therefore, links the critique of elites and elitism to that of (a narrow, homogenous understanding of) identity politics. The latter are portrayed as necessarily and fundamentally opposed to class politics. This discursive articulation is the starting point for my article, in which I want to shed a light on (left) politics of representation and how they structure what we can conceive of the correlations between class and identity politics.

Critiques of identity politics within the left are not entirely new. This topic has been a matter of elaborate debates, especially in the US context. Indeed, a long tradition of materialist feminist, queer and anti-racist critiques of liberal multiculturalism and institutionalised, *white*- and middle-class-dominated women's and LGBT movements exist. The contributions of these writers and activists, among them, Barbara Smith, Audre Lorde, and other authors of the Black Feminist and Socialist Combahee River Collective, Leslie Feinberg, Amber Hollibaugh, Cathy Cohen, Maria Lugones, Cherríe Moraga, Kevin Floyd, Roderick Ferguson or Dean Spade (to name just a few), are hardly considered in the broader left's controversies on identity politics.

Wendy Brown (1993) prominently discussed how the departure from class politics and the critique of capitalism enabled the successes of liberal identity politics – especially gay and feminist politics – since the late 1970s.[3] A decade later, Lisa Duggan traced the cultural politics through which US-neo-liberalism has evolved since the 1970s and began to manifest itself as a new hegemonic project (Duggan, 2003). In the 1990s, a new power bloc slowly took shape, drawing on new alliances between neo-liberal forces and parts of feminist, LGBTIQ and anti-racist movements. However, Duggan also thoroughly engaged with what she saw as skewed critiques of (seemingly) liberal identity politics by many US intellectuals, for example, in Brown (1995, 2002), Fraser (1997a), Fraser in an exchange with Judith Butler (Butler, 1997; Fraser, 1997b) or Naomi Klein (2000). She considered their arguments as over-generalising and "abstracting some overall or general effect of 'identity' politics from its most conservative/neoliberal instantiations" (Duggan, 2004, p.79). Against this trend towards overgeneralisation among left critics of neo-liberalism, Duggan called for not mistaking forms of identity politics "that do in fact not offer any economic critique [...] for the radical critiques informing feminist queer, antiracist political creativity" (2003, p.83). Her observations from nearly 20 years ago can shed light on current debates about the supposedly elitist aspects of emancipatory politics. They offer a vantage point for a much-needed negotiated position in polarised and unproductive debates around class vs. identity politics.

Struggles against racism, for transgender and reproductive rights are generally not understood as class struggles in the left because they are not directly about wages or work (Bhattacharya, 2018). But quite a few of these movements – from *Black Lives Matter* and the feminist strike waves in Latin America and beyond, to smaller initiatives such as *Sisters Uncut* in the UK – connect struggles against racist and gendered violence with class perspectives (Becker, 2018). The Combahee River Collective in the US mentioned previously has pointed to the indivisibility of their struggles as Black, lesbian and working-class women. They coined the term "identity politics", in their now well-known statement, to describe an approach to combat the interlocking systems of oppression that structure their constant marginalisation (Combahee River Collective, 1977, quoted in Taylor, 2017, p.19). This understanding of identity politics finds its continuities in today's Black Feminist (-inspired) thought and political strategising in the US and beyond, for example, in the context of the *Black Lives Matter* movement (Taylor, 2016). It is a lived reality. Such movements point to the importance of struggles over representation: They name and articulate class experiences that are often hard to grasp – for example, female, Black and lesbian class experiences.

I want to argue that one problem that underlies current debates around class vs. identity politics is in the restrictedness of conceptualisations of (working-) class subjects. British Cultural Studies have made it clear since around the 1960s that class is a whole way of life and struggle (Williams, 1989; Thompson, 2002) and have shown how hegemonic rule is based precisely on the relative invisibility, containment and shifting of class conflicts (Hebdige, 1979; Hall, 2004; Hall and Jefferson, 2006; Hall et al., 2013). Precisely for this reason, many of their writings focused on the eclipsed class dimension in cultural conflicts and struggles. (Political) identities – including class identities, or the ways we can conceive of them – are, as Stuart Hall put it, constituted through and "within, not outside representation" (1996, p.4). The writings of the Combahee River Collective and British Cultural Studies have broadened the understanding of class struggle in correlation with the politics of representation and identity.

Such approaches are also helpful in formulating a counterargument in reductionist debates about class vs. identity today. While left critics of identity politics are often quick to dismiss the politics of identity and representation as seemingly detached from redistributive struggles, they tend to ignore that any kind of political subjectivity requires identity and, therefore, representation. British Cultural Studies and researchers in the tradition of these approaches have emphasised the role of articulation, media and aesthetic politics in the (trans)formation of such political identities. Against the backdrop of current neo-liberal austerity policies in Great Britain, for example, cultural studies scholar Rhian E. Jones points out that the crisis of *political* representation of the working classes is also a problem of their *cultural/media* representation. She shows how stereotypical, demeaning representations of a supposed underclass serve to individualise and moralise social "failure". Such portrayals produce an image of a "discarded Other", in contrast to which, a *white*, heteronormative, middle-class lifestyle ideal is co-constituted. Jones (2013, p.20) makes these media representations comprehensible as techniques of neo-liberal governance. They mobilise the old distinction between respectable and immoral working-class subjects. They help restrict the capacity for powerful action "from below" (Jones, 2013, p.20) by ridiculing the poor, reducing them to a mere spectacle.

Instead of scapegoating identity politics and the politics of representation, I suggest that we ask the following questions: How can emancipatory politics of representation help produce images of precarious lives that allow for more resistance? What is their role in times of culture wars? Building on Jones' thesis and the current (discursive) chasm between class and identity politics, I argue that one role of the emancipatory politics of representation is to support and strengthen (self-)representations of "discarded Others" – hitherto largely ignored in common conceptualisations and portrayals of class – thereby making unlikely solidarities across identitarian divides conceivable in ways that do not need to rely on normative images of "respectable" lives. Thus, instead of separating class and identity politics, I propose that we take a relational approach and look at possible and existing intersections and connections between queer and class politics. Arguments for a simple "return to class" must be troubled. Instead, we have to ask, who is (not) represented as affected by neo-liberal politics? This entails not only a critique of mainstream representations but also a thorough reflection of the left's own "politics of representation": How is the ambiguous concept of the elite filled with social content in the current debates?

In this chapter, I will trouble the antagonism of identity vs. class politics. To this end, I will focus on the often overlooked realm of queer aesthetic and subcultural politics, and concrete aesthetic strategies of representation that queer notions of class and identity. I turn to a recent example of such strategies, the Swedish film *Folkbildningsterror* (2014). The film focusses on precarious queer and trans working-class lives. In this regard, it is part of an archive of emancipatory movements, cultures and the knowledge productions emanating from their orbits that are made invisible in the chasm of class vs. identity (as has been the fate of the materialist researchers mentioned previously). It is, for example, in its jarring colours and sense of humour, reminiscent of the films of John Waters, which also, at the core, often focus on (sub-)proletarian queer lives and solidarities. It also takes inspiration from queer Canadian film maker Bruce LaBruce, who often provokingly "queers" and sexualises the figure of the "revolutionary" in his films. Folkbildningsterror has also been compared to the feminist science fiction dystopia Born in Flames (1983), in which an army of women forms to struggle against their continuing racist and patriarchal subjugation in a post-revolutionary US socialist society.

While *Folkbildningsterror* did not screen in big cinemas, it soon caught larger attention beyond the Swedish borders, particularly in queer/feminist and left circles.[4] It has a certain connective potential between those fields because it shows how "discarded others" of class *and* gender/sexuality can become intelligible through strategies of (self-)representation, and how the resulting reworkings of queer and class identities can help transform collective notions of class struggle and solidarities.

My reading of the film is inspired by José Esteban Muñoz' concept of queer utopias. Building on Frankfurt School philosopher Ernst Bloch, the queer theorist points to the concrete utopian potentials in queer aesthetics. The latter "map future social relations", he argues, and by this, they help expand the political horizon: Ideas and relations, which cannot yet be articulated in the present, could already become perceptible in their realm (Muñoz, 2009, p.1). Drawing from interpretive resources such as literary theory and queer-feminist Marxism, I will analyse in more detail which aesthetic strategies are developed in the film for this purpose. I will discuss how *Folkbildningsterror* criticises neo-liberal (cultural) politics and (indirectly) the emergent heteronormative, left-social democratic answers to the neo-liberal crisis and the rise of right-wing populism. Regarding the question of elite-criticism, I argue that the film not only challenges discursive constructions of queer and feminist elitism but also enables a different perspective on the "elitist". I will come back to this point at the end of the text.

Articulated struggles: how the film *Folkbildningsterror* redefines notions of class war

Folkbildningsterror was developed and shot over three years by a group of artists and activists from Gothenburg. It was created as a response to the rise and electoral success of right-wing populist party the Sweden Democrats (*Sverigedemokraterna*). Almost 13% of Swedes voted for them in 2014's parliamentary election. According to director Lasse Långström, the film was directed against the party's slandering and discrimination of minorities, as well as their agenda to dismantle the Swedish welfare state. In Swedish, *Folkbildning* is the term used to describe non-formal forms of adult education, so, the film's cheeky title – the English translation would be "Popular Education Terror" – reveals an attempt to counter and shift popular forms of knowledge. That is, in a drastically different direction than that aimed for by the Sweden Democrats. *Folkbildningsterror*'s screaming aesthetic arguably works to support this claim. In genre terms, the glitter-punk, colourful queer film alternates between musical, fantasy comedy and left-autonomous-queer agit-prop.

As for the plot, the focus is on the trans characters Theo and Kleopatra, and a nameless, tense and angry rabbit. Theo's mother is on benefits. She is chronically ill but, nonetheless, harassed by the employment office. Theo wants to help her, but he also has other problems. In addition to his own troubles with the employment office and a lack of future prospects, he is preoccupied with thoughts of gender transitioning and unfulfilled romantic desires. As the story unfolds, Theo meets trans woman/drag queen[5] Kleopatra Caztrati and the rabbit. Kleopatra offers to support Theo with his plan to help his mother in return for allowing her to move in with him. She is new to the city and has no money.

One of the film's first scenes shows Theo in a small room with two doctors. As in a stereotypical interrogation, they harass him with questions under the cold light of a naked bulb: Did he play with dolls or cars in October 1991? What is his star sign? Theo is shouted at: He must decide "what" he is. The doctors collapse in hysterical laughter, which merges into a mobile ringtone: *The Internationale* anthem. Theo wakes up; fortunately, it was just a dream. But the employment office is on the phone, summoning him to an appointment to make an "action plan" to find a job. If he does not attend, he will lose his benefits. Theo's "problem world", shortly characterised in this scene, consists as much of class issues (problems with the job centre, unemployment) as of "trans issues" that belong to the kind of problems identity politics is expected to take care of: the binary gender order, the discrimination of trans people, legal access to hormones and transition, etc. Yet, he also wants to help his mother: Class solidarity, starting out as kinship solidarity, is a main issue in the movie. This also becomes apparent at the end of the first scene in which Kleopatra Caztrati appears. In a light-pink tulle dress, striding across tarmac, fields and railway tracks, she sings the first electropop song of the musical:

> *The ghost of Europe is shifting shape.*
> *As one of several unknown factors, I have arrived*
> *soft as a southern breeze, heading for a storm.*
> *Because now, the times will shift.*
> *Synchronise your fights, my friends.*
> *Shoulder a piece each of the horizon.*
> *Our Big Bang is now.*

The figure of Europe's ghost in the lyrics of the song is reminiscent of the famous first words of the *Communist Manifesto*: "A spectre is haunting Europe – the spectre of Communism" (Marx and Engels, 1999, p.43). On the one hand, the shifting ghost's shape can be understood as referring to the rise of right-wing and fascist powers in Europe. Kleopatra calls for a turnaround and a change of times. On the other hand, the line "Our Big Bang is now" also marks the dangerous situation as one of opportunity. Her words make clear that the situation calls for quick action ("we must be *swift*"). She demands the coordination and synchronisation of different struggles – in which she *includes* herself.

The *Communist Manifesto* called on the workers of the world to rise, unite and fight for communism together. In a way, they were to *become* the spectre haunting Europe. Instead of bonding with the "progressive" neo-liberal bloc and the ruling elites, Kleopatra describes herself as one of the unknown factors that has arrived in this unity to be built: Unforeseen by Marx and Engels, the trans woman/drag queen also seems to be rising, giving the wretched of the Earth a hitherto less known face.

The post-irony of it all

In a personal conversation, director Lasse Långström described the rather playful aesthetic politics in this musical fantasy comedy as *post-ironic* – meaning an aesthetic approach that constantly shifts between ironic and earnest meanings.[6] Irony is an indirect way of formulating critique. "Stating the opposite of what is meant" is often defined as its basic operating principle (Colebrook, 2004, p.1). More precisely, ironic meanings are located "in the space *between* (and including) the said and the unsaid" (Hutcheon, 1994, p.12). The interpreter has to derive them from the interplay of both spheres, which only "mean in relation to each other" (Hutcheon, 1994, p.12). Post-irony as an aesthetic concept evolved among writers and artists out of a certain weariness with the omnipresence of irony in much of contemporary, "postmodern" art and popular culture but also, for example, out of a critique of irony's frequent conjunctions with pretentiousness, cynicism and its taking a stance that serves as "an excuse for passivity" (Hoffmann, 2016. pp.38, 39). Hoffmann characterises post-ironic approaches instead as attempts to (re-) enchant an audience. The combination of irony and cynicism is refused, and aesthetic irony instead reconnected to its now only seeming counterpart: sincerity (Hoffmann, 2016, p.39).[7]

Constant oscillations between ironic, serious and sincere meanings are specific to the aesthetics of *Folkbildningsterror*. As an effect, viewers are permanently challenged to reflect on what has been said and done. This makes the film indeed a form of popular education (terror). In the following, I will discuss how the film uses post-irony to draw connections between political issues that are usually separated and sorted into the different "camps" of queer identity and class politics. One example is a scene in which Kleopatra and Theo go out to a queer-feminist community centre. There, she does a tarot card reading for him. Theo draws three cards, representing his past, present and future. Kleopatra is deadly serious when she interprets them. Her reading reveals an endless struggle between the haves and the have-nots. The Swedish Social Democrats appear. According to Kleopatra, they have betrayed the lower classes: "They've cleared the way for the right wing to sell out our public welfare", she says.

What is notable for the viewer is that something is "off" in this conversation. While pretending to read Theo's personal past, Kleopatra uses the situation to criticise the Swedish Social Democrats and their political failures that preceded the rise of the Sweden Democrats.[8] Her interpretation of Theo's situation seems too deliberate, even propagandistic, and too far from his personal life. Without explicitly saying so, she has turned the card reading into an opportunity for social criticism.

Theo shrugs it off. "What am I gonna do – save the welfare?", he asks. His eyes wander over to a couple kissing on the other side of the room. He is romantically interested in them. "What about love, can you see anything?", he wants to know. Kleopatra flips the two remaining cards, which reveal the "lovers" upside down and the "devil". Anyone slightly familiar with Tarot knows that this combination bodes very ill. "Oops!", she says, going on to explain that "Love is a bourgeois construct from the 19th century. It was instituted by the patriarchy to keep women subordinate – and drain them of love power." She then looks up from her cards and straight into Theo's eyes to close her interpretation: "You won't find love until you die."

Theo has dismissed the first part of Kleopatra's reading and changed the topic to something more "tarot-like": love. While he also does not say so explicitly, his move, nevertheless, subtly communicates the stance he takes towards her appeal: Even if he opens up to her suggestion and her political-analysis-first approach to the world – what could he do about "the welfare" anyway? *Now what*, his ironic answer seems to suggest – should this problem become a queer person's most important struggle? Kleopatra cannot be serious about this. What about *love* – a matter so central to queer and transgender lives? Theo's (character's) action illustrates that he has internalised irony's widespread fusion with cynicism. Viewers are invited to have a collective laugh about the apparently outlandish idea that he *must*, even just *could* do something about the dismantling of the welfare state. Irony is incorporated in this scene, but in an attempt to overcome the cynicism and passive stance connected to it – since to save the welfare is *exactly* what will become part of Theo and Kleopatra's joint mission and the main story of the film. Post-irony is used to evoke a sense of urgency, to challenge a sense of defeat that has resulted from, but also sustains the neo-liberal logic of no alternative. But also, to rearticulate and expand the horizons of both: Current leftist discourses which cannot comprehend the fact that queer subjects might not be entirely profiteers but also the target of neo-liberal policies and currently dominant queer politics, in which class issues feature too seldom. *Yes we can*, is the subtext of Theo's answer: *Me, and you, fellow queers, we* have to *do something*.

Kleopatra further concludes from her reading of the combined cards of the upside-down lovers and the devil that Theo will not find love until he dies. Her character appears quite cynical when she denounces love as a flexible extension of instituted patriarchy. But, at the same time, her biting remark can also be understood as a sideswipe at the prevailing themes of love, formal equality and gay marriage that have now long been at the heart of mainstream

LGBTIQ politics. Indirectly, by way of irony, she challenges their dominance. The use of irony, thus, allows her to criticise both the heteronormativity of desire and embodiment and the hegemonic homonormative focus of LGBTIQ politics – in which the right to marry can also be understood as the right to assimilate into the heteronormative regulation of love and desire.

Lisa Duggan, in the text mentioned previously, coined the term "the new homonormativity" to describe a shift in LGBTIQ politics in the 1990s. She describes this as a "new neo-liberal sexual politics" (Duggan, 2004, p.50): "it is a politics that does not contest dominant heteronormative assumptions and institutions, but upholds and sustains them, while promising the possibility of a demobilized gay constituency and a privatized, depoliticized gay culture anchored in domesticity and consumption." (Duggan, 2004, p.50) This new kind of neo-liberal sexual politics is what became increasingly adaptable into the "progressive neo-liberal" ruling bloc under Clinton in the 1990s that right- and left-wing populists are attacking as "elitist" today. A strain of conservative LGBTIQ politics increasingly moved forward to adapt to the new chances and formal rights being offered, as I explained earlier.

The ironic subtext of the reading as a whole points to the fact that sexual difference is often perceived as the main dividing line in this strand of LGBTIQ politics. This idea is thoroughly challenged through Kleopatra's dismissal of love as unattainable. Her remarks, however, are neither merely cynical nor sarcastic. They also illuminate a quagmire: The anomality of queer desire and, even more so, transgender embodiment are the basis for the regulation and normalisation of heteronormativity. The only feasible way out seem to be assimilation to what is considered normal, changing what is normal or adapting to the position as other(-ed).

I understand Kleopatra's critique as not only aimed at homonormativity's attempts to normalise queer desire. It is a more general critique of the concept of love in modern "bourgeois" societies. This can be read through a queer/feminist Marxist perspective of the connections between heteronormativity and the gendered relationships of exploitation that form the backbone of patriarchal and capitalist re-/production.

The sexual politics of capitalism: Or how we all learned to only love what we can get

I will now turn to Rosemary Hennessy's theoretical engagement with the entanglements between (the politics of) gender, sexuality and capitalism to situate the discussion of the film's aesthetic strategies in the broader frame of my argument in this text and to further determine their scope for reimagining the connections between class and identity politics. The queer-feminist Marxist argues that capitalism structurally requires that basic human needs are kept unmet, for example, for "love and affection, for education, leisure time, health care, food, and shelter" (Hennessy, 2018, p.22). Only under the condition of this basic inequality can a few extract and accumulate profit from the many (Hennessy, 2018, p.xxi). Hennessy suggests we take this as a vantage point to start questioning "how sexuality as a cultural discourse is entangled in that process" (2018, p.xxi). This approach allows her to take a different look at the construction of gendered identities and heterogendered sexuality in capitalism. She describes how naturalised heterosexuality and (hierarchical) differences between binary genders serve to legitimise and secure relationships of exploitation and domination (Hennessy, 2018, p.19). Such cultural discourses enable, for example, the devaluation of feminised and increasingly racialised reproductive labour by rendering them as a "typical" female capacity or acts of love (Hennessy, 2018, p.25). While she acknowledges the increasing flexibility of heteronormative structures, she characterises heteronormative desire as the prevailing organisational form in which the basic human need for love and affection under capitalism is fulfilled. What connects hetero- and homonormativity is that they are similar ways of organising atomised and privatised social relationships that can fulfil the rejected needs for love and affection under capitalist conditions.

Other writers have pointed to interesting temporal parallels between the neo-liberal restructuring and privatisation of formerly public care (such as "welfare", healthcare) and the increasing flexibility of heteronormative orders and the way they are governed (Engel, 2009; Laufenberg, 2014; Woltersdorff, 2016). Aside from the prominent example of gay marriage and an increased social acceptance of a range of erotic practices, this also entails, for example, the pluralisation of family concepts. These "flexible normalisations" (Engel, 2008) articulate a growing, yet, precarious acceptance for non-normative sexual and gendered lifestyles, promises of freedom and self-realisation, to the parallel normalisation of an increased privatisation of care and social responsibility.

Instead of merely privatised and atomised solutions, Hennessy suggests a reorientation process for sexual politics for *all of us* – no matter what our gendered identities or sexual preferences are. The idea is to put the fulfilment of the need for love and affection into the framework of a broader and collective struggle against capitalism's fundamental denial of basic needs. The reorientation process would link "the human potential for sensation and affect that the discourses of sexual identity organize to the meeting of other vital human needs and calls for a movement for full democracy to begin there" (Hennessy, 2018, p.231f.). However, such a movement would also require a continual process of disidentification with the roles and identities that have allowed us to fulfil parts of these needs. Although such a process is counterintuitive, and laborious, it could enable broader alliances between class and identity politics, Hennessy argues that by displacing

> [...] identity politics with a practice of disidentification that draws attention to the role of human affect in social life and in social movement through the cultivation of what I would call the more comprehensive action potence of 'revolutionary love'
>
> (2018, p.220)

the latter would not be atomised and private but more open in its scope, traversing human connections as an affect that binds and enables a different kind of togetherness, *collective action potence*, and a more democratic fulfilment of basic human needs.

Taking cues from Hennessy's perspective, I would like to re-read the scene in *Folkbildningsterror* discussed previously and reflect the film's aesthetic strategies from another angle. Kleopatra states that Theo will never find love. But this is not exactly true: instead of the romantic love of a couple and the quest for private happiness (the foundation upon which many contemporary LGBTIQ films and the whole world of romantic comedy are based), the film centres on friendship and solidarity. The somewhat absurd situation of the card reading and the political mission that Kleopatra sees for Theo suggest that there is something bigger that is worth fighting for, something that must be done urgently and is apparently more important than romantic love. But this, by no means, forecloses romance and erotic connections. They are woven into the entire film.

The queer politics of representation, as developed in the film, are not elitist. Quite the opposite: Matters of sexual identity, and the quest for affective connection, are not dealt with separately but explicitly intertwined with broader social justice struggles. In this regard, the film can be understood as critical of homonormative, neo-liberal sexual politics *and* of the limited scope of heteronormative versions of social democratic communitarianism.

Do not divide: unite and multiply!

During the film, many characters from different other queer-feminist subcultures and movements join the movement that sets out to reclaim "the welfare": queer straight-edge punks, the BDSM community, among them also a pregnant trans man, bohemian anarcha-feminists, radical lesbians who see the (rather transwomen-exclusive) lesbian struggle as the most important one, a tiny international lesbian militant organisation, etc. Due to their diversity, the group has to constantly negotiate differing views on and tensions over political strategy, identities and disagreements over the importance of particular fights. Instead of digging deeper into the problems of queer/left organising, the film leaves it with magical solutions: Despite their differences, the group members manage each time to rise above and organise. Together, they successfully protest against transport ticket inspections, employment office harassment, machoism and deportations. They aim for an unrestricted access to the good life. Thus, as becomes clear, their notion of a new welfare society far exceeds conventional notions of state welfare politics. This story of smoothly forged alliances might not be convincing in real life, but this is a fantasy comedy. It offers an opportunity to those involved in queer/left organising to laugh at themselves. It also strengthens a hopeful, optimistic perspective. By fighting struggles on "many fronts" in their process of "saving the welfare", the characters and their queer quest for a better life far exceed the narrow political identities and notions of equality suggested by mainstream LGBTIQ politics. However, the film also multiplies and relocates the sites of class struggle to adapt to the lived class experiences of the characters.

While the group prepares for another action, Theo receives a letter from the gender investigation authority. His application for transition has been rejected. This is not represented as a solely private problem. Kleopatra tries to comfort him and offers her solidarity. They go for a walk, and she deconstructs the binary-gendered world using exaggerated, Butler-esque terminology: "All the violence aimed at our bodies is an attempt to uphold the system that categorises us as women and men." As they reach Gothenburg's Museum of Art, she opens her pink handbag and lets packets of tablets rain down upon Theo. "Testosterone, oestrogen. Take it all!", she cries out and runs towards the steps of the building. "Or, you don't need to take anything!"

In an over-the-top way, the scene points to what access to healthcare could and should be: free and easily accessible to all. If class struggle is also understood as a struggle over private vs. collective property and, therefore, for example, (access to) common vs. private goods, then the public healthcare system moves into focus. Over the last few decades, public access to healthcare has shrunk immensely in Europe. More and more services must be paid for privately. A two-tiered health system is the result. However, due to the ongoing pathologisation of trans people, their access to medications and treatments that would allow them to transition – and decide about their bodies – has been restricted in separate and violent ways (Appenroth and Lottmann, 2019; Sauer and Nieder, 2019). This scene, thus, represents a crucial element in the struggle for common goods and a good life for all from a trans perspective. It suggests that fighting for a redistribution of resources also means struggling against pathologisation, against being coerced into attaining the expensive, privately paid-for medical "expertise" to get access to hormones and treatment. The film also makes these fights comprehensible as a class struggle. At the same time, Kleopatra's Butler-esque explanations shed light on the particular reasons for the restricted access for trans people: The capitalist division of labour and exploitation have long been based on the ideological constructions of the heteronormative two-gendered system. Struggles around trans healthcare cannot, thus, be easily placed into the camp of (liberal) identity politics: They must be rooted in the very foundations on which capitalist exploitation thrives.

This is further emphasised as the scene unfolds into one of the film's most spectacular musical numbers. Wrapped in a black and gold cloth, Kleopatra begins to sing as more and more figures, some with bare upper bodies and black angels' wings, emerge from behind the arches of the museum's façade. They stride onto the steps and start to form a kind of choreographed army.

"Hear the angels of trans liberation; your gender is yours, proletarian", sings the choir. "We will combat the identitarian and validational charity. We'll uncover society's torments, not be preserved in minority." The scene suggests a refusal of "queer" politics that aim solely at recognition, preservation and the better regulation of nonconforming gender identities. Instead, the violent effects of gender norms are criticised and identity is unmasked as a kind of force. At the same time, shedding a light on context reveals the complexity of struggles for recognition, representation and redistribution – and why conceiving of them as separate does not suffice as an analytical basis for emancipatory politics. Naming and representing transgender experiences within the current healthcare systems is an important step in developing the forms of collective action potence that are needed to advance the struggle for a just redistribution of common goods for all. Speaking with Muñoz and the notion of queer utopias: The real utopian value of the film is not in its representation of a coming queer revolution but in its mapping out the relationality and intersections of different struggles and social actors as they can be glimpsed in its queer aesthetic.

Conclusion

The growing movement imagined in *Folkbildningsterror* wants everything, for everyone, and for free: love, free choice of gender and access to hormones, a self-determined life instead of exploitation, disciplining, state-sanctioned harassment and pathologisation. It aims at a re-evaluation of the common goods *and* the common struggle. The film represents different forms of precarity, experiences of violence, exclusion and marginalisation, and, thus, makes it possible to reflect collectively upon them. The characters cannot be reduced to one-sided notions of queer or class identities. The movie's representational strategies aim as much at inscribing queer and trans subjects into the figure of those affected by neo-liberalism as they do at a disidentification with hegemonic queer politics and the traditional image of the proletarian. In doing so, the film aids in making it possible to conceive of "the proletariat" as heterogenous. The allocation and imposition of identity can be questioned.

At the same time, a different notion of queer culture emerges: Queer and trans needs are also formulated as class needs, as part of the struggle for the good life for everyone. To put this differently, the film makes queer class experiences accessible and queer struggles also recognisable as class struggles. Queer subcultural practices are not separate from class experiences, but they are often marginalised and split off as a result of intersectional power relationships, lack of access to resources and existing dominant regimes of representation. What do we recognise as class struggle? And what are considered legitimate contributions to the discussion of social utopias in times of neo-liberal crises and neo-right wing "identity politics"?

The film is a stimulating example of an existing cultural practice that is not invisible, but – as an effect of representational regimes – primarily recognisable as queer subcultural practice. To situate the emancipatory potential of queer subcultural practice in films such as *Folkbildingsterror* and the collective spaces in which they are perceived, discussed and appropriated, I would like to recall Stuart Hall's approach to changing hegemony: For Hall, referring to Gramsci, ideologies cannot be "transformed or changed by replacing one, whole, already formed, conception of the world with another, so much as 'renovating and making critical an already existing activity'" (Hall, 1986, p.23). Rather, through complex processes of *de- and reconstruction*, "old alignments are dismantled and new alignments can be effected between elements in different discourses and between social forces and ideas" (Hall, 1986, p.23). This is to say that social forces and the ideas that shape and co-constitute them cannot simply be destroyed and replaced by new ones. In this sense, *Folkbildningsterror* is a small but important part of the "war of positions" that Hall deems the core of politics of hegemony and representation: The film re-articulates the notion of class in queer politics. Its disruptive force lies further in its *de facto* disidentification with the panorama of existing poles in the current constellation of forces. This does not just entail right-wing populism but also progressive neo-liberalism, liberal identity politics and social democratic "communitarian" politics that appropriates an anti-elite rhetoric and articulates it to heteronormative resentment.

In fact, the queer and trans class identities created and represented in the film are discursively marginalised by all of these forces. They also shape the ways in which the film and its narrative can be received. However, the film's intervention is far more than a mere disruption or subversion of identities: It offers a different *standpoint*. The figure of the elite is transferred to the terrain of the critique of hetero- and homonormative identity politics. It is articulated with a critique of neo-liberal attacks on social security systems and – without denying their existing limits and exclusions – a different, inclusive, radical democratic and expanding notion of welfare. Read as a performance with concrete utopian value, the film invites us to imagine class as a necessarily intersectional politics and struggle. A *potential* that might in yet unknown ways contribute to the broader processes of recomposition in social movements, as can currently be observed in other terrains, for example, in the articulations of women's, queer, anti-racist and class struggles in the *Black Lives Matter* and feminist strike movements mentioned previously. If nothing else, the film offers an interesting example of the emancipatory potentials of the politics of representation in times of culture wars. It makes notions of (strategic) alliances conceivable that exceed identitarian divides, thereby expanding the political imagination: Queerly shaped class alliances and solidarities across identitarian divides come into focus, which are, indeed, helpful steps "against the elites!".

Notes

1 All translations from the original German have been provided by the author.
2 In this alliance, Fraser argues, "[i]deals like diversity and empowerment, which could in principle serve different ends, now gloss policies that have devastated manufacturing and what were once middle-class lives" (2017).
3 Brown describes the political impact of US identity politics since the 1970s as having been linked to a re-naturalization of capitalism. In this regard, she explains how the "invisibility and inarticulatedness of class" is not a historical accident but a constitutive moment of identity politics that has remained firmly attached to the

liberal discourse and measures the violations suffered by capitalist violence, racism and patriarchy against the normative ideal of the middle class. In this way, the middle class represents "the normalization rather than the politicization of capitalism, the denial of capitalism's power effects in the ordering of social life, the representation of the ideal capitalism to provide the good life for all [...]" (Brown, 1993, p.395).

4 In the German context, for example, the left party ("Die Linke") has used it for a feminist education camp, while the Rosa-Luxemburg-Foundation, closely linked to "Die Linke", screened it in its "Feminist Futures Festival" in 2018. Numerous women's and queer film festivals, universities, museums and activist spaces have screened it, and the makers of the film are often invited for talks and discussions.

5 According to Långström, the character's gender identity remains open, in-between.

6 Interview with the author, 2019.

7 Therefore, post-irony is sometimes also described as "new sincerity" (Hoffman, 2016, p.11). Such approaches also have a tradition in the contexts of queer art and filmmaking. The films of Bruce LaBruce and Todd Haynes come to mind, also the art of Canadian multimedia artist G.B. Jones.

8 Until the 1970s, Sweden's comprehensive welfare model served as a model for many other countries. It was inextricably linked with the SAP, the Swedish Social Democratic Workers Party. However, since the 1980s, the SAP has started to shift to the right. Neo-liberal ideas found their way into their rhetoric and, especially since the 1990s, increasingly also into their policy (Feld, 2000, pp.306, 314f.). In the 2000s, the Alliance for Sweden sealed many further neo-liberal reforms, which led to the further corrosion of the welfare model.

Bibliography

Appenroth, M. and Lottmann, R., 2019. Altern Trans anders? Empirische Befunde internationaler Untersuchungen zu trans Identitäten, Gesundheit und Alter(n). In: M. Appenroth and M. do Mar Castro Varela, Eds., *Trans & Care. Trans Personen zwischen Selbstsorge, Fürsorge und Versorgung*. Bielefeld: transcript, pp.287–302.

Becker, L., 2018. New queens on the block. Transfeminismus und neue Klassenpolitik. *Luxemburg – Gesellschaftsanalyse und linke Praxis* [online] September. https://www.zeitschrift-luxemburg.de/new-queens-on-the-block/ [Accessed 10 March 2021].

Bhattacharya, T., 2018. Ragpicking through history. Class memory, class struggle and its archivists. *Luxemburg – Gesellschaftsanalyse und linke Praxis* [online] September. https://www.zeitschrift-luxemburg.de/ragpicking-through-history-class-memory-class-struggle-and-its-archivists/ [Accessed 10 March 2021].

Brown, W., 1993. Wounded attachments. *Political Theory*, 21(3), pp.390–410.

Brown, W., 1995. *States of Injury: Power and Freedom in Late Modernity* Princeton: Princeton University Press.

Brown, W., 2002. Moralism as antipolitics. In: R. Castronovo and D.D. Nelson, Eds., *Materializing Democracy: Towards a Revitalized Cultural Politics*. Durham/London: Duke University Press, pp.368–392.

Butler, J., 1997. Merely cultural. *Social Text*, 15(3/4), pp.265–277.

Colebrook, C., 2004. *Irony: The New Critical Idiom*. Oxford/New York: Routledge.

Duggan, L., 2003. *The Twilight of Equality? Neoliberalism, Cultural Politics and the Attack on Democracy*. Boston: Beacon Press.

Duggan, L., 2004. *The Twilight of Equality. Neoliberalism, Cultural Politics, and the Attack on Democracy*. Boston: Beacon Press.

Engel, A., 2008. *Wider die Eindeutigkeit. Sexualität und Geschlecht im Fokus queerer Politik der Repräsentation*. Frankfurt am Main: Campus.

Engel, A., 2009. *Bilder von Sexualität und Ökonomie. Queere kulturelle Politiken im Neoliberalismus*. Bielefeld: transcript.

Eribon, D., 2016. *Rückkehr nach Reims*. Berlin: Suhrkamp.

Feld, R., 2000. Schweden: Vom "Dritten Weg" der achtziger Jahre zur "globalisierten Sozialdemokratie" des 21 Jahrhunderts?. *PROKLA*, 119(2), pp.301–328.

Fraser, N., 1997a. *Justice Interruptus: Critical Reflections on the "Post-Socialist" Condition*. London/New York: Routledge.

Fraser, N., 1997b. Heterosexism, misrecognition, and capitalism: a response to Judith Butler. *Social Text*, 15(3/4), pp.279–289.

Fraser, N., 2017. The end of progressive neoliberalism. *Dissent* [online] 2 January. https://www.dissentmagazine.org/online_articles/progressive-neoliberalism-reactionary-populism-nancy-fraser [Accessed 10 March 2021].

Hall, S., 1986. Gramsci's relevance for the study of race and ethnicity. *Journal of Communication Inquiry*, 10(2), pp.5–27.

Hall, S., 1996. Introduction: who needs "identity"? In: S. Hall and P. du Gay, Eds., *Questions of Cultural Identity*. London: Sage, pp.1–17.

Hall, S., 2004 [1983]. Ideologie und Ökonomie: Marxismus ohne Gewähr. In: J. Koivisto and A. Merkens, Eds., *Ideologie, Identität, Repräsentation*. Ausgewählte Schriften 4. Hamburg: Argument, pp.8–33.

Hall, S., 2014. Popular-demokratischer oder autoritärer Populismus. In: V. R. Diaz, J. Koivisto and I. Lauggas, Eds., *Stuart Hall. Populismus, Hegemonie, Globalisierung*. Ausgewählte Schriften 5. Hamburg: Argument, pp.101–120.

Hall, S., Critcher, C., Jefferson, T., Clarke, J. and Roberts, B., 2013 [1978]. *Policing the Crisis. Mugging, the State and Law and Order*. 2nd ed. London: Palgrave Macmillan.

Hall, S. and Jefferson, T., Eds., 2006 [1975]. *Resistance Through Rituals. Youth Subcultures in Post-War Britain*. 2nd ed. Oxford: Routledge.

Hebdige, D., 1979. *Subculture. The Meaning of Style*. London: Methuen.

Hennessy, R., 2018 [2000]. *Profit and Pleasure. Sexual Identities in Late Capitalism*. London; New York: Routledge.

Hoffmann, L., 2016. *Postirony. The Nonfictional Literature of David Foster Wallace and Dave Eggers*. Bielefeld: transcript.

Hutcheon, L., 1994. *Irony's Edge: The Theory and Politics of Irony*. London; New York: Routledge.

Jones, R.E., 2013. *Clampdown: Pop-Cultural Wars on Class and Gender*. Alresford: Zero Books.

Klein, N., 2000. *No Logo: No Space, No Choice, No Jobs*. New York: Picador.

Laufenberg, M., 2014. *Sexualität und Biomacht. Vom Sicherheitsdispositiv zur Politik der Sorge*. Bielefeld: transcript.

Marx, K. and Engels, F., 1999 [1848]. *Das kommunistische Manifest. Eine moderne Edition*. Hamburg: Argument.

Merkel, W., 2016. Trump und die Demokratie. *ipg* [online] 10 November. https://www.ipg-journal.de/regionen/nordamerika/artikel/trump-und-die-demokratie-1694/ [Accessed 10 March 2021].

Muñoz, J.E., 2009. *Cruising Utopia. The Then and There of Queer Futurity*. New York/London: New York University Press.

Nölke, A., 2017. *Links-Populär. Vorwärts handeln statt rückwärts denken*. Frankfurt am Main: Westend Verlag.

Sauer, A. and Nieder, T.O., 2019. We care. Überlegungen zu einer bedarfsgerechten, transitionsunterstützenden Gesundheitsversorgung. In: M. Appenroth and M. Do Mar Castro Varela, Eds., *Trans & Care. Trans Personen zwischen Selbstsorge, Fürsorge und Versorgung*. Bielefeld: transcript, pp.75–102.

Stegemann, B., 2018. *Die Moralfalle. Für eine Befreiung linker Politik*. Berlin: Matthes & Seitz.

Taylor, K-Y., ed., 2016. *How We Get Free: Black Feminism and the Combahee River Collective*. Chicago: Haymarket Books.

Taylor, K.-Y., 2019. *From #BlackLivesMatter to Black Equality*. Chicago: Haymarket Books.

Thompson, E.P., 2002 [1963]. *The Making of the English Working Class*. London: Penguin.

Wichterich, C., 2019. Die antifeministische Internationale. *Blätter für deutsche und internationale Politik*. 12, pp.103–111.

Williams, R., 1989 [1958]. Culture Is Ordinary. In: R. Williams. *Resources of Hope: Culture, Democracy, Socialism*. London: Verso, pp.3–18.

Woltersdorff, V., 2016. Das gouvernementale Projekt der Prekarisierung von Heteronormativität. In: M.T. Herrera Vivar, P. Rostock, U. Schirmer and K. Wagels, Eds., *Über Heteronormativität. Auseinandersetzung um gesellschaftliche Zugänge und konzeptuelle Zugänge*. Münster: Westfälisches Dampfboot, pp.32–50.

Filmography

Folkbildingsterror, 2014. [film] Directed by Lasse Långström. Sweden.

Born in Flames, 1983. [film] Directed by Lizzie Borden. USA.

Index

Note: Pages in **bold** refer tables and pages followed by 'n' refer notes

For Product Safety Concerns and Information please contact our EU
representative GPSR@taylorandfrancis.com
Taylor & Francis Verlag GmbH, Kaufingerstraße 24, 80331 München, Germany

www.ingramcontent.com/pod-product-compliance
Ingram Content Group UK Ltd.
Pitfield, Milton Keynes, MK11 3LW, UK
UKHW022318250926
475702UK00046B/935

* 9 7 8 0 3 6 7 6 9 2 6 0 5 *